2nd International Conference on Advanced Technologies in Manufacturing and Materials Engineering (ATMME 2018)

IOP Conference Series: Materials Science and Engineering
Volume 389

Harbin, China
4 – 6 May 2018

ISBN: 978-1-5108-6785-7
ISSN: 1757-8981

Printed from e-media with permission by:

Curran Associates, Inc.
57 Morehouse Lane
Red Hook, NY 12571

Some format issues inherent in the e-media version may also appear in this print version.

Printed by Curran Associates, Inc. (2018)

For permission requests, please contact the Institute of Physics
at the address below.

Institute of Physics
Dirac House, Temple Back
Bristol BS1 6BE UK

Phone: 44 1 17 929 7481
Fax: 44 1 17 920 0979

techtracking@iop.org

Additional copies of this publication are available from:

Curran Associates, Inc.
57 Morehouse Lane
Red Hook, NY 12571 USA
Phone: 845-758-0400
Fax: 845-758-2633
Email: curran@proceedings.com
Web: www.proceedings.com

Table of contents

Volume 389

2018 2nd International Conference on Advanced Technologies in Manufacturing and Materials Engineering (ATMME 2018)

4–6 May 2018, Harbin Institute of Technology, Harbin, China

Accepted papers received: 5 July 2018
Published online: 19 July 2018

Preface

Papers

Manufacturing Engineering

IOP Conf. Series: Materials Science and Engineering **389** (2018) 011001 doi:10.1088/1757-899X/389/1/011001

PREFACE

2018 2nd International Conference on Advanced Technologies in Manufacturing and Materials Engineering (ATMME 2018) is an annual conference which offers a place for participants to present and to discuss their innovative recent and ongoing research and theoretical work and/or their applications. The first ATMME was held in Bangkok, Thailand in 2017. This year is the second ATMME annual conference, which was held in Harbin, China, during May 4-6, 2018. ATMME 2018 is co-organized by Harbin Institute of Technology (HIT) and Hong Kong Society of Mechanical Engineering (HKSME), technically sponsored by Swinburne University of Technology (SUT) and The Chinese University of Hong Kong (CUHK). The conference provides a useful and wide platform for researchers and engineers to present their latest research findings and exchange their ideas in a broad range of Advanced Technologies in Manufacturing and Materials Engineering topics. The participants of the conference were from different regions of the world, with background of either academia or industry.

There were 142 submissions including 125 papers and 17 abstracts. After rigorous peer review, 45 papers and 5 abstracts were accepted. Finally, 37 papers are collected in this ATMME 2018 conference proceeding. This proceeding covers a wide range of topics in Material Design of Computer Aided, Manufacturing Processes and Systems, New Materials and Advanced Materials, Surface Engineering/Coatings, Advanced Composites and Hybrid Materials, Materials Forming, Micro / Nano Materials, Materials Machining, etc.

We would like to thank all of those who supported ATMME 2018. Particularly, we would like to thank the organizing committee for their invaluable advices in the organization and the program committee for the peer review of the papers.

We sincerely hope that ATMME will be an excellent forum for discussions that will lead to fruitful research studies and other new scientific ideas and engineering products that will be beneficial to the society and the industry.

Prof. Tao Sun, Harbin Institute of Technology, China
Prof. Du Ruxu, The Chinese University of Hong Kong, Hong Kong

IOP Conf. Series: Materials Science and Engineering **389** (2018) 011002 doi:10.1088/1757-899X/389/1/011002

COMMITTEES

Conference Chairman

Prof. Du Ruxu, The Chinese University of Hong Kong, Hong Kong

Conference Committee Co-Chairs

Prof. Tao Sun, Harbin Institute of Technology, China

Prof. Alan Lau, Swinburne University of Technology, Australia

Program Committee Chair

Prof. Yongda Yan, Harbin Institute of Technology, China

Local Chair

Prof. Junjie Zhang, Harbin Institute of Technology, China

International Technical Committeee

Prof. Yongsheng Ma, University of Alberta, Canada

Dr. Ping Guo, The Chinese University of Hong Kong, HK

Prof. Raghu Prakash, Indian Institute of Technology Madras, India

Prof. Zhijiang Du, Harbin Institute of Technology, China

Prof. Wenjun Zong, Harbin Institute of Technology, China

Prof. Qingliang Zhao, Harbin Institute of Technology, China

Prof. Bo Wang, Harbin Institute of Technology, China

Prof.Feihu Zhang, Harbin Institute of Technology, China

Prof. Hamid Bahai, Brunel University London, UK

Dr. Ştefan Ţălu, The Technical University of Cluj-Napoca, Romania

IOP Conf. Series: Materials Science and Engineering **389** (2018) 011003 doi:10.1088/1757-899X/389/1/011003

Peer review statement

All papers published in this volume of *IOP Conference Series: Materials Science and Engineering* have been peer reviewed through processes administered by the proceedings Editors. Reviews were conducted by expert referees to the professional and scientific standards expected of a proceedings journal published by IOP Publishing.

Ash-and-Slag-Waste-based Porous Glass: Synthesis and Properties

I.S. Grushko[1]

[1] Department "Industrial, civil engineering, geotechnics and foundation engineering", South-Russian State Polytechnic University (NPI) named M.I. Platov

grushkois@gmail.com

Abstract. In recent years, porous glass (sponge-glass) is becoming increasingly popular as an environmentally and hygienically safe non-flammable insulating material in the construction of civil and industrial buildings and structures. Composite materials with sponge-glass can be used for exterior insulation and facade finish. These materials are characterized by sufficiently high insulating effect, mechanical strength and chemical durability in combination with low cost of their production. Ash and slag waste from thermal power plants can be used as one of the main raw materials for the production of sponge-glass for construction purposes. The paper presents the results of research on the development and study of the properties of porous glass on the basis of ash and slag wastes of the Novotcherkasskaya GRES and bottled glass scrap: various compositions of the furnace charge have been synthesized, the influence of the fluxing agent and blowing agent on the functional properties of porous glass (density, porosity, strength and heat conduction) has been studied. Crystalline borax was used as the fluxing agent, and anthracite as the foaming agent. Standard methods were used for the study: electron microscopy methods, measurements of heat fluxes (tests were carried out in two modes: cooling and heating), determination of apparent density, etc., qualified equipment. The following was established in the course of study: the porosity of porous glass samples increases by more than two times with an increase in the anthracite additive; with an increase in the anthracite additive the strength of the porous glass decreases by more than 1.5 times; the optimal composition of the furnace charge, taking into account the effect of anthracite on the most important properties of the synthesized porous glass, is SHPS-B5 composition, which provides a sufficiently high porosity of 37.7% and a density of 466.8 kg/m^3.

1. Introduction

In recent years the porous glass (foam glass) spread more and more due to its being used as an ecologically and hygienically safe incombustible heat-insulating material for building civil and industrial facilities. Composite materials with applied foam glass can be used for outer insulation and furnishing of buildings facades [1]. These materials are characterized by a pretty high heat-insulating ability, mechanical and chemical durability combined with a low manufacturing price. As one of the main raw components of foam glass manufacturing, ash and slag waste of TPSs can be used [2]. Using ash and slag waste could not only decrease foam glass based construction materials costs, but also move forward a solution of the ecological problem of utilizing big amounts of ash and slag waste which is accumulated around thermal power stations (TPS). That's why it's the current aim to develop a technology of foam glass manufacturing with ash and slag waste of the TPSs used as a raw material [3].

IOP Conf. Series: Materials Science and Engineering **389** (2018) 012001 doi:10.1088/1757-899X/389/1/012001

2. Results and discussion

Following parameters have a key value for the ash and slag foam glass being used as a heat-insulating material: apparent density (hereafter density), real porosity, thermal conductivity and compressive strength. For investigating the dependence of amount of the above pore former parameters, the following investigations were performed [4]:

- compositions with different anthracite and borax proportions were received based on an optimal composition of waste foam glass;
- values of density and porosity of received samples were estimated;
- results of the heat conductivity coefficient estimation and compressive strength of the test foam glass samples were received.

For solving the tasks set, following compositions for synthesizing of the waste glass based on charges with borax proportions from 1 to 7% were developed (table 1). All samples were synthesized by a temperature of 825 °C.

Table 1. Waste porous glass charges composition

Composition No.	Proportions of components, %			
	Waste	Cullet	Borax*	Anthracite*
PCG-A1	30	70	1	5
PCG-A2	30	70	2	5
PCG-A3	30	70	3	5
PCG-A4	30	70	4	5
PCG-A5	30	70	5	5
PCG-A6	30	70	6	5
PCG-A7	30	70	7	5

* components' proportion shown over 100% of charge

A visual inspection of the synthesized samples was made. Samples PCG-A1, PCG-A2, PCG-A3, PCG-A4 had clear geometrical edges, the lengthwise cut had a high density, almost no pores seemed to exist. Samples PCG-A5, PCG-A6, PCG-A7 weighed less, the lengthwise cut showed evenly located pores.

Considering the results, the other equal parameters and a rational use of raw materials (especially borax), the composition PCG-A5 was claimed as an optimal one.

Researches were performed on the effect of borax as both a melter and a foaming agent. Results of the investigation showed that under its composition of 5%, borax conditions high melting of the waste porous glass, which consequently can not provide the porous glass synthesis, which will be shown further. This is a convincing argument of lack of pores emergence while using borax. That's why the further waste porous glass synthesis was performed by using anthracite only as a pore blowing agent and borax as a melter. According to these conclusions, waste porous glass compositions with different proportions of anthracite were developed, which are shown in the table 2.

Results of the porous glass synthesized samples microstructure research allow us to conclude the following: sample PCG-B1 (figure 1) has a dense structure, spherical pores. Number of pores is quite small relatively to the volume of the sample. Sample PCG-B2 (figure 2) also has a dense structure, pores have a bigger size compared to PCG-B1, but their number is also insufficient. Interporal partitions are dense. Sample PCG-B3 (figure 3) has a heterogeneous porosity, pores allocation isn't even. Samples PCG-B4 and PCG-B5 (figures 4, 5) have a pretty homogenous structure, pores are closed and isolated from each other, even porosity. Samples PCG-B6 and PCG-B7 (figures 6, 7) have large closed and also connected spherical pores, interporal partitions are thin and can't stand constructional loads. Increasing the anthracite proportion leads to an increase of pores' size and its number, which should provide higher heat insulating properties of the synthesized material. The greatest interest according to results of the microscopic researches represent the samples PCG-B4 and PCG-B5.

Table 2. Waste porous glass charges composition (PCG)

Waste porous glass charges composition	Proportions of components, %			
	Waste	Cullet	Borax*	Anthracite*
PCG-B1	30	70	5	1
PCG-B2	30	70	5	2
PCG-B3	30	70	5	3
PCG-B4	30	70	5	4
PCG-B5	30	70	5	5
PCG-B6	30	70	5	6
PCG-B7	30	70	5	7

* component's proportion shown over 100% of charge

The microstructure of synthesized porous glasses is shown in pictures 1 – 7 (a- is an increase of 20X, b - is an increase of 540X).

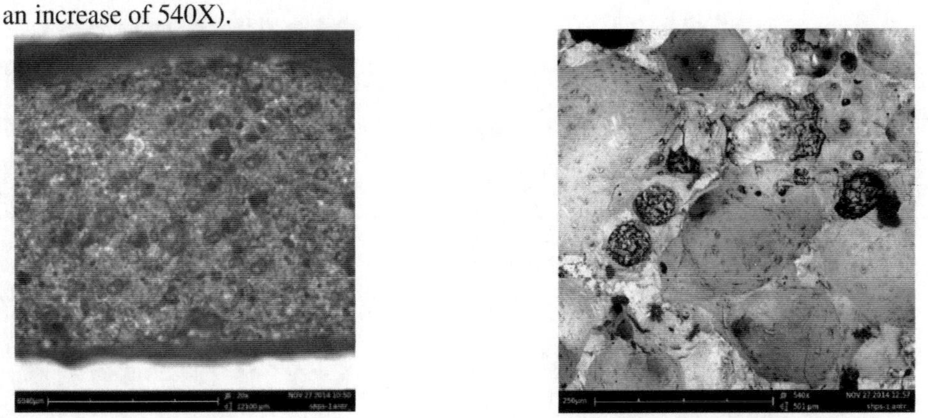

(a) (b)

Figure 1. Microstructure of the sample SHPS-B1

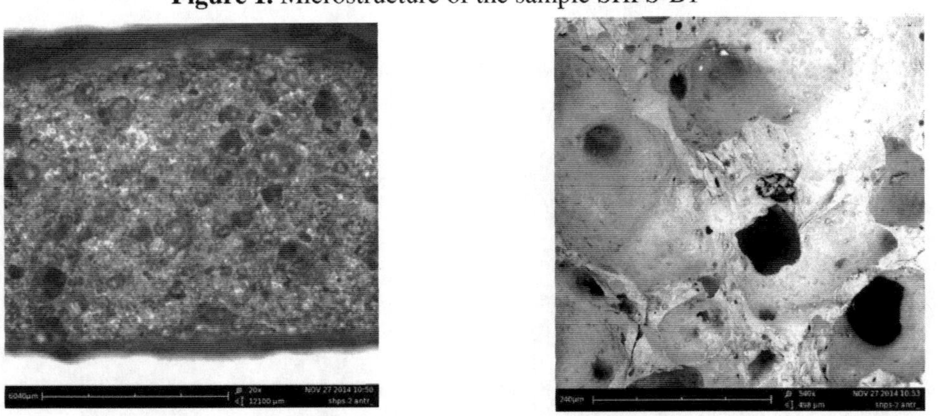

(a) (b)

Figure 2. Microstructure of the sample SHPS-B2

IOP Conf. Series: Materials Science and Engineering **389** (2018) 012001 doi:10.1088/1757-899X/389/1/012001

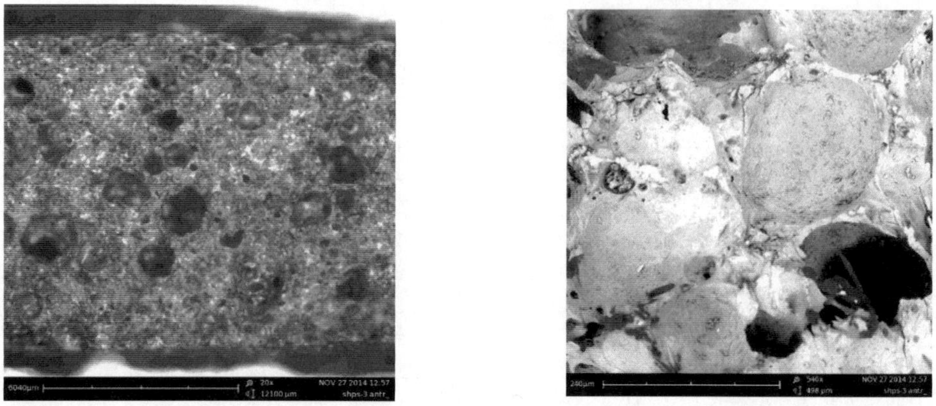

(a) (b)

Figure 3. Microstructure of the sample SHPS-B3

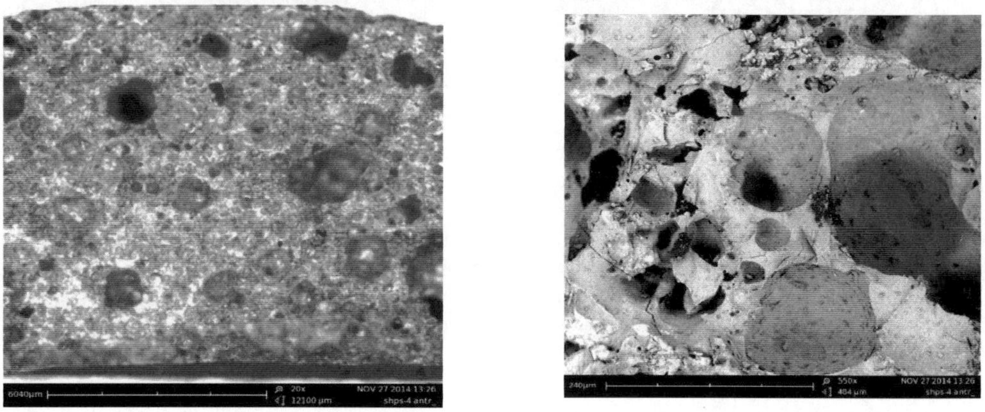

(a) (b)

Figure 4. Microstructure of the sample SHPS-B4

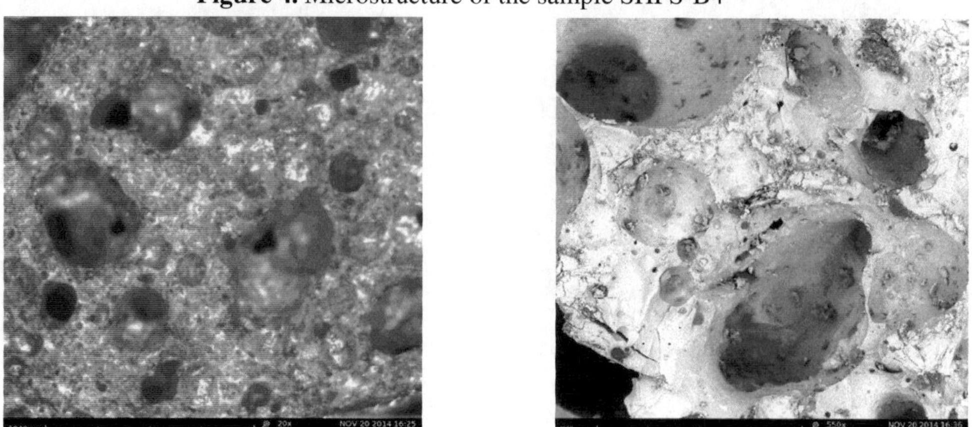

(a) (b)

Figure 5. Microstructure of the sample SHPS-B5

IOP Conf. Series: Materials Science and Engineering **389** (2018) 012001 doi:10.1088/1757-899X/389/1/012001

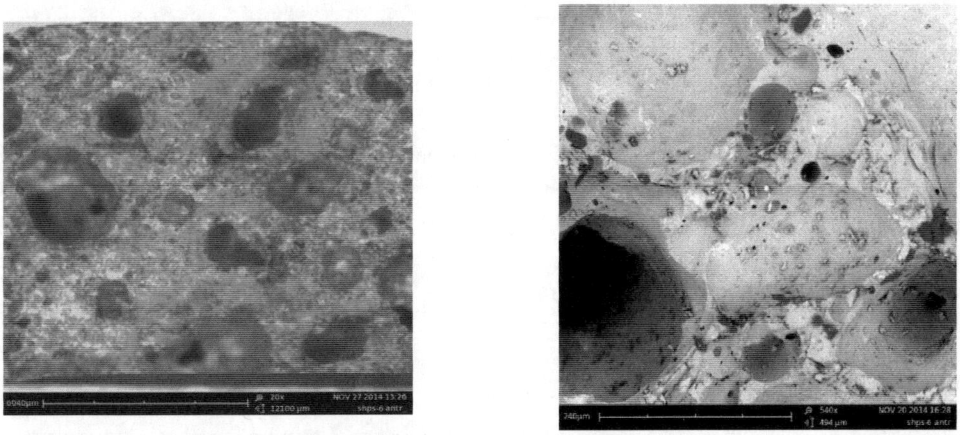

(a) (b)

Figure 6. Microstructure of the sample SHPS-B6

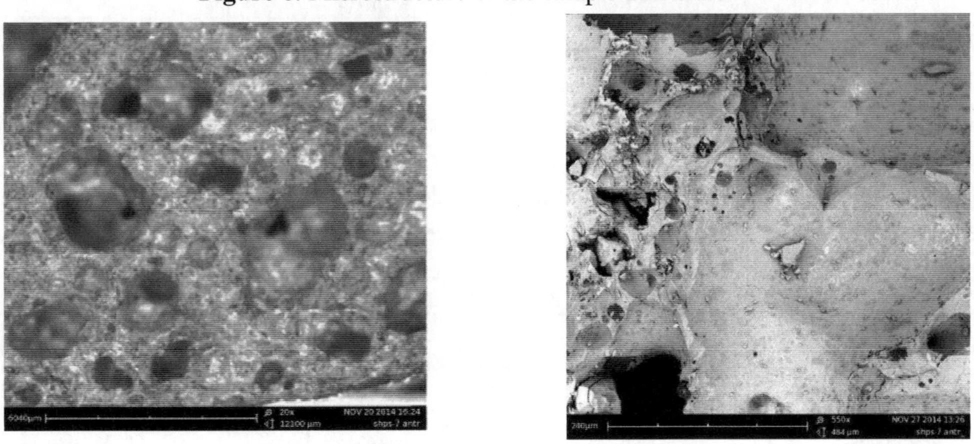

(a) (b)

Figure 7. Microstructure of the sample SHPS-B7

The synthesized glass properties research was also performed based on charges with anthracite proportions of 1-7 % in the charges composition presented in the table 2. The results of calculations are presented in table 3.

The true porosity of the material is considered to be the degree of filling its volume with pores. Its value can be calculated with a following formula [5]:

$$P = (1 - \rho_k/\rho_t)*100 \quad (1)$$

where P is the true porosity, %;

ρ_k–material's density, kg/ m^3;

ρ_t – true material's density, kg/ m^3.

Density of the samples is calculated considering the GOST R EN 1602-2008 «Heat-insulating products, used in construction sphere. Method of estimating the "apparent density" according to the formula [5]:

$$\rho_k = m/V \quad (2)$$

where m is the mass of the materials, kg;

V – the volume of the material, pores and empty space included, m^3.

IOP Conf. Series: Materials Science and Engineering **389** (2018) 012001 doi:10.1088/1757-899X/389/1/012001

Due to the fact that the quality of synthesized porous glass, characterized by its heat-insulating properties determined by the heat conductivity, we launched several researches on the subject of complex borax and anthracite proportions in the samples PCG-B1, PCG-B2, PCG-B3, PCG-B4 and PCG-B5.

For estimating the dependence of materials' density and porosity on its heat-insulating properties, the synthesized samples [6] were investigated on the subject of heat conductivity.

Table 3 . Results of density, porosity and compressive strength of the waste porous glass estimating.

Name	Density, kg/m^3	Porosity, %	Compressive strength, H/m^2
PCG-B1	579	24,7	2,48
	573	26,0	2,44
	577	25,4	2,45
	579	24,7	2,48
	571	25,1	2,46
PCG-B2	537	31,9	2,36
	526	33,1	2,31
	532	32,5	2,35
	539	31,0	2,38
	537	31,9	2,36
PCG-B3	509	37,1	2,23
	521	36,0	2,26
	516	36,5	2,25
	521	36,0	2,26
	529	35,7	2,27
PCG-B4	491	38,0	2,19
	497	37,7	2,21
	497	37,7	2,21
	504	37,3	2,22
	512	37,2	2,23
PCG-B5	475	37,1	2,20
	463	38,0	2,17
	463	38,0	2,17
	470	37,5	2,18
	463	38,0	2,17
PCG-B6	441	44,7	1,75
	417	46,9	1,70
	435	45,2	1,74
	429	46,4	1,73
	435	45,2	1,74
PCG-B7	410	48,9	1,15
	401	49,7	1,13
	410	48,9	1,15
	414	47,4	1,16
	392	54,0	1,07

For measuring the heat fluxes conducted through the samples we used the method of measuring the density of heat flows passing through the enclosing structures. Such a method allows to estimate the heat-technical properties of enclosing structures' and construction material properly, and also to estimate the real heat losses through the external enclosing structures.

This work required a heat flow and temperature density measurer ITP-MG4.03/5(I) «Potok», № 1210, adequate with TC 7648-027-12585810-2008. Heat flow converter № 5191 with a converting coefficient of K=26,94 W/(m^2 x mW), sensors type – diameter of 27 mm, mode - observation.

For maintaining the needed temperature and humidity modes during the research, a climatic cell SM – 60/75-250 TVX was used.

Five waste porous glass samples with a different anthracite proportion were involved into the research, each of them had a porous structure. The research was performed in two modes: «cooling» and «heating» till -30°C and +50°C were reached, according to the requirements of the GOST 25380-82 «Method of measuring the density of heat flows passing through enclosing structures». Each sample had a thickness of 10 mm. Received data is presented below in the table 4.

Table 4. Results of the waste porous glass samples heat conductivity estimation.

SampleNo.	Mode	q, W/m^2	t_v, °C	t_n, °C	λ, W/(m·°C)
PCG-B1	heating +50 °C	80,6	32,6	38,1	0,147
		82,0	32,4	37,9	0,149
		81,2	33,1	38,6	0,148
		81,7	32,9	38,3	0,151
		80,9	32,6	38,1	0,147
	cooling -30 °C	148,7	22,4	6,1	0,091
		150,1	21,7	5,4	0,092
		154,3	21,0	5,1	0,097
		151,8	21,4	5,0	0,093
		149,5	22,1	5,9	0,092
PCG-B2	heating +50 °C	70,9	28,1	34,7	0,107
		71,8	28,2	34,8	0,109
		71,9	28,2	34,8	0,109
		72,4	28,4	34,7	0,115
		70,1	28,5	34,9	0,110
	cooling -30 °C	133,0	18,1	3,5	0,091
		136,5	17,8	3,2	0,093
		136,7	17,9	3,5	0,095
		134,0	17,8	3,6	0,094
		133,7	18,0	3,7	0,093
PCG-B3	heating +50 °C	55,2	24,1	29,2	0,108
		56,3	24,2	29,4	0,108
		55,9	23,9	30,0	0,092
		57,3	23,6	32,0	0,068
		56,2	23,6	31,4	0,072
	cooling -30 °C	54,5	23,3	11,7	0,047
		53,3	23,4	11,8	0,046
		52,8	23,5	12,6	0,048
		48,5	23,5	11,9	0,042
		46,1	23,7	12,3	0,040

PCG-B4	heating +50 °C	36,2	22,6	28,9	0,057
		38,3	22,5	29,0	0,059
		39,8	22,6	29,4	0,059
		40,9	22,7	29,6	0,059
		42,3	22,8	29,9	0,060
	cooling -30 °C	46,3	21,9	14,4	0,062
		45,7	22,0	14,7	0,063
		45,6	22,2	14,7	0,061
		45,6	22,3	15,0	0,062
		45,3	22,5	15,2	0,062
PCG-B5	heating +50 °C	76,9	30,7	40,4	0,079
		80,9	30,5	39,4	0,091
		74,8	29,9	38,3	0,089
		62,6	29	36,3	0,086
		48,4	28,2	34,2	0,081
	cooling -30 °C	46,9	22,6	15,0	0,062
		45,7	22,7	15,6	0,064
		49,3	22,9	14,7	0,060
		47,6	22,8	15,4	0,064
		48,0	23,1	15,2	0,061

Received values were generalized and presented below in the table 5.

Table 5. Results of density, porosity and heat conductivity estimation.

Sample No.	Anthracite's proportion, %	ρ, kg/m^3	P, %	Λ_{av}, W/(m·°C)
PCG-B1	1	575,8	25,2	1,121
PCG-B2	2	534,2	32,1	0,102
PCG-B3	3	519,2	36,3	0,067
PCG-B4	4	500,2	37,6	0,060
PCG-B5	5	466,8	37,7	0,062
PCG-B6	6	431,4	45,7	-
PCG-B7	7	405,4	49,8	-

As can be seen from the table 5, by increasing the amount of anthracite in the charge from 1 % to 7 %, considering even properties, the porosity value increases from 25,2 % to 49,8 %. By decline of the density value and incline of the porosity the heat conductivity decreases. This happens because the heat flow goes not only through the material, but also through its pores filled with air. Due to air having an extremely low heat conductivity there emerges a significant resistanceto the heat flow passing. In the presence of large mutually connected pores, the transfer of heat by the movement of air increases. A lack of connected pores also decreases the heat conductivity value. According to the results of this research we could estimate that the sample PCG-A5 represents the most optimal waste porous glass composition due to its lowest heat conductivity.

3. Conclusion

Based on the results of the complex of studies on the synthesis of porous glass based on ash and slag wastes, the following conclusions can be drawn.

1.The optimal composition of porous glass (SHPS-B5) as a component of the composite with the use of borax as a smelt, anthracite as a blowing agent has been developed.

2. The porosity of the samples increases with an increase in the anthracite additive by more than 2 times (Table 3). This shows that anthracite is an effective pore-former.

IOP Conf. Series: Materials Science and Engineering **389** (2018) 012001 doi:10.1088/1757-899X/389/1/012001

3. Dependence of the strength of glass on compression on the amount of anthracite is completely different. As can be seen from Table 3, with an increase in the anthracite additive, the strength value decreases by more than 1.5 times. Thus, the optimal composition of the charge, taking into account the effect of anthracite on the most important properties of the synthesized porous glass, is the composition of SHPS-B5, which provides a sufficiently high porosity of 37.7% and a density of 466.8 kg / m^3.

Acknowledgments
The article is prepared based on the results of a study carried out with the financial support of the Russian Foundation for Basic Research in the framework of the scientific project No. 16-33-60177 Mol_a_dk.

References

[1] Minko N.I., Puchka O.V., Evtushenko E.I., Nartsev V.M., Sergeyev S.V. Foamglass - a modern effective inorganic thermal insulation material // Fundamental research. - 2013. - No. 6.
[2] Shikhova V.A., Yatsenko E.A. Production of heat-insulating materials for construction purposes on the basis of waste from the fuel and energy complex. // Izvestiya Vysshikh Uchebnykh Zavedenii. The North Caucasus region. Series: Engineering sciences. - 2013. - No. 4.
[3] Smoliy V.A., Kosarev A.S., Yatsenko E.A. Cellular heat-insulating building glass materials based on waste from thermal power plants and ferrous metallurgy // Glass and ceramics. - 2017. - No. 2.
[4] Grushko I.S. Dependence of foam-slag glass parameters on furnace charge and technological additives // Scientific Review. - 2015. - № 6.
[5] GOST 2409-80 "Method for determination of apparent density, open and total porosity, water absorption".
[6] Research of the physicochemical regularities in the process of thermoplastic sintering and the formation of a cellular structure as a function of temperature-time regimes [Text]: report on R&D (concl.), RFBR project No. 13-03-90756 / 13 // Pr. manager. Grushko I.S. - St. Petersburg: 2015 - 230 pp.

IOP Conf. Series: Materials Science and Engineering **389** (2018) 012002 doi:10.1088/1757-899X/389/1/012002

Experimental Study and Matrix Effect Correction of Pseudo-binary Samples in XRF Analysis

Hefan LIU, Xiaoling ZHOU, Xin ZHANG, Keyin WU and Chengwei LU

Chengdu Research Academy of Environmental Sciences,Chengdu,China

Email: lhf@cdaes.cn

Abstract. This basic study conducted an X-ray fluorescence (XRF) analysis on pseudo-binary samples, which emphasis on the binary influence coefficient and correction for matrix effects. The Ti-V, V-Ti, Ti-Fe, Fe-Ti, V-Fe and Fe-V samples were confected and measured by XRF analysis using a Si-PIN detector. These coefficients were calculated using mathematical models proposed through the Lachance-Traill algorithm, which can be applied in practice via an efficient calibration procedure. For the content measurement, the maximum absolute error decreased from 5.92% to 0.27%, and the relative error decreased from 14.84% to 0.66% via fitting correction for matrix effects. This study describes and emphasizes the application of the binary influence coefficient and fitting correction method.

1. Introduction

X-ray fluorescence (XRF) analysis technology has beenwidely applieddue to its advantageofnon-destructive and fast analysis, and easy application. For instance, energy dispersive fluorescence analysis equipment has beenwidely used in chemical, geological, environmental and other fieldsdue to its portability. XRF has been utilized to analyse unknown samples for several decades. Under excitement through X-ray sources, the characteristics of the X-ray fluorescence elements measuredwereemitted through the energy level transition intheinner-shell electron. By measuring the energy and intensity characteristics of X-ray fluorescence, the element type and concentration for unknown samples could be determined. However, adifferenceexists between the standard sample and unknown sample, such as physical structure, chemical composition and so on. Therefore, there would be some deviations in the quantitative results. This is called the matrix effect. The variety of chemical composition and physical property in these samples would affect these analytical results, which presented as an enhancement or absorption. Due to the complex ingredients in these samples, the matrix effects would bring in an unacceptable error in XRF measurements. With the proposal of the Sherman Equation, many scholars have conducted a number of working XRF analyses [1-4]. Ying Xiaohu corrected the matrix effectsof Cr-Fe-Niby directlyusing theoretical alpha coefficients in stainless steel measurements[5]. This method can achieve good results in analyzing the element content, which varies greatly in unknown samples. Furthermore, this method may also appropriately extend the calibration curve. Li Siwei corrected the matrix effects of lead-copper alloy samples through the r-α coefficient [6].This method reduced the number of required standards, and their analysis results are better than those through the fundamental parameter method or theoretical alpha coefficients method. Li Zhe measured the element Ti and Fe in ore samples by EDXRF [7], andcorrected the non-linear relationships between the characteristics X-ray fluorescence counts and itscontent by neural network technology. Compared with chemical analysis methods, 65.4% of the sample's relative error was less than 1%, and the rest were less than 3%. This result could meet the requirements for the industrial production equipment analysis, which requires a relative error of less than5%.Most of these methods limit the specific range, as well as the range of applications [8-11]. This study attempted to conductamatrix effects analysis in EDXRF by establishing pseudo-binary samples and designing experiments to obtain two-dimensional coefficients. In addition, it also discusses the binary influence coefficient and correction for matrix effects in pseudo-binary samples, such as samples of Ti-V, V-Ti, Ti-Fe, Fe-Ti, V-Fe and Fe-V. The Lachance-Traillalgorithm mand fitting correction method were applied.

2. Binary influence coefficient

In these pseudo-binary samples, the elements comprise of *j* and *i*. These samples can be expressed as the composition (C_i, C_j), and

$$C_i + C_j = 100\% \tag{1}$$

With monochromatic excitation sources, β_{ij} is the absorption factor, and δ_{ij} is the enhancement factor. According to the Sherman Equation [12],

$$\beta_{ij} = \frac{\mu_i(\lambda_k) + \mu_i(\lambda_i)}{\mu_j(\lambda_k) + \mu_j(\lambda_i)} - 1 \tag{2}$$

$$\delta_{ij} = \frac{1}{2} D_j(\lambda_k) D_i(\lambda_k) k_j \mu_j(\lambda_k) \frac{\mu_i(\lambda_j)}{\mu_i(\lambda_k)} P_{ij}(\lambda_k) \tag{3}$$

The parameter, $P_{ij}(\lambda_k)$, presented in these two previous equations was defined by

$$P_{ij}(\lambda_k) = \frac{1}{\mu_s^{'}(\lambda_k)} In[1 + \frac{\mu_s^{'}(\lambda_k)}{\mu_s(\lambda_k)}] + \frac{1}{\mu_s^{''}(\lambda_i)} In[1 + \frac{\mu_s^{''}(\lambda_i)}{\mu_s(\lambda_j)}] \tag{4}$$

In Equations (2) to (4), the meaning of each parameter is as follows:
$\mu_i(\lambda_k)$ The mass absorption coefficient of the *i*element on *X*-ray, in which the wavelength is λ_k;
$D_j(\lambda_k)$ 1 if the incident radiation $\lambda_k \leq \lambda_{abs\ edge}$ of the *j* element,
0 in all other cases;
$D_i(\lambda_k)$ 1 if the fluorescence radiation $\lambda_j \leq \lambda_{abs\ edge}$ of the *i*element,
0 in all other cases;
k_jThe absorption mutation factor;
μ_s, μ_s', μ_s'' The mass absorption coefficient, $cm^2 \cdot g^{-1}$
and

$$\mu_s^{'}(\lambda_k) = \sum_{n=1}^{N} \mu_n(\lambda_k) C_n \csc\phi^{'} \tag{5}$$

$$\mu_s^{''}(\lambda_i) = \sum_{n=1}^{N} \mu_n(\lambda_i) C_n \csc\phi^{''} \tag{6}$$

Where N is the amount of elements in the sample, $\phi^{'}$ is the incidence angle and $\phi^{''}$ is the exit angle.

In addition, with polychromatic excitation sources, α_{ij} is the absorption factor, and$\varepsilon_{ij}$is the enhancement factor. According to the Rousseau Equation [13],

$$\alpha_{ij} = \frac{\sum_k W_i(\lambda_k) \beta_{ij}(\lambda_k)}{\sum_k W_i(\lambda_k)} \tag{7}$$

$$\varepsilon_{ij} = \frac{\sum_k W_i(\lambda_k) \delta_{ij}(\lambda_k)}{\sum_k W_i(\lambda_k)} \tag{8}$$

The weighting factor, $W_i(\lambda_k)$, presented in the two previous equations was defined by

IOP Conf. Series: Materials Science and Engineering **389** (2018) 012002 doi:10.1088/1757-899X/389/1/012002

$$W_i(\lambda) = \frac{\mu_i(\lambda_k) I_O(\lambda_k) \Delta\lambda_k}{\mu_i^*[1 + \sum_j C_j \beta_{ij}(\lambda_k)]} \tag{9}$$

In Equation (9), μ_i^* is the effective mass absorption coefficient, which is defined by

$$\mu_i^* = \mu_i^{'}(\lambda_k) + \mu_i^{''}(\lambda_i) \tag{10}$$

According to Equation (1) and the Lachance-Traill algorithm [14-15], the binary influence coefficient A_{ij} can be calculated by

$$A_{ij} = \frac{\alpha_{ij} - \varepsilon_{ij}}{1 + \varepsilon_{ij}(1 - C_i)} = \frac{\alpha_{ij} - \varepsilon_{ij}}{1 + \varepsilon_{ij} C_j} \tag{11}$$

Actually, the measurement results were determined through numerous factors, such as absorption effect, enhancement effect, and so on. This study used the binary influence coefficient A_{ij}, which is a comparative statistic, in order to discuss the matrix effects in pseudo-binary samples.

3. Sample preparation

This study configured a series of samples using a chemical analytical reagent, such as Fe_2O_3 powder (the content of Fe was 70%), TiO_2 powder (the content of Ti was 60%), and V_2O_5 powder (the content of V was 56%). For the sample's composition (C_i, C_j) in the pseudo-binary system, the *i* element is the primary base, and the *j* element is the so-called impurity. The content of these pseudo-binary samples is presented in Table 1.

Table 1. Content of the pseudo-binary samples

Fe-V Samples		V-Fe Samples		V-Ti Samples		Ti-V Samples		Fe-Ti Samples		Ti-Fe Samples	
C_{Fe}	C_V	C_V	C_{Fe}	C_V	C_{Ti}	C_{Ti}	C_V	C_{Fe}	C_{Ti}	C_{Ti}	C_{Fe}
0.05	0.52	0.05	0.66	0.05	0.63	0.05	0.52	0.05	0.56	0.05	0.66
0.10	0.49	0.10	0.60	0.10	0.59	0.10	0.48	0.10	0.52	0.10	0.61
0.15	0.44	0.15	0.52	0.15	0.52	0.15	0.42	0.15	0.48	0.15	0.58
0.20	0.39	0.20	0.45	0.20	0.45	0.20	0.38	0.20	0.43	0.20	0.50
0.25	0.36	0.25	0.41	0.25	0.40	0.25	0.32	0.25	0.39	0.25	0.42
0.30	0.33	0.30	0.34	0.30	0.32	0.30	0.28	0.30	0.35	0.30	0.35
0.35	0.29	0.35	0.26	0.35	0.27	0.35	0.25	0.35	0.30	0.35	0.34
0.40	0.24	0.40	0.21	0.40	0.20	0.40	0.20	0.40	0.26	0.40	0.24
0.45	0.21	0.45	0.16	0.45	0.16	0.45	0.16	0.45	0.22	0.45	0.17
0.50	0.16	0.50	0.08	0.50	0.09	0.50	0.09	0.50	0.18	0.50	0.14
0.55	0.12	0.55	0.03	0.55	0.02	0.55	0.03	0.55	0.15	0.55	0.10
0.60	0.09	0.56	0.00	0.56	0.00	0.60	0.00	0.60	0.10	0.60	0.06
0.65	0.04	-	-	-	-	-	-	0.65	0.04	0.65	0.03
0.70	0.00	-	-	-	-	-	-	0.70	0.00	0.70	0.00

Each sample was made into sheets using the pressing method. The best pressure was 25 KN. Recovery time was 90 seconds. Every sample was placed in a dryer for one hour at a temperature of 105-110°C. This sample preparation can reduce its impact on measurement results, such as the uniformity effect, particle distribution effect, moisture effect, and so on.

The detector of the X-ray fluorescence analyzer was the Si-PIN detector, which was produced by AMPTEK INC. Its main technical indicators were as follows: the energy resolution was 150-180 eV (^{55}Fe, 8.90 KeV), the scope of analysis was 0.001%-99.99%, the measuring range was 2-20 KeV, measuring time was <200 seconds, repeatability was <5%, stability was <0.05%, and power consumption was 300 W.

IOP Conf. Series: Materials Science and Engineering **389** (2018) 012002 doi:10.1088/1757-899X/389/1/012002

4. Sample preparation

The mass absorption coefficient $\mu_s$is the sum of the optical absorption coefficient, coherent mass absorption coefficient, and incoherent mass absorption coefficient. The primary X-ray intensity E_o was 12.5 *KeV*, and the content of the sample (C_i,C_j) was 100%. The X-ray characteristic energy, wavelength and mass absorption coefficient are presented in Table 2 [16].

Table 2. Mass attenuation coefficients of Ti, V and Fe elements

X-ray's Energy and Wavelength (*KeV*, *Å*)	Mass attenuation coefficient(cm^2·g^{-1})		
	Ti	*V*	*Fe*
Ti_{KI}(4.510, 2.749)	120.00	132.00	199.00
V_{KI}(4.952, 2.503)	89.00	98.20	148.00
Fe_{KI}(6.403, 1.936)	386.00	427.00	75.50

Absorption mutation factor k_{Ti}=0.8842, k_V=0.8828, k_{Fe}=0.8786[17].

The calculation and measurement results are shown in Figures 1-3. These figures represent the visual analysis of the relationship between C_i and A_{ij}. When $A_{ij}$was positive, the *i*element was absorbed and the *j* element was enhanced in the sample (C_i,C_j). However, the consequences reversed when A_{ij} was negative.

For Fe-V samples (Fig. 1), the binary influence coefficients are quite negative. $A_{Fe\text{-}V}$ was -0.015 to -0.724. The V element shows the enhancement effects for the Fe element. The greater the Fe content, the larger the enhancement effects. For V-Fe samples, $A_{V\text{-}Fe}$ was -0.674 to 0.006. The Fe element shows the enhancement effects for the V element. The greater the V content, the smaller the enhancement effects. When Fe content was greater than 65%, the Fe element exhibited a slight absorption effect for the V element.

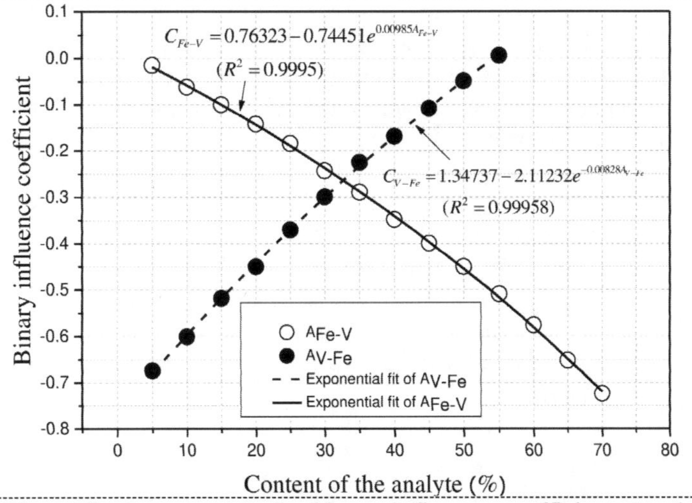

Figure 1. The ordinate represents the binary influence coefficient $A_{V\text{-}Fe}$(•), and the abscissa represents the V content; while the ordinate represents the binary influence coefficient $A_{Fe\text{-}V}$(○), and the abscissa represents the Fe content.

In Table 2, μ_V was 427.00 cm^2·g^{-1} for the Fe element. In this case, the fluorescence of Fe was obviously aborted by the V element. In addition, the fluorescence intensity of V increased through its second fluorescence. Similarly, for the V element, μ_{Fe} was 148.00 cm^2·g^{-1} and μ_V was 98.20 cm^2·g^{-1}. In this case, the Fe element produces scattering effects for the V element. Moreover, these slight enhancement effects were exhibited for the Fe element.

For the Ti-V samples (Fig. 2), the binary influence coefficients were quite negative. $A_{Ti\text{-}V}$ was -1.200 to -0.990. The V element exhibited enhancement effects for the Ti element. The greater the Ti content, the smaller the enhancement effect. For V-Ti samples, the binary influence coefficients were also quitenegative. $A_{V\text{-}Ti}$ was -1.113 to -1.315. The Ti element exhibited enhancement effects for the V element. The greater the V content, the larger the enhancement effects.

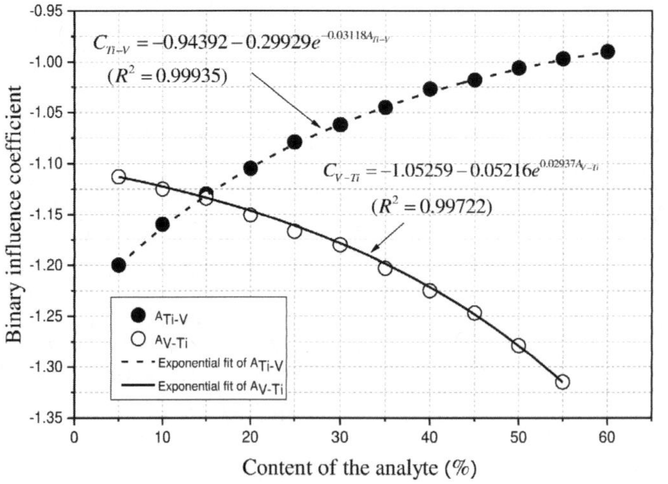

Figure 2. The ordinate represents the binary influence coefficient $A_{Ti\text{-}V}$(●), and the abscissa represents the Ti content; while the ordinate represents the binary influence coefficient $A_{V\text{-}Ti}$ (○), and the abscissa represents the V content.

According to Figure 2, with the gradient change in the primary base, the binary influence coefficients appear in a slightly convex curvature. The reason for this is because the initial emitted fluorescence of the Ti element had strong absorption for coherent and incoherent scattering.

For the Fe-Ti samples (Fig. 3), the binary influence coefficients are quite negative. $A_{Fe\text{-}Ti}$ was -1.138 to -1.082. The Ti element exhibited enhancement effects for the Fe element. The greater the Fe content, the smaller the enhancement effects. For the Ti-Fe samples, the binary influence coefficients were also quite negative. $A_{Ti\text{-}Fe}$ was -1.112 to -1.094. The Fe element also exhibited enhancement effects for the Ti element. The greater the Ti content, the smaller the enhancement effects.

In Table 2, for the mass absorption coefficient, the Fe element was 386.00 $cm^2 \cdot g^{-1}$, which was greater than the Ti element (120.00 $cm^2 \cdot g^{-1}$). For these reasons, the probability of photoelectric effects induced by the Fe element was less than that induced by the Ti element. Therefore, the secondary fluorescence of the Ti element was not obvious. Similarly, for the Ti element, the impact of Fe was mainly the scattering effect, rather than the photoelectric absorption effect. Therefore, the curve in Figure 3 had no obvious concave form.

IOP Conf. Series: Materials Science and Engineering **389** (2018) 012002 doi:10.1088/1757-899X/389/1/012002

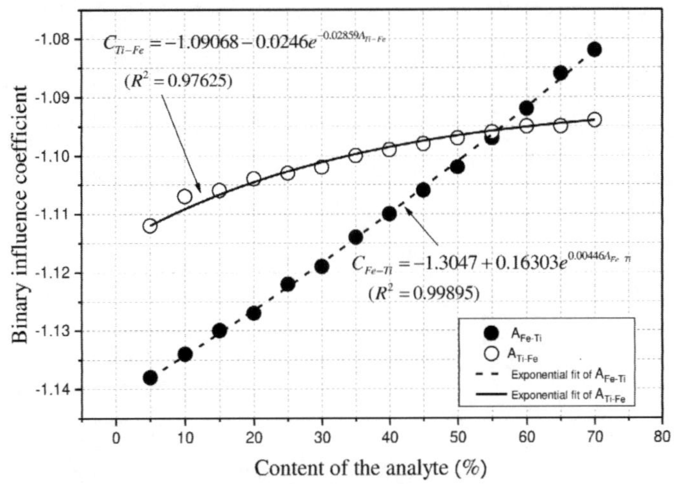

Figure 3. The ordinate represents the binary influence coefficient $A_{Fe\text{-}Ti}$(●), and the abscissa represents the Fe content; while the ordinate represents the binary influence coefficient $A_{Ti\text{-}Fe}$ (○), and the abscissa represents the Ti content.

5.Correction

In Figures 1, 2 and 3, A_{ij}, C_i and μ_s were closely correlated. This shows that A_{ij}cannotbe regarded as a constant when analyzing the wide range content of samples. The relationship between A_{ij} and C_i was nonlinear. This study applied a mathematical fitting model to correct the matrix effects in these pseudo-binary samples.

In Section 3, $C_{ij}$and A_{ij}satisfied the following exponential relationship, as presented in Equation (12).

$$C_{ij} = A_1 - A_2 \exp(-kA_{ij}) \qquad (12)$$

In Equation (12), the parameters of the fitting equation such as A_1, A_2, k and R^2 are listed in Table 3.

Table 3. Parameters of Equation (12) and goodness-of-fit

A_{ij}	Coefficient of the Eq. (12)			Goodness of fit
	A_1	A_2	K	R^2
$A_{V\text{-}Fe}$	1.34737	2.11232	0.00828	0.99958
$A_{Fe\text{-}V}$	0.76323	0.74451	-0.00985	0.99950
$A_{V\text{-}Ti}$	-1.05259	0.05216	-0.02937	0.99722
$A_{Ti\text{-}V}$	-0.94392	0.29929	0.03118	0.99935
$A_{Ti\text{-}Fe}$	-1.09068	0.02460	0.02859	0.97625
$A_{Fe\text{-}Ti}$	-1.30470	-0.16303	-0.00446	0.99895

This study corrects the matrix effects of the Fe-Ti samples using Equation (12). The results are shown in Figures 4 and 5.

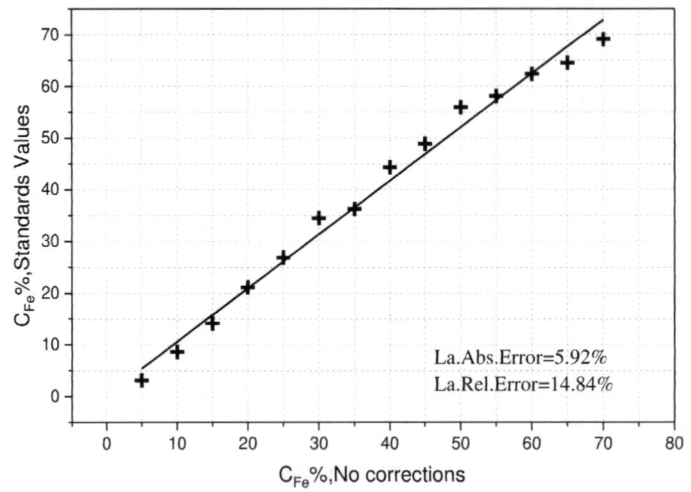

Figure 4. Fe element concentrations in Fe-Ti sample alloys plotted against concentrations calculated from uncorrected Fe intensities (%). Note the maximum absolute and relative errors.

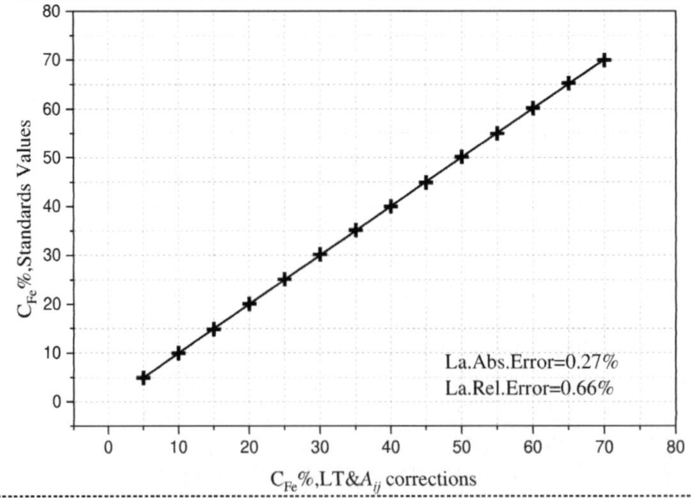

Figure 5. Fe element concentrations in Fe-Ti sample alloys plotted against concentrations calculated using $A_{Fe\text{-}Ti}$ and Equation (12). Note the maximum absolute and relative errors.

In Figures 4 and 5, the maximum absolute error decreased from 5.92% to 0.27%, and the relative error decreased from 14.84% to 0.66%. This means that the fitting method of the correction for matrix effects in pseudo-binary samples was available. Similarly, this fitting method may be applied to other samples.

5. Conclusions

The method in this study can be an effective way to correct matrix effects in pseudo-binary samples. The binary influence coefficient is a significant parameter to analyze the absorption or enhancement effects. The mathematical fitting model can be used to correct matrix effects. This analytic method can be applied to othersamples, and become an effectual manner for matrix effects correction in XRF analysis. However, the exponential fitting correction method is only a simple application for the matrix effect correction in pseudo-binary samples, and needs to be improved for practical applications.

Acknowledgements
The presented results receive funding from the national key research and development project of China under contract number 2016YFC0201506.

References

[1] A. N. Smagunova, N. F. Losev, A. G. Revenko and A. N. Mejevich, General Chart of Development the X-Ray Spectral Analysis Methods, ZavodskayaLaboratoriya, 1974, 40(12): 1461-1463.
[2] K.Mauser. Workshop on quantitative techniques and errors in XRF, 47th Annual Denver X-ray Conference, 1998.
[3] D. Hampai, S.B. Dabagov, C. Polese, A. Liedl, G. Cappuccio. Laboratory total reflection X-ray fluorescence analysis for low concentration samples, Spectrochimica Acta Part B: Atomic Spectroscopy, 2014, 101: 114-117.
[4] Injuk J, Van Grieken R. Literature trends in X-ray emission spectrometry in the period 1990-2000-A review, X-Ray Spectrometry, 2003, 32(1): 35-39.
[5] YING Xiao-hu,WANGQian,CAO Guo-Zhou,YU Ming-fang. The Application of Variable Theoretical α InfluenceCoefficent to the Matrix Correction in Cr-Fe-Ni Stainless Steel. Spectroscopy and Spectral Analysis, 2004, 24(12): 1681-1683.
[6] LI Siwei,LAIWanchang,ZHAIJuan,JIAXuehui. Determination of Major Elements in Lead Brass Alloy by R-α Coefficient Method for Matrix Effect in XRF. Journal of Sichuan Nonferrous Metals (In Chinese), 2013, 4: 62-64.
[7] LI Zhe,TUOXianguo,MUKeliang,WANGHonghui,ZHONGHongmei. Matrix effect and ANN correcting technique in EDXRF analysis Of Ti and Fe in core samples. NUCLEAR TELCHNIQUES (In Chinese), 2009, 32(1): 35-40.
[8] Ethan C. Geil, Steven A. LeBlanc, Darren S. Dale, Robert E. Thorne. Application of X-ray fluorescence imaging to ceramics from the American Southwest, Journal of Archaeological Science, 2013, 40(12): 4780-4784.
[9] Rafet Yilmaz. Measurements of L2,3 subshells X-ray production cross-sections and average L2,3 subshells fluorescence yield values of some elements in the atomic number range 70⩽Z⩽78, Radiation Physics and Chemistry, 2012, 81: 1539-1542.
[10] Henryk Drozdowski, Anna Romaniuk, ZdzisławBłaszczak. Structure and intermolecular interactions in selected binary solutions studied by X-ray methods, Radiation Physics and Chemistry, 2013, 93: 47-53.
[11] Rousseau R M. Some consideration on how to solve the Sherman equation in practice, Spectrochemca Acta Part B, 2004, 59: 1491-1502.
[12] Sherman J. The theoretical derivation of fluorescent X-ray intensities from mixture, Spectrochim Acta, 1955, 7: 283-306.
[13] Rousseau R M. Corrections for matrix effects in X-ray fluorescence analysis-A tutorial, Spectrochimica Acta Part B, 2006, 61: 759-777.
[14] G.R.Lachance, R.J.Traill, A practical solution to the matrix problem in X-ray analysis, Can.J.Spectrosc., 1966, 11: 43-48.
[15] Rousseau R M. Concept of the influence coefficient, Rigaku J., 2001, 18: 8-21.
[16] Liang Y. The basis of X-ray fluorescence spectroscopy. (Science Press, Nanjing), 2007, 193-224.
[17] Broll N. Quantitative X-Ray Fluorescence Analysis, Theory and Practice of the Fundamental Coefficient Method X-ray spectroscopy, 1986, 15(4): 271-285.

IOP Conf. Series: Materials Science and Engineering **389** (2018) 012003 doi:10.1088/1757-899X/389/1/012003

Molecular Dynamics Simulation Study on the Possible Factors Affecting Stability of ODS Steel

M.Mustafa Azeem, Zhongyu Li , Qingyu Wang and Abid Hussian

College of Nuclear Science & Technology, Harbin Engineering University, 150001, P.R China.

lizy820601@163.com (Zhongyu Li); mustafa@hrbeu.edu.cn (M.Mustafa Azeem)

Abstract. Oxide-dispersed-strengthened (ODS) steel has excellent mechanical, thermodynamic and radiation resistant properties, which makes it an important candidate material for high-temperature reactors applications. Radiation stability of oxide is very important to study at the atomic level. In present case molecular dynamics (MD) simulation is used to study stabilities of Y_2O_3 particles in ODS steel as a function of particle sizes. The size dependence is obvious for particle sizes larger than 2.8 nm. Furthermore, the stabilities depend on charges of Y and O atoms.

1. Introduction

The structural materials in nuclear reactors have degradation caused by irradiation damage. Research on advanced fission and fusion reactors put forward higher requirements for the safety and economy of reactors, which lead to critical needs for materials of better performance [1,2,3,4]. So far candidate materials for reactor-core components include nickel-based alloys, austenitic stainless steels, ferrite / martensitic steels, and oxide dispersion strengthening (ODS) steels [5]. ODS steel is unique for high temperature applications due to their high creep resistances from embedded oxide particles of nanometre sizes [3,5,6,7,8]. In order to further optimize ODS steels, MD simulations are needed to study atomic scale details of the structural evolution of oxide particles under high temperatures. The thermal stabilities of particles are important for their applications in high temperature fast reactors. In present study some possible factors affecting the stability of ODS are discussed by MD simulation method.

2. Modeling procedure

The dispersoids in ODS steels are typically Y-Ti-O oxide. As a model material, Y_2O_3 embedded Fe is simulated by using LAMMPS (Large scale Atomic Molecular Massively Parallel Simulator)[9]. The interatomic potentials are described by hybrid potentials from Hammond et al [10]. The potential is linked to Ziegler-Biersack-Littmark (ZBL) potential for short-range interactions [11]. Simulation cell has dimensions of 22nmx22nmx22nm and has one oxide particle of size 5 nm embedded at the center. Periodic boundary conditions are applied on all sides. The cell is relaxed initially by isobaric-isothermal ensemble NPT for 100ps, followed by microcanonical ensemble NVE for 100ps at 300K. Figure 1 shows the cell structure. The red color represents O atoms, and green color for Y atoms, and the purple color for Fe atoms. The cell contains 1015597 atoms including 21550 O atoms and 14086 Y atoms. The Open Visualization Tool (OVITO) [12] is used for visualization of LAMMPS output.

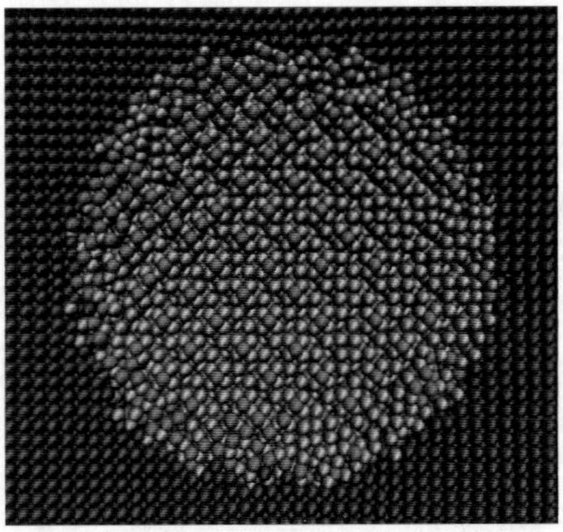

Figure 1. Structure of model after thermal relaxation with Fe embedded 5nm oxide at 300K.

3. Results and discussion

Figure 2 shows the temperature changes of the system as a function of time. The initial temperature is 300K. With increasing time, the temperature rises up and then drops down to reach a saturated, value at a time longer than about 20 ps. Figure 3 shows a variation of total energy vs. time for a thermal relaxation at 300K. The trend of the total energy changes is roughly the same as the change of temperature as shown in Figure 2.After relaxation longer than 20ps, the energy fluctuations are very small, which suggests the system has reached an equilibrium state. We find that thermal stability of the oxide particle at a longer time depends on the factors other than temperatures, as explained below.

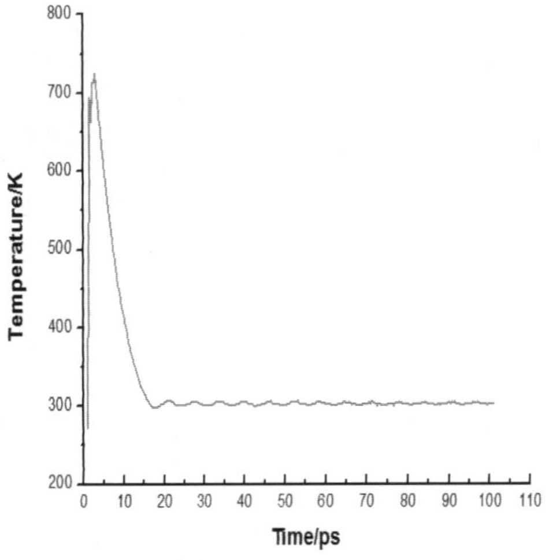

Figure 2. Temperature changes of the cell as a function of time for the thermal relaxation at 300K.

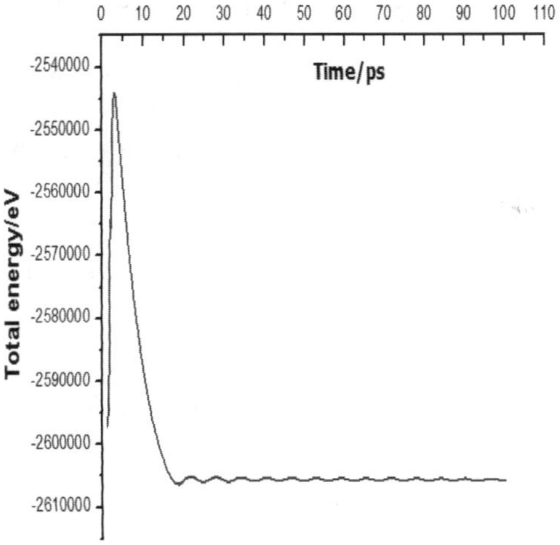

Figure 3. Energy changes as a function of time for the cell relaxed at 300K.

3.1. Factors affecting the stability

3.1.1. Dependence on size of the oxide. MD simulations are performed by changing the radii of oxide particles as shown in Figure 4. For small oxide particle sizes of 1.5 nm and 2 nm, MD simulations show non-spherical configurations and amorphous-like structures after thermal relaxation at 300K. At a radius of 2.8 nm, atomic arrangement starts to become spherical and ordered. At larger sizes crystallinity of oxide particles is further improved. Only at the interfaces of oxide particle and Fe matrix highly disorder regions exist. Clearly, the size plays an important role to influence stabilities of oxide particles. This can be understood by a large interface-to-volume ratio of small oxide particles. A defective interface has high energy [13-16]. Large interface-to-volume ratios may increase the system energy of the oxide particle to a level exceeding the critical energy for amorphization [17-19].

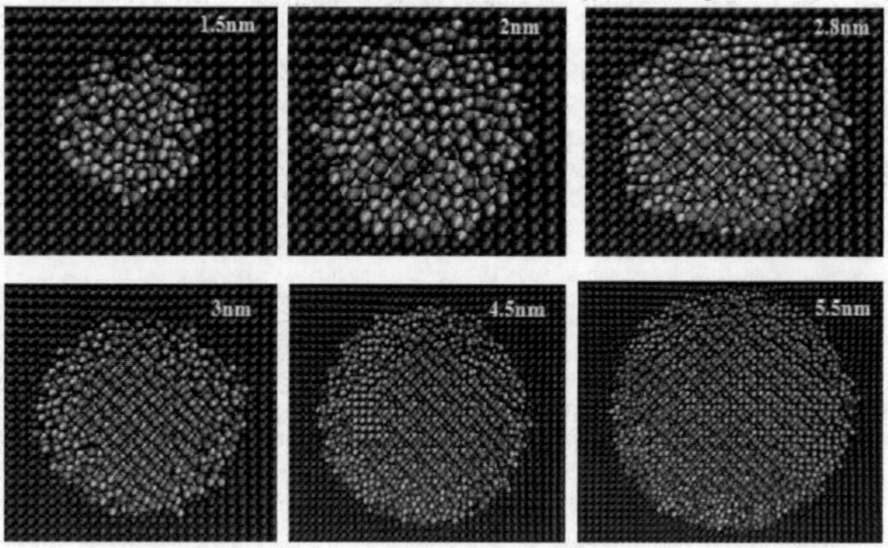

Figure 4. Snapshot of oxide particles of different radii at 300K.

Figure 5 plots the atomic energy per atom as a function of oxide particle sizes. When oxide sizes are 1nm and 1.5nm, the total numbers of Y and O atoms are 287 are 943 and average atomic energies of Y and O atoms are -15.67 eV and -15.37 eV respectively. When oxide size is 2.0 nm, the total number of Y and O atoms is 2227, and the average atomic energy of Y and O is -0.0023 eV. At 2.5nm, the total number of Y and O atoms is 4,392, and the average atomic energy of Y and O is -0.0028 eV. At 2.8nm, the total number of Y and O atoms is 6132, and the average atomic energy of Y and O atoms is -14.93 eV. At 3nm, the total number of Y and O atoms is 7584, and the average atomic energy of Y and O atoms is -14.96 eV. Our modeling shows that at sizes at 2.8 nm and larger, average atomic energy reaches saturation.

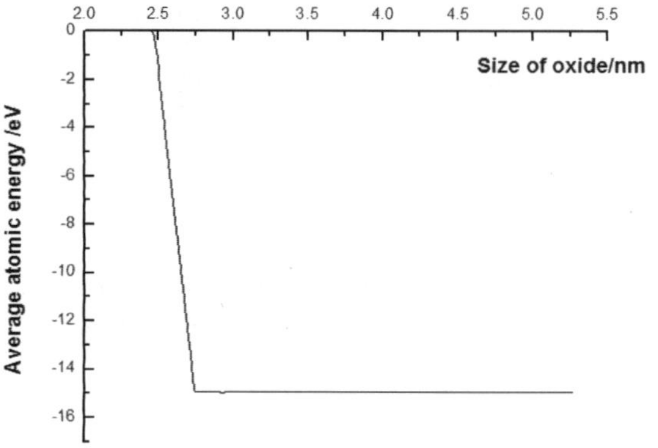

Figure 5. Variation of average atomic energy as a function of oxide particle sizes.

IOP Conf. Series: Materials Science and Engineering **389** (2018) 012003 doi:10.1088/1757-899X/389/1/012003

3.1.2. Influence of charge. Under the same relaxation temperatures, we further tested the effect of Y/O charge ratios to the oxide particle stability. For charges of Zy=2.55 (for Y) and Zo=-1.7 (for O), the oxide particle shows instability and the generation of cavity near the interface (as shown in Figure. 6a). In a comparison, for charge values of Zy=1.66 and Zo=-1.11,relaxed structure looks stable without any cavity as shown in Figure. 6b [20,21].

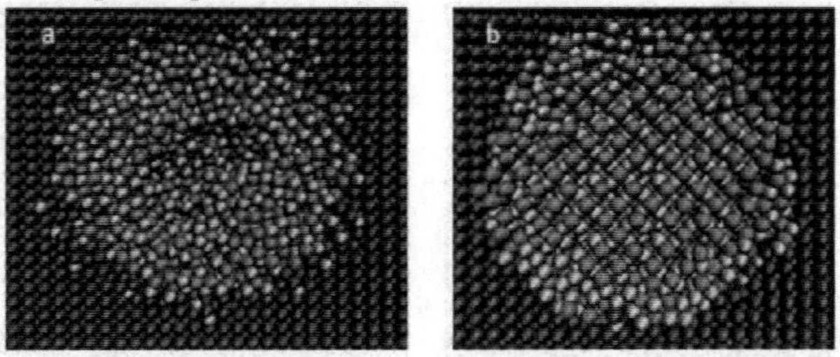

Figure 6. Snapshot of a 2.8nm oxide at 300Kwith different Y/O charges.

3.1.3. Influence of temperature. Due to the importance of temperature in reactor application, we further tested stabilities of oxide particle of 2.8 nm size at 300K, 900K, 1200K, and 1500K. The charge values of Zy=1.66 and Zo=-1.11 are used for the best stabilities. Figure 7 compares the oxide particle morphologies as a function of temperatures.

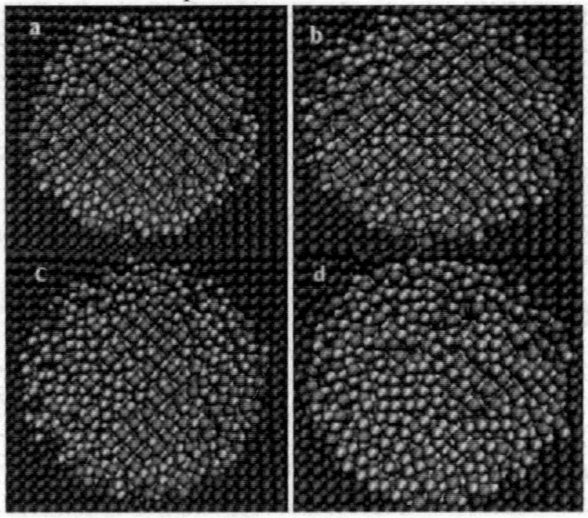

Figure 7. Structural morphologies of oxide particle of 2.8nm size at temperatures of (a) 300K (b) 900K (c) 1200K and (d) 1500 K respectively.

The critical stability sizes are found to increase with increasing temperature. Figure 8 shows structural morphologies of oxide particles of 2.5nm, 2.8nm, 3nm respectively, at 900K. It shows that at the size of 3nm (Figure. 8c) oxide particle still maintain crystallinity.

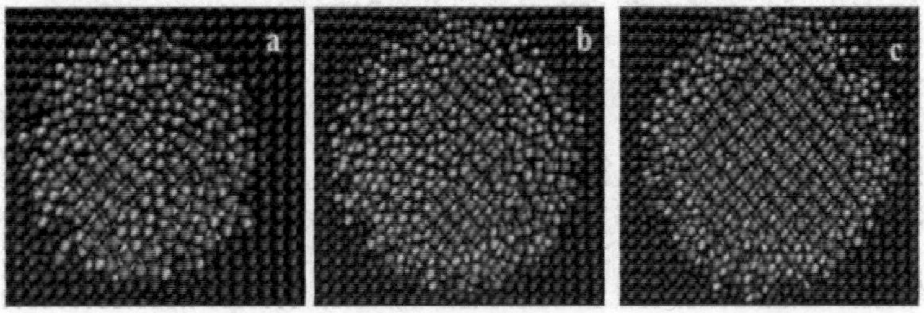

Figure 8. Structural morphology for oxide with sizes (a) 2.5nm, (b) 2.8nm, (c) 3 nm at 900K.

Figure 9 plots the critical stability sizes as a function of temperatures. The higher the temperature, the larger the size required to reach stability. For realistic applications which involves either high temperatures during the materials processing or high temperatures during operations, it is needed to consider this size effects and introduce stabilize particles. On the other hand, we hypothesize that amorphous or unstable oxide particles may get dissolved into the matrix and promote nucleation of new oxide particles and growth of these particles into the favorable size ranges.

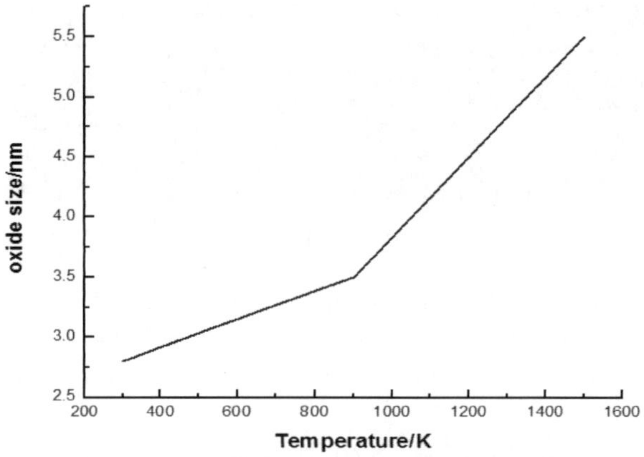

Figure 9. Critical sizes for structural stability as a function of temperature.

4. Conclusion

Through systematic modeling of oxide particles in Fe under various temperatures and as a function of sizes, we found that there is an existence of critical sizes below which oxide particles will become unstable and amorphized. This critical stability size increases with increasing temperatures. The oxide particle stability is sensitive to charge/ionization levels of Y and O atoms. There is an existence of optimized charge ratios for obtaining structural stability.

Acknowledgments

The authors greatly appreciate the financial support from the International Exchange Program of Harbin Engineering University P.R China & Chinese Scholarship Council (CSC) for their international scholarship program. Qingyu Wang acknowledges the support from National Science Foundation of China (Grant No.11505037).

References

[1] Muroga T, Gasparotto M and Zinkle S J 2002 *Fusion Eng. and Des.* 61–62 13–25.
[2] Yvon P and Carré F 2009 *J. Nucl. Mater*.385 217–22.
[3] Zinkle S J and Busby J T 2009 *Mater. Today* 12 12–19.
[4] Yvon P, Le Flem M, Cabet C and Seran J L 2015 *Nucl. Eng. Des*.294 161–9.
[5] Azevedo C R F 2011 *Eng. Fail. Anal*.181943–62.
[6] Murty K L and Charit I 2008 J. *Nucl. Mater*. 383 189–95.
[7] Klueh R L,Shingledecker J P,Swindeman R W,Hoelzer D T 2005 *J. Nucl. Mater*.341 103–14
[8] Klueh R L and Nelson A T 2007 *J. Nucl. Mater*.371 37–52.
[9] Plimpton S 1995 *J. Chem. Phys*. 1171–19.
[10] Hammond K D,Lee Voigt H-J,Marus L A,Juslin N and Wirth B D 2013 *J.Phys:Condens.Matter* 25 55402–13.
[11] Biersack J P and Ziegler J F 1982 *Nucl. Instruments Methods* 194 93–100.
[12] Stukowski A 2010 *Simul.Mater. Sci. Eng.*18 15012.
[13] Lu C et al 2016. *Sci. Rep*. 6 19994.
[14] Brodrick J, Hepburn D J and Ackland G J 2013 *J. Nucl. Mater* 445 291–7.

IOP Conf. Series: Materials Science and Engineering **389** (2018) 012003 doi:10.1088/1757-899X/389/1/012003

[15] Backhaus-Ricoult M 2003 *Physical Chemistry Chemical. Phys*. 5 2174–82.
[16] Becquart C S and Domain C 2011 *Metall. Mater. Trans. A*42 852–70.
[17] S. Rogozhkin et al 2016 *Nucl. Mater. Energy* 9 66–74.
[18] Beyerlein I J, Demkowicz M J, Misra A and Uberuaga B P 2015 *Prog. Mater. Sci.*74 125–210
[19] Demkowicz M J, Bellon P and Wirth B D 2010 *MRS Bull*. 35 992–8
[20] A M Mustafa, Z Li, Lin Shao 2017 *ASME 25th Int. Conf.on Nuclear Eng. ICONE-25* (China) 5 1-5.
[21] M M Azeem, Z Li, Q Wang, Q M N Amjad, M Zubair and O M H Ahmed 2018 *IEEE* 15th*Int. Bhurban Conf. on Applied Sci. & Tech.IBCAST* (Pakistan) 12–15.

Thermal Properties Analysis of Compact Motorized Spindle Considering Fluid-Solid Thermal Coupling

Li Cui and Qingsheng Wang

College of Engineering, Shanghai Polytechnic University, Shanghai, China

cuili@sspu.edu.cn

Abstract. A new model for thermal properties analysis of compact motorized spindle bearing system is constructed. Heat generation model of ball bearing and compact spindle are given. Considering fluid-solid thermal coupling, a new model of temperature analysis of compact motorized spindle bearing is constructed by using moving heat sources method. Calculation model considering fluid-solid thermal coupling is given. Based on the model and finite element method, heat generation and temperature rise of a high-speed compact motorized spindle bearing system is studied. Temperature rising analysis have been done under different conditions, results show that temperature decreases when coolant flow becomes faster, preload of bearing decreases. It can be found that the maximum heat generation and temperature rise of the spindle system located on the rotor surface. When fluid-solid thermal coupling effect is considered, temperature rising becomes lower than without coupling, and the difference becomes more obvious with the increase of rotation speed of bearing. Experiment proves the accuracy of temperature calculation results considering fluid-solid thermal coupling effect. Fluid-solid thermal coupling should be considered in the calculation for predicting the thermal characteristics of high speed compact spindle bearing system. The study will provide evidence for analyzing thermal characteristics of high speed compact spindle bearing system.

1.Introduction

Compact motorized spindle is used widely in high speed machine center, which has a built-in motor rotor and it is directly involved in processing, therefore, temperature rise in the operation process will affect dynamic and thermal characteristics of spindle, which will directly affect the machining quality of products. Compare with traditional machine tool spindle, heat production and heat transfer of high speed compact motorized spindle system are more complex, therefore, thermal properties analysis of high speed electric spindle has attracted substantial attention because of heat stability effects.

Jeng[1] studied a infinite surface heat transfer method to analyze temperature field of bearing system. But the study only considered the gyro heat. Tu[2] investigatedoperating temperatures of a rolling bearing which combines dynamics calculation with finite element thermal analysis method, which can predict transient temperature responses of the bearing but the calculation cost too much time. Kim[3] reviseda friction heat equation of ball bearing through the temperature rising test of high speed ball bearings, results showed that temperature prediction by the transient model is much more accurate than the steady model. Bossmanns[4] constructed a thermal analysis model of the high speed motorized spindle, the heat transfer mechanism of the motorized spindle was studied. Creighton[5] proposed a method of thermal displacement compensation for spindle, in which was used to reduce the machining error caused by heat. Mizuta[6] studied heat transfer characteristics of the angular contact ball bearing using finite element method. However, it cannot consider the local heat in the contact area of the bearing.In order to obtain more accurate temperature results, this paper will present a local friction model to construct the heat generation model of the roller bearing.

This paper present a new model of thermal analysisof high speed motorized spindle bearing system under fluid-solid thermal coupling.The influence of fluid-solid thermal coupling has been contrasted. The study will provide evidence for analyzing thermal characteristics of high speed compact spindle bearing system.

2. Heat generation model

A new heat generation of high-speed ball bearing is proposed by analyzing the local friction in the bearing. The friction power of high-speed ball bearings include: sliding friction power between balls and raceway, sliding friction power between cage and guide surface of rings, flow resistance power caused by balls move in the lubricant. Quasi-dynamic equations of bearing can be constructed based on Hertz contact theory. Displacements of bearing is obtained from the dynamic model of spindle bearing system, Newton-Raphson method and Runge-Kutta method are used to solve these equations, then, contact results and kinematic parameters of ball bearing can be obtained[7].

Contact force acting on dAcan be defined as

$$\sigma = \frac{3Q}{2\pi ab}\left[1-\left(\frac{x}{a}\right)^2-\left(\frac{y}{b}\right)^2\right]^{0.5} \tag{1}$$

Friction force and friction moment can be given as

$$dF = \frac{3Q\mu}{2\pi ab}\left[1-\left(\frac{x}{a}\right)^2-\left(\frac{y}{b}\right)^2\right]^{0.5} dA \tag{2}$$

$$M_s = \frac{3Q\mu}{2\pi ab}\iint_{\Omega}(x^2+y^2)^{0.5}\left[1-\left(\frac{x}{a}\right)^2-\left(\frac{y}{b}\right)^2\right]^{0.5} \cos(\phi-\theta)dA \tag{3}$$

Power loss of the bearing can be calculated based on the friction force and friction moment, and the whole power consumption can be obtained by the sum of every local heat source[8].

The compact motorized spindle is different from the traditional machine spindle, the motor of the electric spindle is built in the inner of the spindle housing. There are three main losses in the loss of the motor, mechanical loss, electric loss andmagnetic loss.

The mechanical power loss are caused by the friction of stator, rotor of the motor and air. The power loss can be written as

$$P_n = \pi C\rho\omega^3 R_r{}^4 L_r \tag{4}$$

Where, P_n is mechanical power loss, C is friction coefficient, ρ is density of air, ω is rotating speed of rotor, R_r , L_r are radius and length of the rotor.

The electric power loss of the motor is caused by coil of the stator, which can be obtained from

$$P_e = I^2 R_c \tag{5}$$

Where, P_e is electric power loss, I is electric current, $R_c$is electric resistance.

Magnetic power loss is caused by the alternation of the magnetic field in the motor core. It includes eddy current power loss and hysteresis power loss.

Hysteresis power loss can be written as

$$P_t = C_t f_t B_{max}{}^2 \tag{6}$$

Where, $P_t$is hysteresis power, $C_t$is a constant value, $f_t$is magnetization frequency, B_{max} is maximum magnetic induction intensity.

Eddy current power loss can be calculated as

$$P_o = \frac{\pi^2 t_s{}^2 (f_t B_{max})^2}{6\rho_s r_c} \tag{7}$$

Where, $P_o$is eddy current power, $t_s$is thickness of silicon steel sheet, ρ_s is resistivity of iron core, $r_c$is density of iron core.

Heat generation rate of the motorized spindle is defined as power loss every unit volume, which can be obtained by

$$q_s = \frac{p_n + p_e + p_t + p_o}{\frac{\pi}{4}\left(D_o^{\,2} - D_i^{\,2}\right)L_s} \tag{8}$$

Where, D_o, D_i,L_s are outside diameter, inner diameter, and length of the stator.

3. Modeling and Calculating method

Taking an compact spindle bearing system as an example, the rotating speed is 10000rpm, power of the spindle is 30kW, 7010contact angle ball bearing is used as front bearing, and 7008 is used as back bearing. The power of the motor is 30kW, the power loss of each part of the motorized spindle is calculated as Table 1.

Table 1. Power loss of the spindle bearing system

Location	Power loss (W)
Front bearings	903
Back bearings	293
Stator	1033
Rotor	1866

The results of the heat transfer coefficient of the main parts of the motorized spindle are shown in Table 2.

Table 2. Heat transfer coefficient of high speed motorized spindle

Location	Heat transfer coefficient w/(m^2·k)
Spindle and air	205
Housing and air	10
Rotor and air	228

3D model of compact spindle ball bearing system is established in Solidworks software, fluid and solid regions of bearing are defined and meshed in ICEM software. Convection heat transfer coefficient are calculated and set in FLUENT[9]. Figure 1 gives calculation flow chart considering fluid-solid thermal coupling.

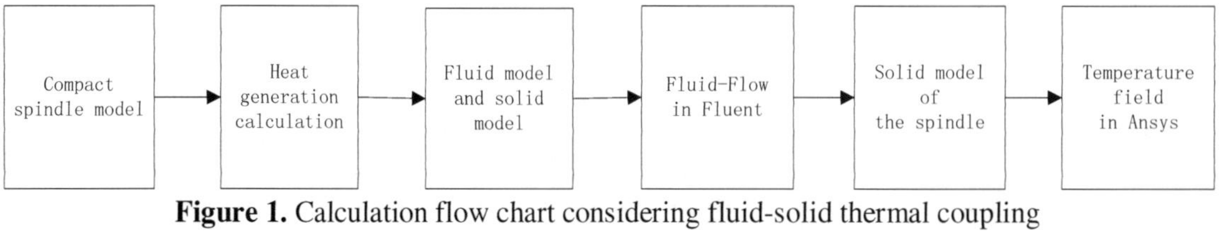

Figure 1. Calculation flow chart considering fluid-solid thermal coupling

Contacting surface between the internal fluid and solid parts is set as default coupling-surfacein order to consider fluid-solid thermal coupling effect, and the convection heat transfer is calculated automatically. Velocity boundary is set at the inlet and velocity is chosen as 50 m/s. Finite difference method is used to discrete continuous equation. The RNG two-equation model is used in the calculation[10]. SIMPLEC algorithm is used to simulate the coupling of pressure and velocity. Solid model of compact spindle bearing system is constructed in ANSYS software, heat transfer coefficient is applied to the model, which is shown in Table 3.

Table 3. Heat transfer coefficient of high speed motorized spindle

Location	Heat transfer coefficient w/(m^2 ·k)
Stator and cooling sleeve	1353.1~1732.5
Front bearing and cooling	1685.8~1998
Rear bearing and cooling	761.6~808.8

IOP Conf. Series: Materials Science and Engineering **389** (2018) 012004 doi:10.1088/1757-899X/389/1/012004

4. Results and Discussions

According to the above modelling, thermal properties considering considering fluid-solid thermal coupling are analyzed based on FLUENT and ANSYS software. Fluid field and temperature field distribution of the spindle bearing system are obtained.

Assuming rotating speed is 10000r/min, fluid field of coolant and temperature filed results of the spindle are given in Figure 2. The temperature near the entrance of coolant is relatively low, while near the outlet temperature is higher. It is shown that when the coolant passes through the channel, the heat of the stator is absorbed by the coolant, and the coolant acts as a cooling agent.

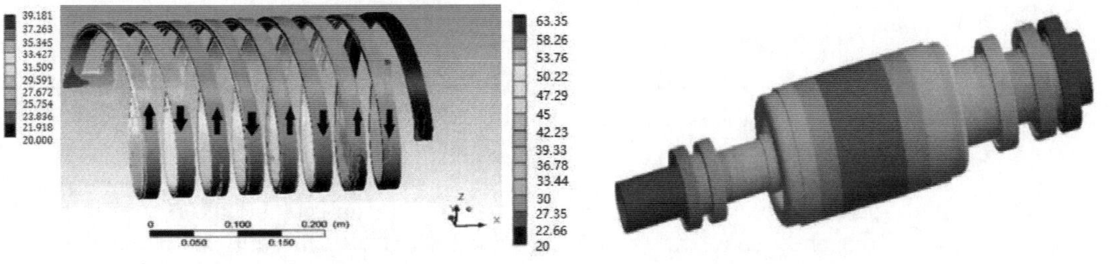

Figure 2. Temperature distribution of the compact spindle

Figure 3 shows the influence of rotating speed on temperature rising of spindle bearing system.

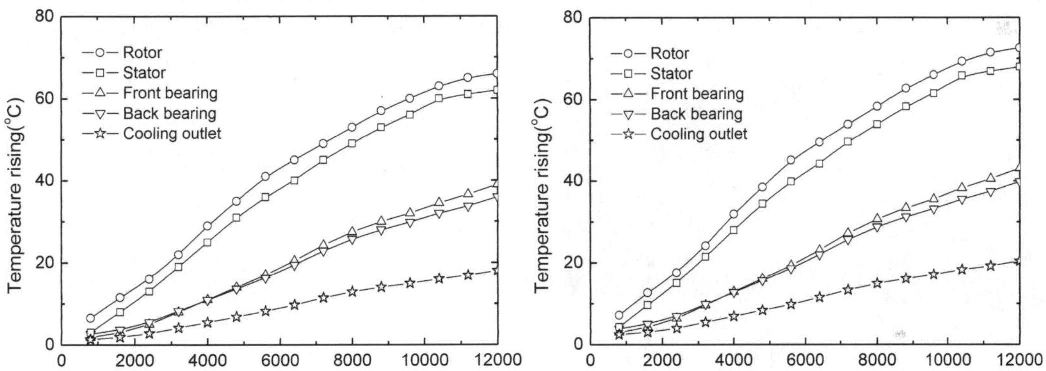

a) with coupling b) without coupling

Figure 3. Effect of rotating speed on temperature rising of spindle bearing system

It can be seen that rotating speed has important influence on the temperature rising. When rotating speed changes from 1000rpm to 12000rpm, temperature rising of rotor, stator, front bearing, back bearing and cooling outlet increase to 65°C, 62°C, 39°C, 37°C, 19°C when considering coupling. While temperature rising of rotor, stator, front bearing, back bearing and cooling outlet increase to 72°C, 68°C, 43°C, 41°C, 21°C when coupling is not considered. It is found that the temperature rising is lower when considering the fluid-solid thermal coupling. The reason is that convective heat transfer is accelerated by the convection and conduction when fluid-solid thermal coupling effect is considered.

Figure 4 and Figure 5 give effect of coolant flowand bearing preload on temperature rising of rotor, front bearing and back bearing. It shows effect of coolant flow on temperature rising is obvious, and effect of preload is small.

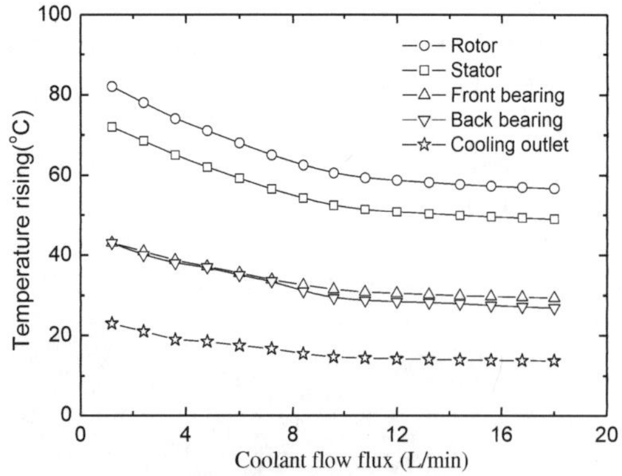

Figure 4. Effect of coolant flow

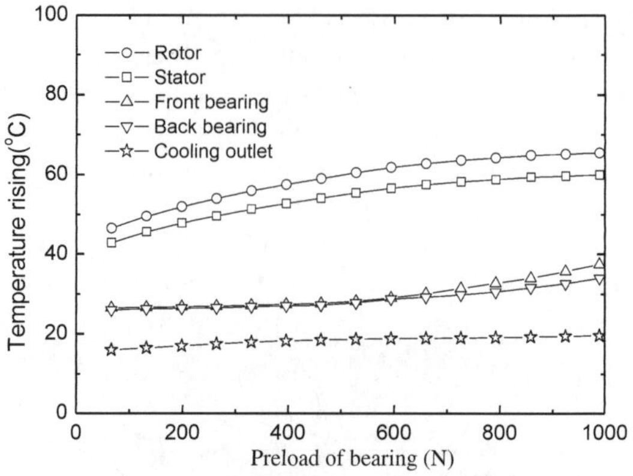

Figure 5. Effect of preload

Figure 6 is a high speed compact motorized spindle test rig under air/oil lubrication, which includes the electric spindle, frequency converter, water cooling device, oil/ air lubrication generator, air compressor, and temperature and vibration test system. The temperature rising of roller bearing can be measured by temperature sensors arranged on the outer sleeve of the electric spindle.

Temperature rising test of high speed compact spindle bearing system have done using this test bench. Figure 7 gives contrast between experimental results and calculation results of temperature rising on outer ring surface of bearing.

Figure 6. Compact motorized spindle test rig

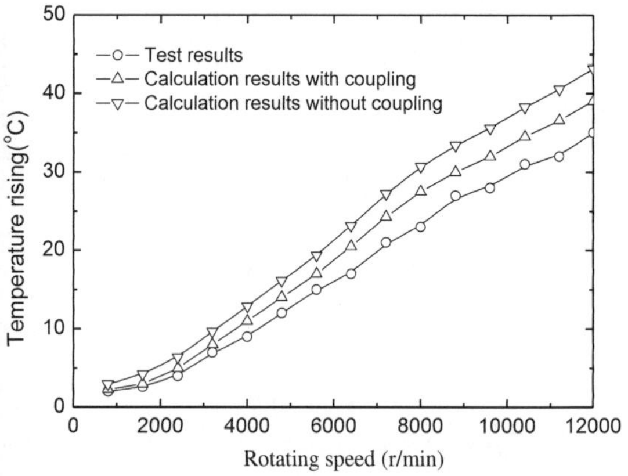

Figure 7. Temperature rising contrast of bearing

It can be seen that bearing surface temperature rising increases gradually with the speed increase. When rotating speed changes from 1000rpm to 12000rpm, the difference between result with coupling and test result changes from 2.3% to 8.7%.While the difference between result without coupling and test result changes from 4.9% to 15.5%. The test results demonstrate that results with fluid-solid thermal coupling are more close to the experimental results, which verify the reliability of the method in this paper.

5. Conclusions

Thermal properties model of the high speed motorized spindle bearing system considering fluid-solid thermal coupling is constructed. Temperature rising analysis have been done under different conditions, results show that temperature decreases when coolant flow becomes faster, preload of bearing decreases. It can be found that the maximum heat generation and temperature rise of the spindle system located on the rotor surface.

When fluid-solid thermal coupling effect is considered, temperature value is lower than that without coupling, and the difference becomes more obvious with the increase of rotation speed of bearing. Experiment proves the accuracy of temperature calculation considering fluid-solid thermal coupling effect. Fluid-solid thermal coupling should be considered in the calculation for predicting the thermal characteristics of high speed compact spindle bearing system.

Acknowledgment
This work was financially supported by the National Natural Science Foundation of China(No. 51675323), Shanghai Natural Science Foundation(No.14ZR1416800), The key subject of Shanghai Polytechnic University (Material Science and Engineering, XXKZD1601).

References
[1] Y. R. Jeng, Prediction of Temperature Rise for Ball Bearings, Tribology Transaction, 46, pp.49-56, 2003.
[2] J. F. Tu, Thermo-Elastic Instability Monitoring for Preventing Spindle Bearing Seizure, Tribology Transaction, 38, pp.11-18, 1995.
[3] Sun Min Kim and Sun Kyu Lee, Prediciton of Thermo-Elastic Behavior in a Spindle-Bearing System Considering Bearing Surroundings, International Journal of Machine Tools & Manufacture, 41, pp. 809-831, 2001.
[4] Bossmanns B, Tu Jufeng, A thermal model for high speed motorized spindles, International Journal of Machine Tools & Manufacture, 39(9), pp.1345-1366, 1999.
[5] Creighton E, Honegger A, Tulsian A, et al. Analysis of thermal errors in a high-speed micro-milling spindle, International Journal of Machine Tools and Manufacture, 50(4), pp.386-393, 2010.
[6] J. R. Brown and N. H. Forster, Operating Temperatures in Lubricated Rolling Element Bearings for Gas Turbines, ASME Journal of Tribology, 12, pp.1268-1275, 2000.
[7] Li Cui, Jianrong Zheng, Nonlinear vibration and stability analysis of a flexible rotor supported on angular contact ball bearings, Journal of Vibration and Control, 20(12), pp. 1767-1782, 2014.
[8] S. Y. JiangandH. B.Mao. Investigationof thehigh speed rolling bearing temperature risewith oil-air lubrication. ASME Journal of Tribology, 2011, 133, 021101-021101.
[9] Li Cui, Yafei He. A new logarithmic profile model and optimization design of cylindrical roller bearing. Industrial Lubrication and Tribology. 2015, 67(5): 498-508.
[10] Li Cui, Wenxia Wang, Zhang Yanlei. Bubble Flow Analysis of High Speed Cylindrical Roller Bearing under Fluid-Solid Thermal Coupling. International Journal of Rotating Machinery. 2017, 8372176

Scheduling on Parallel Machines with Mold Constraints

H D Zhao[1,*], J Gao[1] and F Zhu[1]

[1]China State Shipbuilding Corporation System Engineering Research Institute, Beijing, China

*Email: haidan.zhao@outlook.com

Abstract. This paper deals with makespan minimization scheduling problem on two identical parallel machines with mold constraint. The mold constraint in this problem is described as a kind of resource-constrained. In the production of wafer fabrication, two jobs with same mold cannot be processed on two parallel machines at the same time. This problem is quite common in practice, but there are few literatures relevant to this problem. Since the problem is NP-hard, a mathematical model is discussed to describe and solve the problem and a heuristic is proposed. Computational results show that the proposed heuristic can efficiently obtain good solutions for medium and large size problems.

1. Introduction

This problem arises from a real application in the production of wafer fabrication in semiconductor plant. There are silicon wafers starting out blank and pure. In wafer fabrication, the production process consists of five stages: cleaning of the wafer's surface, applying of the photoresist, exposing of the wafer, striping the resist and etching the pattern. The exposing wafer is one of the most important stages because this step is to fulfill customer needs and the final wafer quality is decided in this step. At the stage of exposure, a kind of patterning tool known as mask usually severs as a mold. They are dismountable and contain patterns which can be transferred to an entire wafer in just a single exposure. According to the different customer orders, each wafer has a specific mold. In the wafer exposing process, the machine with molds is very expensive and therefore the number of machines is fewer than the number of molds in the semiconductor plant. Thus, molds are not fixed on the machines and should be changed all the time. To increase the efficiency of this stage, the scheduling problem of molds and jobs should be considered.

In this paper, we consider a scheduling problem of a set $N=\{1,2,...,n\}$ where there are n jobs on two identical parallel machines named M_1 and M_2. There are q mold types in the problem. Each job with a job processing time p_i where $i=1,2,...,n,$ and corresponding mold m_i where $i=1,2,...,n,$ should be processed on one of two identical parallel machines. Let S_k ($k=1,2,...,q$) be the set where jobs belong to the same mold type and V_k be the total processing time of S_k. It is noted that $V_1 \le V_2 \le ... \le V_q$ in this paper. Each machine can install at most two molds and remove them when necessary. Any mold on the machine cannot be replaced when job is processing and two jobs with the same mold cannot be processed at the same time as well. The mold replacement time is not considered in this paper. Each machine can process at most one job at a time. The job preemption, division, or cancellations are not allowed. The objective is to minimize the makespan.

Following the three-field notation, the problem can be denoted as $P_2 | m_i | C_{\max}$, where P_2 designates the two identical parallel machines, m_i represents the mold constraint for job i, and $C_{\max}$ denotes the maximum completion time (or makespan). The $P_2 | m_i | C_{\max}$ problem is NP-hard since the problem without mold constraint is known to be NP-hard and proved by Sethi[1] and Garey and Johnson[2].

In what follows, we review the literature related to the $P_m || C_{\max}$ problem. There are several exact algorithms which can obtain the optimal solutions. Dell'Amico and Martello [3] presented a branch and bound algorithm based on sophisticated lower and upper bound computations and the dominance properties. An exact cutting plane algorithm based on the identification of valid inequalities was proposed by Mokotoff[4]. Dell'Amico *et al.*[5] presented an exact algorithm, which is based on a specialized binary search and a branch-and-price scheme. However, the exact algorithm can only be used for the small or medium scale problems. Many heuristics and meta-heuristics have developed for the $P_m || C_{\max}$ problem. Graham[6] conducted the worst case analysis of the Longest Processing Time first (LPT) rule. In addition, they also provided a worst case analysis of an arbitrary list schedule for $P_m || C_{\max}$. A more sophisticated heuristic for $P_m || C_{\max}$, called MULTIFIT, was presented by Coffman et al[7]. The MULTIFIT algorithm based on techniques from bin-packing has a bound of 1.22. Then Friesen[8] improved the bound of this algorithm upon to 1.2. In addition, Lee and Massey[9] analyzed a heuristic combined by the LPT rule and the MULTIFIT algorithms. They used the result of LPT rule as the incumbent and then applied MULTIFIT algorithms with less iteration. Gupta and Ruiz-Torres[10] presented a new heuristic based on bin-packing and list scheduling, called LISTFIT.

Besides, metaheuristics have been applied in the $P_m || C_{\max}$ problem. Liu and Cheng[11] proposed a kind of genetic algorithm (GA) to minimize makespan for an identical machine scheduling problem. They provided several different scale numerical examples to demonstrate that the genetic algorithm is efficient. Lee *et al.*[12] proposed a Simulated Annealing (SA) approach with hill-climbing moves to tackle the makespan problem on identical parallel machines. Davidović *et al.*[13] studied the static scheduling of independent tasks on homogeneous multiprocessor systems and proposed a Bee Colony Optimization (BCO) algorithm. The BCO algorithm belongs to the class of stochastic swarm optimization methods inspired by the foraging habits of bees in nature. Tian *et al.*[14] presented a Discrete Particle Swarm Optimization (DPSO) algorithm to solve the two stage assembly scheduling problem where the first stage is a workstation of several identical parallel machines and the second stage is a single assembly machine workshop. The results show that DPSO is an effective and efficient for assembly scheduling problem. The DPSO algorithm is also extended to solve our proposed problem due to its good ability to search solution.

We also have reviewed some related references. According to the description of limited resource in the relevant literature, the mold constraints in this problem can be described to a kind of resource constraints. In practice, the resource constraints can be defined as the constraints for jobs handling and processing, such as automated guided vehicles, machine operators, tools, pallets, fixtures, industrial robots and so on. In this paper, the mold used in the wafer process can be as a tool which precludes the job processed with freedom. Hence, we can take this problem as a kind of Resource Constrainted Parallel Machine Scheduling (RCPMS) problem. The RCPMS problem was first studied by Garey and Graham[15] who show that greedy list schedulers can obtain an approximation ratio of $3-m/3$. Garey and Johnson[16] examined the computational complexity of scheduling problems associated with a model. They have proved the RCPMS problem as a NP-hard problem. Blazewicz *et al.*[17] considered the same problem with nonpreemptable jobs and an $O(n^3)$ algorithm was given. Blazewicz *et al.*[18] and Reklaitis[19] gave a comprehensive review of the resource constrainted problem, including applications in the chemical processing industry, semiconductor manufacturing and so on. Daniels *et al.*[20] provided mathematical formulations for static and dynamic versions of the same problem. The characteristic of the problem is the dependence of the processing time on the additional resource allocated. Daniels and Hua[21] extend the formulation of the RCPMS problem for the case where the

IOP Conf. Series: Materials Science and Engineering **389** (2018) 012005 doi:10.1088/1757-899X/389/1/012005

job assignment to the machines is unspecified. They referred to this problem as the RCPMS problem and stated it as a NP-hard problem. More recently, Emrah *et al.*[22] reviewed the literature related RCPMS problem and presented integer programming models for two problems. Niemeier and Wiese[23] studied the identical parallel machine problem with an orthogonal resource constraint and present an algorithm with an approximation ratio of $2+\in$.

This paper is organized as follows. In next section, a mathematical model and a lower bound are presented for the $P_2 | m_i | C_{\max}$ problem. In Section 3, a heuristic is developed. The computational results are summarized in Section 4. Finally, we conclude the paper with a summary discussion on the further research directions in Section 5.

2. A mathematical model for $P_2 | m_i | C_{\max}$

The proposed problem is to schedule jobs with the mold constraints on two identical parallel machines to achieve the aim of minimizing the makespan. We now discuss a mathematical model to describe and solve the $P_2 | m_i | C_{\max}$ problem. We assume here that the mold types are described as different positive integers a_i, i.e. if the mold types of two jobs are the same, the difference between them is zero. The decision variables are defined as follows:

$$y_h = \begin{cases} 1, & \text{if job } h \text{ is the first job on one of the machines,} \\ 0, & \text{otherwise,} \end{cases}$$

$$x_{ih} = \begin{cases} 1, & \text{if job } i \text{ is scheduled directly before job } h, \\ 0, & \text{otherwise,} \end{cases}$$

$$x_{in+1} = \begin{cases} 1, & \text{if job } i \text{ is the last job on one of the machines,} \\ 0, & \text{otherwise,} \end{cases}$$

Let R is a sufficiently large number and the $P_2 | m_i | C_{\max}$ problem can be formulated as the follows:

$$\text{Min} \quad C_{\max} \tag{1}$$

s.t.

$$\sum_{h=1}^{n} y_h \le 2 \tag{2}$$

$$y_h + \sum_{i=1, i\ne h}^{n} X_{ih} = 1 \qquad \forall h = 1,\dots,n \tag{3}$$

$$\sum_{i=1, i\ne h}^{n+1} X_{hi} = 1 \qquad \forall h = 1,\dots,n \tag{4}$$

$$C_h \ge p_h y_h \qquad \forall h = 1,\dots,n \tag{5}$$

$$r_{ih} \ge a_i - a_h \qquad \forall i,h = 1,\dots,n \tag{6}$$

$$r_{ih} \ge a_h - a_i \qquad \forall i,h = 1,\dots,n \tag{7}$$

$$C_h \ge C_i + p_h - R r_{ih} \qquad \forall i,h = 1,\dots,n, i \ne h \tag{8}$$

$$C_h \ge C_i + p_h - R(1 - x_{ih}) \qquad \forall i,h = 1,\dots,n, i \ne h \tag{9}$$

$$\sum_{i=1}^{n} X_{in+1} \le 2 \qquad \forall i = 1,\dots,n \tag{10}$$

$$x_{ih} + x_{hi} \leq 1 \qquad \forall i,h = 1,\ldots,n, i \neq h \tag{11}$$

$$y_i + \sum_{h=1, i\neq h}^{n+1} X_{ih} \leq 2 \qquad \forall i = 1,\ldots,n \tag{12}$$

$$C_{\max} \geq C_h \qquad \forall h = 1,\ldots,n \tag{13}$$

Equation (1) represents the objective function to minimize makespan. Constraint (2) ensures that at most two jobs may be the first job on the two machines. In constraint (3), each job must either start on one of the machines or be preceded by some other jobs. Similarly, constraint (4) presents that each job must either be succeeded by another job or be the last job on one of the machines. Constraint (5) guarantees that the completion time for the first job of each machine must be equal to or greater than its processing time. Constraints (6-8) ensure that two jobs with the same mold type cannot be processed at the same time. In constraint (9), the completion time is defined as the completion time of the predecessor job and its processing time. Constraint (10) states that at most two jobs can be the last jobs on the two machines. Constraint (11) presents that sequence of jobs is fixed by decision variables. Constraint (12) ensures that a job start on one of the machines should also be preceded by some other jobs. Constraint (13) defines $C_{\max}$ as the makespan.

2.1. Lower bound

A lower bound is introduced and evaluated with the following heuristics for the $P_2 \mid m_i \mid C_{\max}$ problem. Mokotoff [4] provided a mixed integer programming (MIP) formulation for the $P_m \parallel C_{\max}$ problem which is slack for $P_2 \mid m_i \mid C_{\max}$ problem. Thus, the optimal solution of the $P_m \parallel C_{\max}$ problem is used as a lower bound, named LB, for the proposed problem. The MIP formulation for the $P_m \parallel C_{\max}$ problem is as follows:

Min $C_{\max}$

s.t.

$$\sum_{i=1}^{m} x_{ij} = 1 \qquad 1 \leq j \leq n \tag{14}$$

$$C_{\max} - \sum_{j=1}^{n} p_j x_{ij} \geq 0 \qquad 1 \leq i \leq m \tag{15}$$

$$x_{ij} \in \{0,1\} \qquad 1 \leq i \leq m, 1 \leq j \leq n \tag{16}$$

where,

$$x = \begin{cases} 1 & \text{if job } j \text{ is assigned to machine } i \\ 0 & \text{otherwise} \end{cases}$$

Example 2. Consider a $P_2 \parallel C_{\max}$ problem with job processing time given in Table 1. Applying the MIP formulation in a C++ application with Cplex and the optimal solution is 22.

Table 1: The processing time and mold type of Example 1

Job	1	2	3	4	5	6	7	8	9	10
m_i	B	D	A	A	D	B	B	D	C	C
a_i	2	4	1	1	4	2	2	4	3	3
p_i	1	7	2	5	7	1	8	2	2	8

3. The proposed heuristic

In this section, two conditions are provided and one of them is a special case which can be solved easily. For the other conditions, a heuristic based on the Longest Processing Time (LPT) rule are proposed to solve the $P_2 \mid m_i \mid C_{\max}$ problem.

The first condition is a special case presented as follows.

Lemma 1. If $V_1 + V_2 + \cdots + V_{q-1} \leq V_q$, there exists an optimal schedule for the $P_2 \mid m_i \mid C_{\max}$.

Proof. As we have emphasized, two jobs with the same mold cannot be processed on the machines at the same time. Thus, V_q is the maximum completion time, shown in Figure 1.

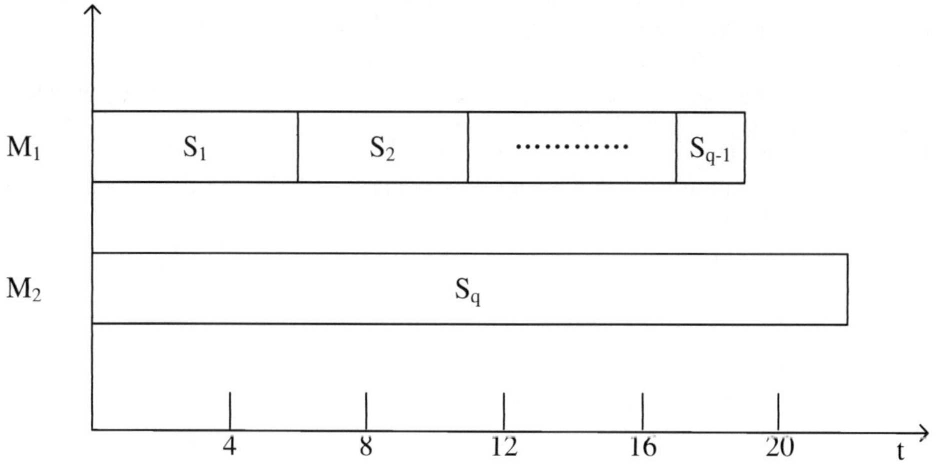

Figure. 1. Gantt chart for Lemma 1.

If $V_1 + V_2 + \cdots + V_{q-1} \geq V_q$, two proposed heuristics are proposed in second condition. The basic idea of two heuristics is extended by LPT rule as a list scheduling method where n jobs are sorted in descending order of processing time. Whenever a machine is available, an unscheduled job with largest processing time will be assigned to this machine. The heuristic is presented as follows.

3.1. Heuristic TLPT

The proposed heuristic can be introduced as a two stage procedure. At the first stage, jobs with the same mold can be grouped as a job set S_k where $k = 1, 2, ..., q$. The job sets, except set S_q with largest total processing time, can be assigned by LPT to the machines. At the second stage, the unscheduled jobs belonging S_q can be arranged by LPT rule to the front of one machine and on the back of the other, alternatively. Let MT_i where $i = 1, 2,$ be the completion time of two machines.

The steps of the Two-stage LPT (TLPT) algorithm can be given as follows.

Step 1. Let jobs with the same mold group into a job set S_k. Calculate the total processing time V_k where $k = 1, 2, ..., q$.

Step 2. Select all the job sets except S_q and assign the job sets by LPT rule to the machine with earliest start time. Obtain MT_1 and MT_2.

Step 3. A dummy can be calculated by $|MT_1 - MT_2|$. Arrange remaining jobs by LPT to the machine with shortest machine completion time until the total processing time of those jobs is equal or larger than dummy.

Table 2: Computational results of PD and CPU time

		Percentage Deviation						CPU time	
		(1,10)			(1,100)			(1,10)	(1,100)
q	n	TLPT	VNS	DPSO	TLPT	VNS	DPSO	TLPT	TLPT
4	20	2.24	0.69	0.69	2.03	0.53	0.53	0.07	0.07
	50	0.59	1.25	2.42	0.86	0.66	2.49	0.17	0.16
	100	0.14	4.14	5.26	0.25	4.12	5.88	0.32	0.31
	150	0.19	6.70	9.11	0.11	6.65	8.05	0.51	0.47
	Ave.	0.79	3.19	4.37	0.81	2.99	4.24	0.27	0.25
6	20	1.08	0.00	0.00	1.92	0.04	0.04	0.07	0.08
	50	0.19	0.06	0.18	0.42	0.01	0.35	0.18	0.17
	100	0.22	1.35	2.19	0.04	1.35	2.15	0.34	0.31
	150	0.10	2.68	3.90	0.06	3.09	4.31	0.51	0.42
	Ave.	0.40	1.02	1.57	0.61	1.12	1.71	0.27	0.24
8	20	0.86	0.00	0.00	0.59	0.02	0.02	0.09	0.07
	50	0.15	0.00	0.04	0.07	0.02	0.06	0.18	0.17
	100	0.07	0.52	1.06	0.08	0.41	1.03	0.29	0.29
	150	0.00	1.35	2.08	0.02	1.41	2.11	0.46	0.46
	Ave.	0.27	0.47	0.80	0.19	0.46	0.80	0.25	0.25
10	20	1.20	0.21	0.21	0.53	0.09	0.09	0.08	0.07
	50	0.07	0.00	0.00	0.08	0.02	0.04	0.18	0.16
	100	0.08	0.12	0.59	0.02	0.13	0.40	0.34	0.31
	150	0.00	0.62	1.07	0.03	0.76	1.41	0.52	0.46
	Ave.	0.34	0.24	0.47	0.17	0.25	0.48	0.28	0.25
	Agg.	0.45	1.23	1.80	0.45	1.21	1.81	0.27	0.25

Step 4. Assign the unscheduled job with largest processing time to the first position on the machine with shortest machine completion time.

Step 5. Assign the unscheduled job with largest processing time to the last position on the machine with shortest machine completion time.

Step 6. If $S_q = \varnothing$, obtain a makespan value and then stop; otherwise, repeat Steps 4 and 5.

Example 3. Consider the same problem as Example 1.

Step 1. $S_1 = \{j_3, j_4\}$, $S_2 = \{j_1, j_6, j_7\}$, $S_3 = \{j_9, j_{10}\}$ and $S_4 = \{j_2, j_5, j_8\}$. $V_1 = 7$, $V_2 = 10$, $V_3 = 10$, $V_4 = 16$.

Step 2. S_1, S_2 and S_3 are on M_1 and M_2, respectively.

Step 3. $|MT_1 - MT_2| = 7$. Assign job 5 and job 8 on M_2.

Step 4. Assign job 2 to the first position on M_1 and $S(q) = \varnothing$.

Step 5. $S_q = \varnothing$.

Step 6. $S_q = \varnothing$. The makespan value is 24 and stop

The solution of TLPT is 24 while the optimal solution obtained by MIP is 22. The ratio of ω'/ω is 1.09. The sequences on two identical parallel machines are $\{7,6,1,3,4,2\}$ and $\{5,8,10,9\}$. The Gantt chart is presented as Figure 2.

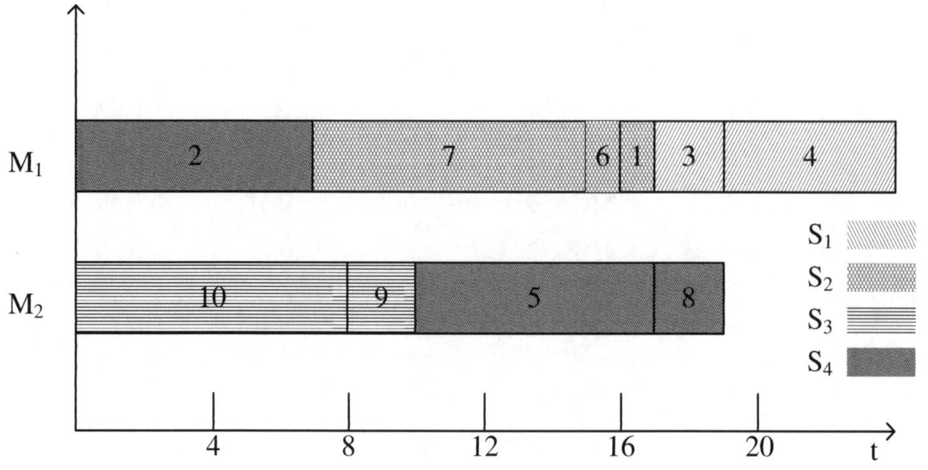

Figure 2. Gantt chart for TLPT.

4. Computational results

To verify the proposed heuristics, extensive computational experiments were conducted. The lower bound was coded in combination of C++ and CPLEX 9.0, and all other algorithms were coded in C++. The algorithms were run on the Intel Core 2 CPU (2.2 GHz) with 2.0 GB RAM.

The test problem was generated with $n = 20, 50, 100, 150$ and $q = 4, 6, 8, 10$. The processing time p_i was given randomly from uniform distributions $U(1,10)$ and $U(1,100)$ respectively. The Discrete Particle Swarm Optimal (DPSO) algorithm by Liao *et al.*[24] and Variable Neighborhood Search (VNS) algorithm by Mladenović and Hansen[25] are used as a comparison. Each of the problems was run ten times. The percentage deviation (PD) is computed as follows:

$$PD = 100 \times \frac{C_{\max}^{H} - LB}{LB} \tag{22}$$

where $C_{\max}^{H}$ is the makespan obtained by the respective heuristics and LB is the lower bound acquired by the MIP formulation. It is noted that the stop criterion of DPSO and VNS is time limited within 10 seconds.

Examining the Table 2, we can observe that TLPT have the performance with $PD \approx 0.45$. Moreover, TLPT has better performance when $q = 4$ and 6. The average PD is getting smaller as n is increasing. Whenever p_i is from uniform distributions $U(1,10)$ or $U(1,100)$, the same situation can be generalized and the heuristic can obtain the solution within 1.

In the Table 2, We can also observe that only when the job number is small, i.e., $n = 20$, VNS and DPSO perform better than the proposed heuristics. It is because the proposed heuristics based on LPT rule perform worse as the job decrease. We also can observe that VNS performs better than DPSO for all problems and the percentage deviation tends to decrease as the mold number and the job number increases. Thus, only VNS is used in the following comparison.

5. Conclusions and future research

IOP Conf. Series: Materials Science and Engineering **389** (2018) 012005 doi:10.1088/1757-899X/389/1/012005

In this paper, a two-identical parallel machine scheduling problem with the mold constraint, i.e., $P_2 \mid m_i \mid C_{\max}$, is considered. This problem is a real problem from the production of wafer fabrication. The objective function is to minimize makespan. A mathematical model and a lower bound are proposed to descript and evaluate the problem. A heuristic is provided for finding the solution. In addition, the Discrete Particle Swarm Optimal (DPSO) algorithm and Variable Neighborhood Search (VNS) algorithm are used as a comparison. Extensive experiments with different size problems show that the proposed heuristic can efficiently yield better solutions.

The mold constraint is a common practice in industry, but it is seldom discussed in the scheduling literature. Extension of the results to other machine environments will be worthwhile in the future research.

References

[1] Sethi R 1977 Mathematics of Operations Research **2** 320–30
[2] Garey M R and Johnson D S 1979 Computer and intractability: a guide to the theory of NP-completeness San Francisco W H Freeman
[3] Dell'Amico M and Martello S 1995 Operations Research Society of America Journal on Computing **7** 191–200
[4] Mokotoff E 2004 European Journal of Operational Research **152** 758–769
[5] Dell'Amico M, Iori M, Martello S and Monaci M. 2008 *INFORMS Journal on Computing* **20** 333–344
[6] Graham R L 1966 *Bell System Technical Journal* **45** 1563–1581
[7] Coffman E G, Jr, Garey M R and Johnson D S 1978 Society for Industrial and Applied Mathematics Journal on Computing **7** 1–17
[8] Friesen D K 1984 Society for Industrial and Applied Mathematics Journal on Computing **13** 170–181
[9] Lee C Y and Massey J D 1988 *Discrete Applied Mathematics* **20** 233–242
[10] Gupta J N D and Ruiz-Torres A J 2001 *Production Planning & Control* **12** 28–36
[11] Liu M and Cheng W. 1999 Artificial Intelligence in Engineering **13** 399–403
[12] Lee W C, Wu C C and Chen P 2006 The International Journal of Advanced Manufacturing Technology **31** 328–334
[13] Davidović T, Šelmić M, Teodorović D and Ramljak D 2012 *Journal of Heuristics* **18** 549–569.
[14] Tian Y, Liu D, Yuan D and Wang K 2012 The International Journal of Advanced Manufacturing Technology **66** 481–499
[15] Garey M R and Grahams R L 1975 Society for Industrial and Applied Mathematics Journal on Computing **4** 187–200
[16] Garey M R and Johnson D S 1975 Society for Industrial and Applied Mathematics Journal on Computing **4** 397–411
[17] Blazewicz J, Kubiak W, Rock H and Szwarcfiter J 1987 *Acta Informatica* **24** 513–524
[18] Blazewicz J, Dror M and Weglarz J 1991 *European Journal of Operational Research* **51** 283–300
[19] Reklaitis G V 1996 Batch Processing Systems Engineering **143** 660–705
[20] Daniels R L, Hoopes B J and Mazzola J B 1996 *Management Science* **42** 1260–1276
[21] Daniels R L and Hua S Y 1999 *Computers & Operations Research* **26** 143–155
[22] Emrah B E, Ceyda O and Irem O 2013 *European Journal of Operational Research* **230** 449–463
[23] Niemeier M and Wiese A 2013 *Approximation and Online Algorithms* **7846** 242–256
[24] Liao C J, Tjandradjaja E and Chung T P 2012 *Applied Soft Computing* **12** 1755–1764
[25] Mladenović N and Hansen P 1997 *Computers & Operations Research* **24** 1097–1100

IOP Conf. Series: Materials Science and Engineering **389** (2018) 012006 doi:10.1088/1757-899X/389/1/012006

Finite Element Simulation of Pre-Heating Effect on Melt Pool Size During Micro-Plasma Transferred Arc Deposition Process

Sagar H Nikam[1,*] **and Neelesh K Jain**[2]

[1]Research Scholar, Discipline of Mechanical Engineering, Indian Institute of Technology Indore, Simrol 453 552 (MP) India.
[2]Professor, Discipline of Mechanical Engineering, Indian Institute of Technology Indore, Simrol 453 552 (MP) India.

*Corresponding author: sagariiti@gmail.com

Abstract. This paper reports on simulating the effect of substrate pre-heating on melt pool size during deposition of titanium alloy (Ti-6Al-4V) on the similar substrate material using micro-plasma transferred arc (μ-PTA) deposition process. It involved (i) three-dimensional finite element simulation (3D-FES) of pre-heating of Ti-6Al-4V substrate material within the range of 400-450 K using furnace, (ii) simulating the formation of melt pool on the pre-heated and non-heated substrate material along μ-PTA torch movement and (iii) image processing of the melt pool to predict its width and depth obtained by μ-PTA deposition process. Obtained results revealed that (i) pre-heated substrate exhibited good distribution of temperature as compared to substrate exposed to ambient temperature and (ii) increase in melt pool width and depth has been observed for pre-heated substrate as compared to melt pool formed at ambient temperature which reveals that pre-heated substrate can be effectively used for depositing titanium alloy.

1. Introduction

Titanium alloys (Ti-6Al-4V) is high strength, high toughness, excellent temperature toughness and high resistance to corrosion material used for manufacturing aircraft structures, engines and bio-medical components. Due to complexity in manufacturing these components, alternative bottom-up approach of manufacturing known as Additive layer manufacturing (ALM) process can be used. However, titanium alloy possesses relatively lower thermal conductivity as compared to other metals, which make it difficult to use in ALM processes. Therefore, most of the researcher have used titanium alloy material with high energy beam (i.e. laser and electron) and arc (i.e. gas metal arc, metal inert gas etc.) based ALM processes.

Titanium alloy (Ti6-6Al-4V) material in wire form was used for manufacturing components using Nd:YAG laser based deposition process [1]. While its mechanical properties such as tension and fatigue tests were done on the layer manufactured components [2]. Lasers were further used for depositing titanium alloy in closed container using selective laser sintering method. In this process, bed of titanium alloy powder was spread inside the sealed container and then the beam was moved over the selected areas doing multi-layer deposition [3]. Electron beam melting has been used for depositing wrought Ti-6Al-4V alloy material for biomedical applications. Performance of manufactured bio-medical components were analysed by conducting tensile test and Vickers hardness test [4]. In arc based processes, tungsten inert gas welding process was used for deposition of titanium alloy. In this study, effect of process parameter on deposition morphology has been investigated [5]. Further, plasma arc welding based ALM was used to build a 3D structure using titanium alloy in wire form. Suitability of this deposition process was investigated by using this process for manufacturing large aerospace components [6].

It has been observed from past literature that, to obtain good strength for layer manufactured components using Ti-6Al-4V, good bonding between each successive layer is required which is mainly dependent on the melt pool formed by the heat source during deposition. However, formation of melt pool is easy in high energy beam and arc based processes, but it becomes difficult when titanium alloy is used for micro-energy arc based processes like micro-plasma transferred arc based deposition process. Therefore, present work aims to develop (i) 3D-FES of pre-heating of Ti-6Al-4V substrate material within the range of 400-450 K using furnace, (ii) Simulating the formation of melt pool along μ-PTA torch movement on the non-heated and pre-heated substrate material and (iii) predict the melt pool width and depth using image processing technique.

2. Simulation

Simulation of melt pool for μ-PTA process was done using ANSYS (version 17.2) software [7].

2.1. Assumptions

- Constant distance is maintained between micro-plasma torch and substrate.
- Geometry of the melt pool remains constant along μ-PTA torch movement direction.

2.2. Governing equation

Micro-plasma transferred arc is a thermal process in which heat conduction is the dominant mode of heat transfer. Therefore, heat conduction governing equation written as follows has been used in present work [8].

$$\rho_s C_{ps}\left(\frac{\partial T}{\partial t}\right) = q + \nabla \cdot (K_s^* \nabla T) + \nabla Q_f \tag{1}$$

In which, ρ_s is the substrate material density (Kg/m^3); C_{ps} is the substrate material specific heat (J/Kg K); t is the micro-plasma arc time (s); q is the actual density of heat flux of micro-plasma arc (W/m^3); T is melt pool temperature (K); K_s* is the modified substrate material thermal conductivity for Marangoni flow (W/m K) and Q_f is the density of heat flux of furnace (W/m^2).

- **Initial condition:** Ambient temperature condition. i.e.

$$T(x, y, z, t) = T_i \tag{2}$$

Where, T_i (K) is the ambient temperature.

- **Pre-heating Condition:** This condition was given to the elements associated with external areas of the substrate material in the form of heat flux. Where the external surface of substrate material was constant heated using furnace operated at 500 W for 240 secs until the substrate material gets heated between 400 to 450 K.
- **Heat source model:** In the present work, Gaussian heat source model modified by considering current density *'J'* (A/m^2) in μ-plasma arc and electric field *'E'* (V/m) has been used [8]. Theoretic density of heat flux *q(x,y)* value obtained at *x* and *y* coordinates considering center of micro-plasma arc radius *'r_o'* is calculated as follows

$$q(x, y) = J\,E\,exp^{\frac{-(x^2+y^2)}{r_o^2}} \tag{3}$$

Here, the micro-plasma current density *'J'* is calculated by multiplying electrical conductivity 'σ_{es}' of the substrate material as 5.8 x 10^5 ohm^{-1}m^{-1} with electric field on the substrate material 'E_{es}'. It is considered approximately equal to electric field *'E'* (i.e. =V/L, Where, V is the arc voltage of micro-plasma in volts and L is the space maintained between nozzle and substrate)

$$J \approx \sigma_{es}\, E_{es} \approx \sigma_{es}\frac{V}{L} \quad Where\ E_{es} \sim E \tag{4}$$

Actual density of heat flux of micro-plasma arc *'q'* (W/m^3) was obtained by multiplying theoretic density of heat flux *'q(x,y)'* with its thermal efficiency *'η'*

$$q = \eta\, q(x,y) \tag{5}$$

Thermal efficiency *'η'* value as 60% is considered [8].

- **Marangoni flow:** Its influence was considered by modifying substrate material thermal conductivity [9] given as follows

$$K_s^* = CK_s \quad if\ T > T_{ms} \tag{6}$$

Where, C is the factor of correction (= 2.5 is used in present study); K_s* is the substrate material thermal conductivity modified for Marangoni flow (W/m K) and K_s is the substrate material thermal conductivity (W/m K).

- **Heat loss from the substrate material:** It can be described as follows

$$(K_s^* \nabla T)\cdot n = \varepsilon_s\, \sigma_{sbc}(T_{ms}^4 - T_i^4) + h_{conv}(T_{ms} - T_i) \tag{7}$$

Here, ε_s is the emissivity of the substrate material; σ_{sbc} is the Stefan-Boltzmann constant (5.67 x 10^{-8} W/m^2 K^4); T_{ms} is melting temperature of the substrate material (K); h_{conv} is the coefficient of convective heat transfer (W/m^2 K).

- **Temperature-dependent material properties:** Figure 1 depicts the distribution of specific heat (Figure 1a); thermal conductivity (Figure 1b) and density (Figure 1c) of the substrate material (i.e. titanium alloy Ti-6Al-4V) with temperature.

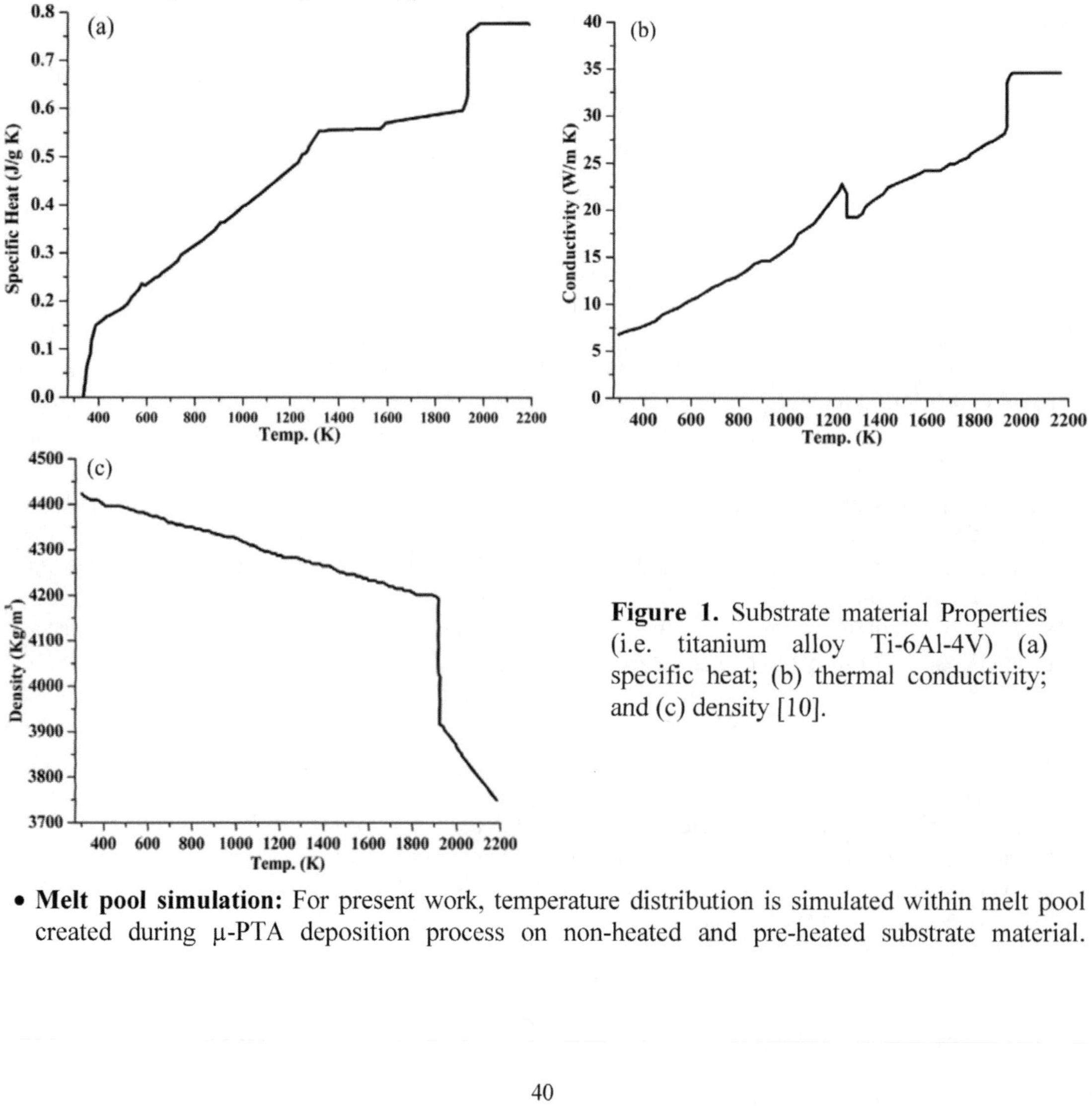

Figure 1. Substrate material Properties (i.e. titanium alloy Ti-6Al-4V) (a) specific heat; (b) thermal conductivity; and (c) density [10].

- **Melt pool simulation:** For present work, temperature distribution is simulated within melt pool created during μ-PTA deposition process on non-heated and pre-heated substrate material.

Simulation was done on substrate material of Ti-6Al-4V, its dimensions as 50 mm × 25 mm × 3 mm were considered. Micro-plasma power as 460 W and torch travel rate as 100 mm/min was used in present simulation. Substrate material geometry was discretized with 8-node cubic elements with an edge length of 0.025 mm as shown in Figure 2. In the ANSYS [7] software, element type SOLID70 was used for 3D-FES of melt pool. Micro-plasma torch movement direction is along y-axis.

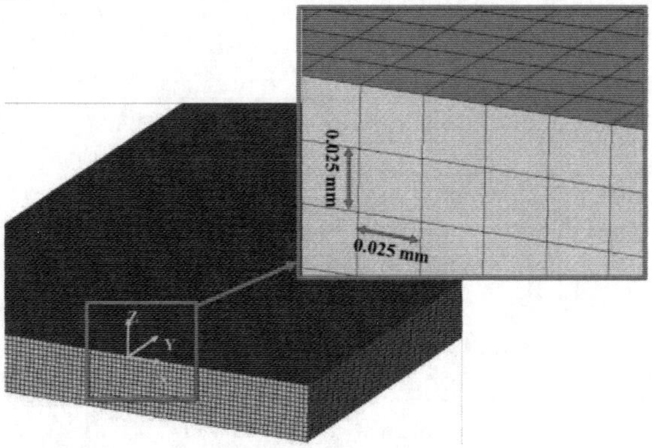

Figure 2. Substrate geometry meshed for 3D-FES of melt pool.

3. Result and discussion

3.1. Temperature distribution analysis

Figure 3 presents distribution of temperature in the melt pool formed in non-heated (Figure 3a) and pre-heated (Figure 3b) substrate material of Ti-6Al-4V during μ-PTA deposition process. It can be observed from Figure 3a that, the temperature distribution is constrained within the substrate material when it is at ambient temperature. While Figure 3b depicts unconstrained temperature distribution within the substrate material. The results revealed that, due to low thermal conductivity of titanium alloy the flow of heat within non-heated substrate material is restricted while due to pre-heated substrate the flow of heat from substrate to atmosphere is easily relieved. Also, the heat affected zone (HAZ) around the area of melt pool in pre-heated substrate is gradually expanding as compared HAZ around melt pool in non-heated substrate. This is due to increase in heat losses by convection and radiation and increase in heat accumulation in the molten pool zone for pre-heated substrate material. From obtained results it can be concluded that pre-heated substrate exhibited good distribution of temperature as compared to substrate exposed to ambient temperature.

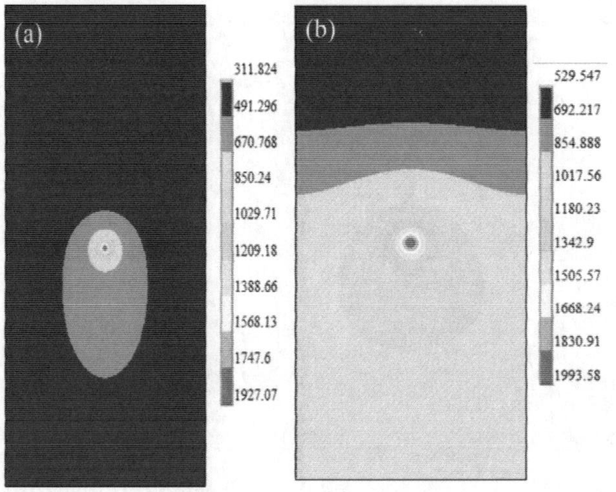

Figure 3. Distribution of temperature in the melt pool formed in (a) non-heated and (b) pre-heated substrate material.

3.2. Measurement and comparison of melt pool width and depth

Melt pool width and depth formed in the substrate material were predicted by image processing technique [11]. Images of temperature distribution obtained by finite element simulation were captured and exported to MATLAB for processing the melt pool images. Processing of the images were done by using following steps:

i) Pixels were counted for width and height of the substrate material as shown in table 1.

ii) Pixels were counted within melt pool to determine its width and depth as shown in table 1.

Table 1. Parameters used for image processing of melt pool.

Parameters	**Non-heated substrate material**	**Pre-heated substrate material**
Width of the substrate material 's_w'	25 mm	25 mm
Height of the substrate material 's_h'	3 mm	3 mm
Pixels within width of melt pool 'P_{mw}'	10.98	27.95
Pixels within depth of melt pool 'P_{md}'	4.97	13.88
Pixels within width of the substrate material 'P_{sw}'	616	616
Pixels within height of the substrate material 'P_{sh}'	73	73

Therefore, using values from table 1, melt pool width 'w_m' for non-heated and pre-heated substrate material were calculated by using following equation

$$w_m = \frac{P_{mw} \times s_w}{P_{sw}} \tag{8}$$

Where, P_{mw} is the pixels counted for melt pool width; P_{sw} is the pixels counted for substrate width and s_w is the width of substrate material (mm). While melt pool depth 'd_m' has been calculated by using following equation

$$d_m = \frac{P_{md} \times s_h}{P_{sh}} \tag{9}$$

Where, P_{md} is the pixels within the melt pool depth; P_{sh} is the pixels within substrate height and s_h is the height of substrate material (mm).

Figure 4 depicts the processing of the images done for predicting melt pool width (using equation 8) and depth (using equation 9) for non-heated (Figure 4a) and pre-heated (Figure 4b). Its value calculated for non-heated substrate material are 0.45 mm and 0.21 mm respectively and for pre-heated substrate material are 1.13 mm and 0.57 mm respectively. Therefore, results revealed that, increase in the melt pool width and depth has been observed for pre-heated substrate as compared to melt pool formed at ambient temperature. Therefore, pre-heated substrate material of titanium alloy can be effectively used for micro-plasma based additive layer manufacturing.

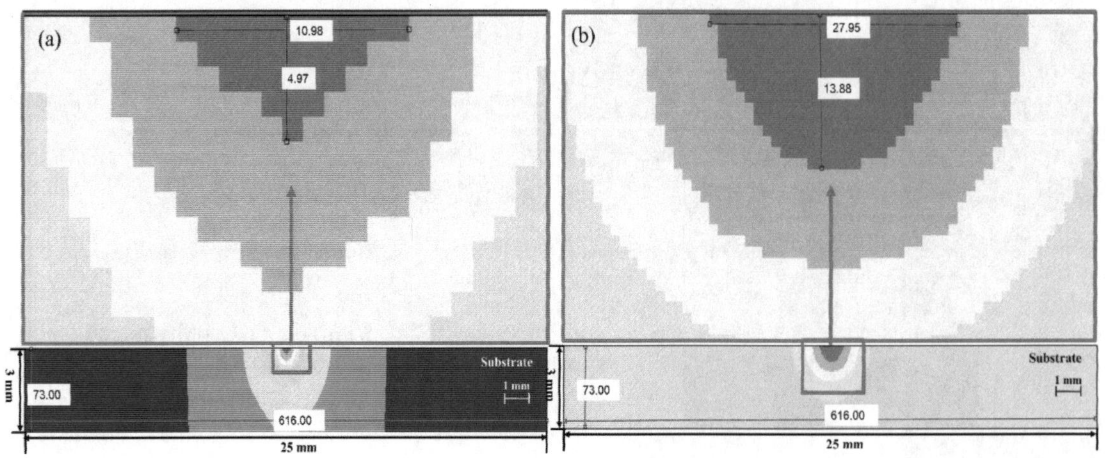

Figure 4. Image processing of melt pool formed in (a) non-heated and (b) pre-heated substrate.

IOP Conf. Series: Materials Science and Engineering **389** (2018) 012006 doi:10.1088/1757-899X/389/1/012006

4. Conclusion

This paper reported on simulating the effect of pre-heating on melt pool size using finite element simulation developed for micro-plasma transferred arc (μ-PTA) process. Temperature distribution within melt pool were studied for pre-heated and non-heated substrate material of Ti-6Al-4V. Images of simulated melt pool for both the cases were captured and exported to MATLAB for image processing to predict melt pool width and depth. Following main conclusions can be drawn from present work:

- Developed simulation can investigate the effect of pre-heating on temperature distribution within the melt pool.
- Width and depth of melt pool were predicted with good accuracy by using image processing technique.
- Pre-heated substrate of Ti-6Al-4V exhibited good distribution of temperature as compared to substrate exposed to ambient temperature.
- Increase of melt pool width and depth has been observed for pre-heated substrate as compared to melt pool formed at ambient temperature which reveals that pre-heated substrate can be effectively used for depositing titanium alloy.

References

[1] Baufeld B, Brandl E and Van der Biest O 2011 Wire based additive layer manufacturing: Comparison of microstructure and mechanical properties of Ti–6Al–4V components fabricated by laser-beam deposition and shaped metal deposition *J. Mater. Process. Tech.* **211** 1146-1158

[2] Brandl E, Baufeld B, Leyens C and Gault R 2010 Additive manufactured Ti-6Al-4V using welding wire: comparison of laser and arc beam deposition and evaluation with respect to aerospace material specifications *Phy. Proce.* **5** 595-606

[3] Roberts I A, Wang C J, Esterlein R, Stanford M and Mynors D J 2009 A three-dimensional finite element analysis of the temperature field during laser melting of metal powders in additive layer manufacturing *Int. J. Mach. Tool. Manu.* **49** 916-923

[4] Murra L E, Esquivel E V, Quinones S A, Gaytan S M, Lopez M I, Martinez E Y, Medina F, Hernandez D H, Martinez E, Martinez J L, Stafford S W, Brown D K, Hoppe T, Meyers W, Lindhe U and Wicker R B 2009 Microstructures and mechanical properties of electron beam-rapid manufactured Ti–6Al–4V biomedical prototypes compared to wrought Ti–6Al–4V *Mater. Charact.* **60** 96-105

[5] Wang F, Williams S and Rush M 2011 Morphology investigation on direct current pulsed gas tungsten arc welded additive layer manufactured Ti-6Al-4V alloy *Int. J. Adv. Manuf. Technol.* **57** 597-603

[6] Martina F, Mehnen J, Williams S W, Colegrove P and Wang F Investigation of the benefits of plasma deposition for the additive layer manufacture of Ti–6Al–4V *J. Mater. Process. Tech.* **212** 1377-1386

[7] ANSYS17.2 2016© *ANSYS Inc. Canonsburg Pennsylvania (USA)*

[8] Nikam S H and Jain N K 2017 Three-dimensional thermal analysis of multi-layer metallic deposition by micro-plasma transferred arc process using finite element simulation *J. Mater. Process. Tech.* **249** 264-273

[9] Alimardani M, Toyserkani E and Huissoon J P 2007 Three-dimensional numerical approach for geometrical prediction of multilayer laser solid freeform fabrication process *J. Laser Appl.* **19** 14-25

[10] Mills K C 2002 *Recommended values of thermo-physical properties for selected commercial alloys* (Cambridge: Woodhead Publishing series) p 211

[11] Nikam S H and Jain N K 2018 3D-finite element simulation and image processing based prediction of width and height of single-layer deposition by micro-plasma-transferred arc process *Int. J. Adv. Manuf. Technol.* 10.1007/s00170-017-1472-x

IOP Conf. Series: Materials Science and Engineering **389** (2018) 012007 doi:10.1088/1757-899X/389/1/012007

On Productivity of WSEM Process for Manufacturing Meso-Sized Helical and Bevel Gears

Sujeet K Chaubey[1] and Neelesh K Jain[2]

[1]Research Scholar, Discipline of Mechanical Engineering, Indian Institute of Technology Indore, Simrol 453 552 (MP) India.
[2]Professor, Discipline of Mechanical Engineering, Indian Institute of Technology Indore, Simrol 453 552 (MP) India.

Corresponding author: sujeetiiti@gmail.com

Abstract. This paper reports on productivity of wire spark-erosion-machining (WSEM) process for manufacturing meso-sized helical gear (MHG) and meso-sized bevel gear (MBG). Productivity of WSEM process is measured by volumetric gear cutting rate (*VGCR*). Variation in *VGCR* with four WSEM parameters such as servo voltage 'S_V' pulse-on time 'T_{on}', pulse-off time 'T_{off}', and feed rate of wire 'W_F' is also studied. Box-Behnken design approach of Response Surface Methodology (RSM) was used to design and conduct 29 experimental runs with replicates twice for analysis of *VGCR*. Analysis of variance (ANOVA) was used to determine the significant parameters and their interactions. It was observed that all considered WSEM parameters are significant and considerably affect *VGCR*. Desirability function analysis was used to find the optimum WSEM parameters so as to enhance the productivity of WSEM process to manufacture the better quality of MHG and MBG. Confirmation experiments were performed to validate the optimized values of *VGCR* corresponding to MHG and MBG and achieved much closer to it. Scanning electron micrographs (SEM) of MHG and MBG revealed uniform and burr-free tooth profile, bore and smoother cracks-free flank surfaces.

1. Introduction

Globally, demands of miniaturized parts and components growing continuously due to rapid development and application of MEMS and microsystems devices. Meso-sized gears (i.e. tip diameter ranging from 1 to 10 mm) are core constituents of these devices and commonly used for actuating and driving purposes. Meso-sized motors, pumps, gearboxes and actuators are some typical applications of these gears which are extensively used in chemical industries, dental and surgical instruments, precision instruments, prototype models, and domestic appliances [1, 2]. These gears are become more popular due to their light weight and small volume, stable superior performance, longer service life, minimum materials and manufacturing cost. Operating performance and service life of these gears are governed by its overall quality (i.e. aspects of gear quality and surface quality) [2]. Manufacturing processes and materials considerably affect the overall quality, manufacturing time and cost of meso-sized gears. Therefore, their proper selection is necessary to manufacture the high quality meso-sized gears with higher productivity and minimum manufacturing cost. Selection of appropriate materials and manufacturing processes of these gears mainly depends on their shape, size and types of applications. Stainless steel, low alloy steel, brass, bronze, copper, aluminium and plastic are commonly used meso-sized gears materials. Manufacturing processes of meso-sized gears are classified as (i) subtracting processes: hobbing, milling; (ii) formative process: die casting, injection molding; (iii) deformative process: extrusion, rolling; (iv) micro-machining processes: LIGA, hot embossing, micro-injection and compression molding, bio etching; and (v) advanced manufacturing process: abrasive water jet machining; laser ablation, ion-beam machining. Poor gear quality (DIN 9-12), sharp edge definition, burr, microcracks, tool marks, applicable for specific types of gears and materials, requirement of appropriate finishing and hardening processes are some limitations of these processes. These inherent limitations of these manufacturing process and demands of high quality meso-sized gears at reasonable price in global market motivate the researchers to develop and explore another alternative process to manufacture high quality of meso-sized gears with higher productivity and minimum manufacturing cost [3, 4]. Wire spark-erosion machining (WSEM) is emerging advanced machining process to manufacture high quality of meso-sized cylindrical (i.e. spur and helical) and conical (i.e. bevel) gears. Generally, various input parameters govern the productivity of WSEM process to manufacture MHG and MBG. Therefore, selection of optimum WSEM parameters plays important role to manufacture high quality MHG and MBG with higher productivity and minimum manufacturing cost. Its offers several advantages to manufacture meso-sized cylindrical and conical gears (MCCG): (i) ability to manufacture MCCG from any electrically conductive materials of any hardness and thickness; (ii) manufacture different size and shape of gears from same gear plate without changing tool electrode (very thin brass wire) thus minimize total setting time; (iii) higher dimensional accuracy with smooth and burr-free flank surfaces; (iv) achieved gear quality of MCCG up to DIN 6-7 along with higher wear resistance on flank surfaces without using any subsequent finishing and hardening process; and (v) unattended manufacturing of MCCG [2, 3].

Review of past works revealed that WSEM is very emerging and promising advanced manufacturing process to manufacture closed to net-shape meso-sized gears from hard and difficult-to-machine materials. Regrettably, very limited work has been stated on spark-erosion based processes to manufacture meso-sized gears, improving gear quality and productivity. Most of the past work dedicated on fabrication of meso-sized spur gears (MSG), dimensional accuracy, surface roughness, and microstructural studies of meso-sized spur gears [5, 6, 7]. Productivity of WSEM process was improved to manufacture closed to net-shape MSG made of brass [8]. Extensive review of relevant past works revealed that no works have been conducted on manufacturing of MHG and MBG by WSEM process and improving its productivity. Therefore, the major objectives of this study are summarized as follow: (i) improve the productivity of WSEM process for manufacturing MHG and MBG; (ii) study variation in *VGCR* with WSEM parameters; (iii) identify significant WSEM parameters and their interaction on *VGCR*; (iv) optimize WSEM parameters with an objective to maximize *VGCR* to manufacture MHG and MBG; and (v) validate the optimized values of *VGCR* through confirmation experiments using optimum combination of WSEM parameters.

2. Experimental details

2.1. Details of Materials and Meso-sized Gears

In this study, austenitic stainless steel (grade SS 304) was chosen as gear materials to manufacture MHG and MBG due to their non-magnetic nature, higher resistance to corrosive atmosphere, very good wear resistance and perform satisfactorily for longer duration. It consists 8.3% Ni; 18% Cr; 0.05% C; 1.08% Mn; 0.48% Si; 0.017% S; 0.038% P; and balance Fe (iron) by weight. MHG and MBG made from SS 304 are commonly used in chemical, bio-chemical, pharmaceuticals, food processing industries and domestic appliances. The detailed design specifications of these gears are as follow: (i) for MHG: involute profile, 8.35 mm as tip diameter, 0.66 mm as module, 5 mm as face width, 10 teeth, 20^0 as helix angle (right hand) and pressure angle; and (ii) for MBG: involute profile, 9.8 mm as major diameter, 7.0 mm as minor diameter, 0.7 mm as module, 5 mm as face width, 12 teeth, 32^0 as cone angle, 20^0 as pressure angle.

2.2. Details of Manufacturing Process

Computerized Numerical Control (CNC) four axes (X, Y and U, V) Sprincut win WSEM machine from Electronica Machine Tools Ltd. Pune, India was used to manufacture MHG and MBG. A very fine brass wire of 250 μm diameter was used as tool electrode to cut the MHG and MBG from gear plate (having dimensions of 100 mm as length; 50 mm as width and 5 mm as thickness). Following is the sequences of manufacturing steps used to manufacture MHG and MBG by WSEM and measurement of volumetric gear cutting rate as shown in Figure 1: (i) workpiece preparation followed by grinding and buffing process to make top and bottom surface perfectly flat and edges perpendicular to each other; (ii) drilling of microholes on workpiece (i.e. gear plate) by μ-electro-discharge drilling process; (iii) mounting of workpiece on main worktable of WSEM through adjustable clamps and using dial gauge to ensure its proper mounting on worktable with respect to direction of feed rate of wire; (iv) using ELCAM software to prepared part program of MHG and MBG along with bore; (v) trepanning of microhole up to 3 mm diameter which served as bore of MHG and MBG and its center also served as a center of the meso-sized gears; (vi) run the WSEM in dry mode to bring upper and lower guide adjacent microhole which served as wire passage to cut MHG and MBG according to their corresponding part program; and (vii) microstructure studies of flank surfaces and calculating volumetric gear cutting rate using weighing machine and stop watch.

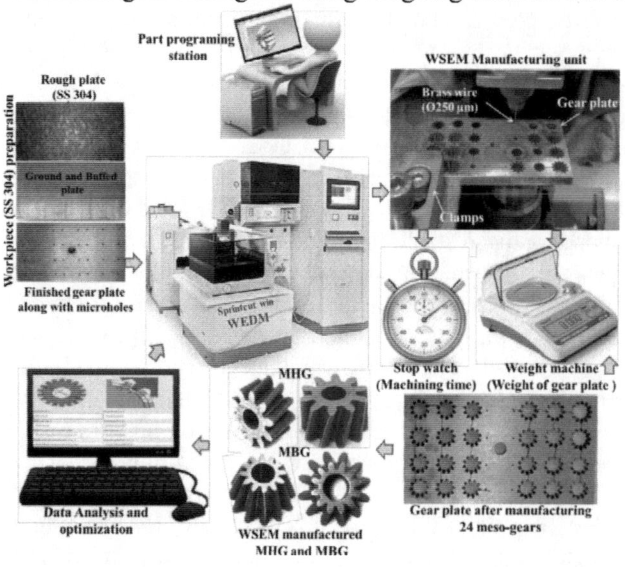

Figure 1. Sequences of different manufacturing steps used to manufacture MHG and MBG by WSEM process and measurement of considered response.

2.3. Methodology

Twenty nine experiments were designed and conducted (for MHG and MBG separately) according to Box–Behnken design (BBD) of response surface methodology (RSM) and each experiments repeated twice thus total 58 MHG and 58 MBG were manufactured by varying four WSEM parameters namely servo voltage (S_V), pulse-on time (T_{on}), pulse-off time (T_{off}) and feed rate of wire (W_F) at three levels each. Values of the fixed parameters, coded levels and their corresponding values of the variable WSEM parameters are depicted in Table 1. These values were selected on the basis of preliminary and pilot experiments conducted using one-variable-at-time approach of design of experiment keeping the view of wire breakage, gear cutting rate, gear quality and surface quality of MHG and MBG.

Table 1. Identified values of the fixed parameters, code levels and their equivalent values of the variable parameters of WSEM used in the experiments.

Variable parameters				Fixed parameters
Parameters (units)	Code levels			
	-1	0	1	Peak current (I_P): 12 A; Cutting speed (C_S):
Pulse-on time 'T_{on}' (μs)	1.0	1.3	1.6	75%; Wire: Soft plain brass; Wire diameter:

Pulse-off time 'T_{off}' (μs)	44.5	46.5	48.5	250 μm; Tensile strength: 470-510 N/mm^2;
Servo voltage 'S_v' (Volts)	10	15	20	Wire tension (W_T): 1,260 g; Dielectric: deionized water; Dielectric flow rate (W_P): 15 kg/cm^2

2.4. Measurement of Responses

Productivity of WSEM process is measured by volumetric gear cutting rate (*VGCR*). It defines as the amount of the material removed per unit time and expressed by the following equation:

$$VGCR = \frac{W}{\rho \times t} \quad \left(\frac{mm^3}{min.}\right) \tag{1}$$

Where, W (*i.e.* $W_{before} - (W_{after} + W_{gear})$ is total material lost during manufacturing of MHG and MBG obtained by subtracting weight of meso-gear (W_{gear}) and gear plate (W_{after}) after machining from the total weight of gear plate (W_{before}) before machining, ρ stands for density of gear materials and t stands for total manufacturing time of MHG and MBG.

Weight of gear plate was measured on precision weighing machine having accuracy up to 0.01g. Manufacturing time of MHG and MBG was directly revealed on the monitor of WSEM or can be calculated by a digital stop watch having a least count of 0.01 seconds. Supra 55 from Carl Zeiss was used for scanning electron microscopic (SEM) images of MHG and MBG.

3. Result and discussion

Following tasks were performed for analyses of *VGCR* of MHG and MBG and their corresponding 29 experimental run with replicates using 10th version of Design Expert software [9]:

(i) Using sequential model sum of squares (SMSS), lack-of-fit and model summary statistics (MSS) to identify the best suitable model for *VGCR* of MHG and MBG. It was concluded from these tests that the quadratic model is adequate to express the best fit polynomial equations for modelling *VGCR* of MHG and MBG.

(ii) Modelling equations for the responses using coded values of the variable parameters in the regression analysis. Then identifying significance of these equations along with variable parameters and their interactions involved in these models at 95% confidence interval using analysis of variance (ANOVA). This led to the following modeling equations in which S_V, T_{on}, T_{off}, and W_F and their squared terms found to be significant parameters and interactions between T_{on} and T_{off}; T_{on} and S_V; and T_{off} and S_V were found to be significant.

For MHG

$$VGCR_{MHG} = 3.86 + 0.97T_{on} - 0.41T_{off} + 0.58S_V + 0.24W_F + 0.25T_{on}^2 + 0.52T_{off}^2 + 0.46S_V^2 + 0.39W_F^2 - 0.28T_{on}T_{off} - 0.28T_{on}S_V + 0.32T_{off}S_V \tag{2}$$

For MBG

$$VGCR_{MBG} = 3.56 + 1.07\,T_{on} - 0.42T_{off} + 0.74S_V + 0.24W_F + 0.30T_{on}^2 + 0.38T_{off}^2 + 0.58S_V^2 + 0.21W_F^2 - 0.4T_{on}T_{off} + 0.32T_{on}S_V + 0.33T_{off}S_V \tag{3}$$

(iii) Identifying the variation in *VGCR* with four WSEM parameters. Figures 2(a) and 2(b) shows variation of volumetric gear cutting rate '*VGCR*' with servo voltage, pulse-on time, pulse-off time, and with feed rate of wire for WSEM manufactured MHG and MBG respectively. Figure 2 depicts main effect plots showing the influences of four WSEM parameters on the VGCR. It can be seen from these figures that (i) *VGCR* increases with increase in servo voltage and pulse-on time because higher values of these parameters produce violent spark which accelerate the formation of craters on flank surfaces results in higher material removal rate; (ii) *VGCR* decreases with increase in pulse-off time due to fact that at higher value of pulse-off time decrease frequency of sparks between wire and workpiece results in minimum material removal from workpiece per unit time; (iii) *VGCR* slightly increases with increase in feed rate of wire because higher feed rate ensure that fresh wire always available for sparking phenomenon in WSEM process. Fresh wire has better dimensional accuracy and ensures stable spark generation between inter electrode gap as compare to exposed wire to the spark results in higher material removal at higher feed rate of wire.

(iv) Identifying the optimum values of the WSEM parameters to maximize the *VGCR*.

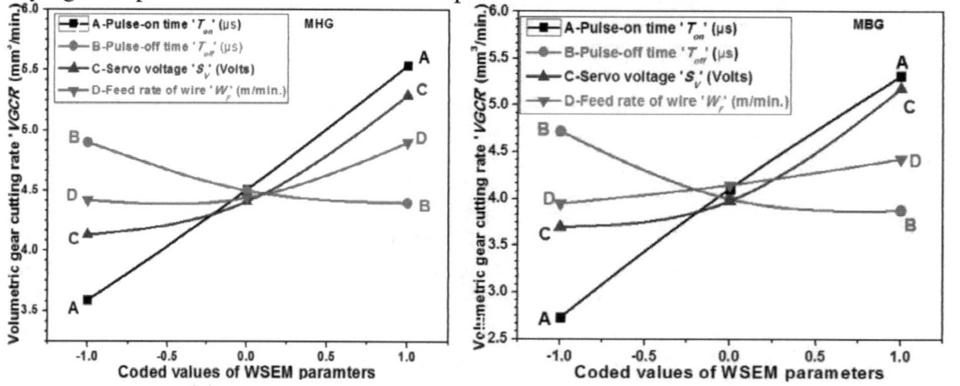

(a) **(b)**

Figure 2. Effect of WSEM parameters on *VGCR*: (a) MHG; and (b) MBG.

4. Optimization

Desirability functional analysis (DFA) was used for optimization of chosen WSEM parameters to maximize the *VGCR* of MHG and MBG. Considered response (y_i) is transformed into an individual desirability function 'd_i' ($0 \leq d_i \leq 1$). This study consist single response *VGCR* with overall objective to maximize it for value of desirability to 1. Therefore, the overall desirability function analysis of considered response of MHG and MBG for i^{th} data set can be expressed by following equation:

$$D_i = \left(\prod_{i=1}^{m} d_i \right)^{\frac{1}{\Sigma w_i}} \tag{4}$$

Or

$$D_i = [(d_{VGCR})_i]^1 \tag{5}$$

Where, w_i represents the importance of each variable relative to the others; d_i represents discrete desirability of each response y_i and($0 \leq d_i \leq 1$). Table 2 presents the optimum WSEM parameters and their corresponding values of *VGCR* for MHG and MBG obtained from desirability analysis.

4.1. Validation of optimized results

Confirmation experiments were performed to validate the optimized values of *VGCR* of MHG and MBG using optimum WSEM parameters closed to their available values in the machine. It was concluded from Table 2 that values of *VGCR* of MHG and MBG obtained from confirmation is much closer to their optimized values.

Table 2. Results of the confirmation experiments.

Results	Optimum vaues of WSEM parameters				Optimized '*VGCR*'
For MHG	T_{on} *(μs)*	T_{off} *(μs)*	S_V *(Volts,*	W_F *(m/min)*	*(mm³/min.)*
By DFA	1.5	44.6	10.3	14.8	6.8
By confirmation	1.5	44.5	10	15	6.7
For MBG					
By DFA	1.58	46.03	19.12	12.17	6.5
By confirmation	1.6	45.5	20	12	6.3

4.1.1. SEM images of optimized MHG and MBG. **Figure 3** depicts the SEM micrographs of tooth profile, flank surfaces and bore of MHG and MBG respectively. **Figure 3(a)** and **Figure 3(b)** showing the tooth profile, bore, and flank surfaces of MHG and MBG at 50X and 31X magnifications respectively. It can be seen from these micrographs that MHG and MBG have accurate involute tooth profile and bore free from burr and sharp edge definition without undercut at root. While, flank surfaces of MHG and MBG are smooth, uniform and cracks free.

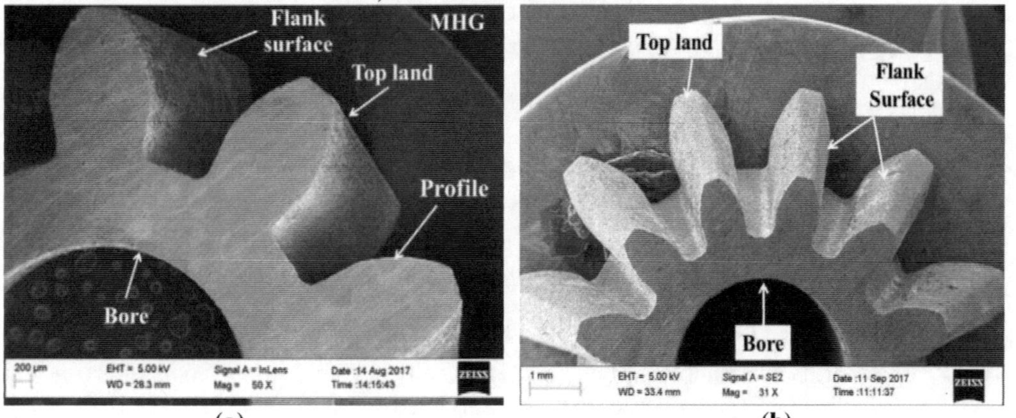

(a) (b)

Figure 3. SEM micrographs of bore, tooth profile, and flank surfaces of WSEM manufactured meso-sized gears: (a) MHG at 50X magnification; and (b) MBG at 31X magnification.

5. Conclusions

This paper aimed to maximize the productivity of WSEM process for manufacturing MHG and MBG by WSEM process. Following are the major conclusions of this study:

- Enhanced productivity of the WSEM by achieving optimum parametric combination by DFA.
- S_V, T_{on}, T_{off}, and W_F found as the significant parameters for both MHG and MBG which significantly affect the volumetric gear cutting rate. The significant interaction between T_{on} and T_{off}; T_{on} and S_V; and T_{off} and S_V on *VGCR* have been observed.
- *VGCR* of MHG and MBG increase with longer pulse-on time and higher servo voltage and decrease with longer pulse-off time. While slightly increase with feed rate of wire.
- Validate the optimized values of *VGCR* through confirmation experiments.
- SEM micrographs of MHG and MBG obtained by confirmation experiments revealed smoother and accurate tooth profile, bore, and flank surfaces free from burr and cracks.

Therefore, the results of the present study will be very useful for the industry users and researchers to manufacture the high quality of meso-sized cylindrical and conical gears with higher productivity.

References

[1] Jain N K and Chaubey S K 2016 *Review of Miniature Gear Manufacturing* In: Comprehensive Material Finishing-vol-1 (Ed. M.S.J. Hashmi) Elsevier Inc. UK p 504-538

[2] Chaubey S K and Jain N K 2017 Investigations on microgeometry of meso bevel and meso helical gears manufactured by WEDM process *Int. J. of Adv. Manuf. Technol.* **93(9-12)** (2017) 4217–4231

[3] Chaubey S K and Jain N K 2017 Investigations on surface quality of WEDM-manufactured meso bevel and helical gears Mat. and Manuf. Proc. 1-10 DOI: 10.1080/10426914.2017.1415440

[4] Chaubey S K and Jain N K 2018 , State-of-art review of past research on manufacturing of meso and micro cylindrical gears Prec. Engg. 51 702-728

[5] Ali M Y, Karim A N M, Adesta E Y T, Ismail A F, Abdullah A A and Idris M N 2010 Comparative study of conventional and micro WSEM based on machining of meso/micro sized spur gear Int. J. of Prec. Eng. and Manuf. 11(5) 779-784

[6] Ali M Y and Mohammad A S 2008 Experimental study of conventional WSEM for micro-fabrication Mat. Manuf. Proc. 23(7) 641–645

[7] Benavides G L, Bieg L F, Saavedra M P and Bryce, E A 2002 High aspect ratio meso-scale parts enabled by wire micro-EDM Microsystem Technol. 8 395–401

[8] Gupta K and Jain N K 2013 On micro-geometry of meso gears manufactured by wire electrical discharge machining Mat. and Manuf. Proc. 28(10) 1153-1159

[9] Design Expert (Trial Version) Stat-Ease, Inc. Minneapolis, MN (USA)

IOP Conf. Series: Materials Science and Engineering **389** (2018) 012008 doi:10.1088/1757-899X/389/1/012008

A Study of Relation between Tool Wear and Dynamic Cutting Force Ratio in CNC Turning

Phathara Chaiyawong and Somkiat Tangjitsitcharoen*

Department of Industrial Engineering, Faculty of Engineering, Chulalongkorn University, Payathai Road, Patumwan, Bangkok, 10330, Thailand

*Somkiat.ta@eng.chula.ac.th

Abstract. Since tool wear directly affects the manufacturing cost in CNC turning process, the aim of this research is to study the effects of cutting conditions on tool wear and dynamic cutting force ratio. It also describes relation of tool wear and dynamic cutting force ratio. The results of experiment show that cutting speed is the most significantly influential factor followed by feed rate, depth of cut and nose radius respectively. In addition, effect of dynamic cutting force on the tool wear analysis show that trend of the dynamic cutting force is not changed even though the cutting condition is changed. Accordingly, the tool wear can be predicted by using the dynamic cutting force ratio in turning process.

1. Introduction

One of the material removal process is the turning process where a cutting tool with a single cutting edge is used to remove material from a rotating work piece to make a cylinder shape. The friction occurs at contacting surface between a cutting tool and chips during turning process by turning for a while. The heat accumulation is generated which affects the tool wear and the surface roughness. These directly reflect to the manufacturing cost. Due to the cutting tool necessary frequently replace and has highly price. Also defects occur from a poor quality of surface finishing. Base on economic production, the tool wear is primary factor to consider the way to extend the tool life as long as possible.

Base on cutting tool wear and cutting force theory show that wearing rate depends on various factors as follow: spindle speed, feed rate, depth of cut, rake angle, nose radius, work material, cutting fluid and etc. Previous researches have been studied the influence of cutting conditions on the tool wears [1] .After that, researchers found out that the tool wear has significantly associated with the cutting force [2, 3] and cutting force monitoring can estimate the tool wear. Due to the difficulty regarding influence of work material and cutting conditions on the cutting force, the cutting ratio has been developed [4] for surface roughness estimation. Similarly, the cutting force ratio can be utilized for estimating the tool wear which is better than that the real time monitoring. Therefore dynamic cutting force measurement was proposed [5]. In term of cutting force analysis, previous research has applied the fast flourier transform [6]. But in actual turning operation, there was many cutting force signal combination such as broken chip signal, chatter signal and tool wear signal and etc. The wavelet transform analysis is a better alternative for filter and separate their cutting force signal from each other[7].

According to previous researches review, there was limitation in term of experiment cutting condition and cutting force analysis method. Thus, this study will provide more experiment cutting

IOP Conf. Series: Materials Science and Engineering **389** (2018) 012008 doi:10.1088/1757-899X/389/1/012008

conditions including cutting speed, feed rate, depth of cut and nose radius for giving more understanding about effects of various cutting conditions on the tool flank wear by utilizing wavelet transform analysis as cutting force analysis method. This study will lead to solving the tool wear problem, extending tool life and developing intelligent manufacturing in the future.

2. Experimental step and cutting conditions

This study utilizes MAZAK Quick turn NEXUS 200MY/MSY, which is 4 axis CNC turning machine with dry cutting condition. The dynamometer is set up to measure the dynamic force signal as shown in figure 1. The experiment was perform on a rod steel S45C with hardness 200 HB and using an insert tool of the KYOCERA turning insert coated carbide.

Figure 1. The dynamometer set up for the dynamic force measurement.

During experiment as condition list shown in table 1 by Box-Behnken design of experiment with 4 cutting conditions and 3 levels, The dynamic cutting force data was transferred to the oscilloscope for record as real-time and also the tool flank wear (V_b) was measured by using microscope of NIKON MM-60. These results were always collected at the same time until the tool flank wear equal to 2 mm or failure. After that a new insert was replaced then the next condition was started. The dynamic cutting force (dFy, dFz) analyzed by using wavelet transform then calculated its ratio by using the dynamic main force (dFy) divided by the dynamic feed force (dFz). The dynamic cutting force ratio (dFy/dFz) and the tool flank wear (V_b) were selected at 15 min of cutting time as the response.

Table 1. Experiment conditions and levels

Cutting conditions	**-1**	**0**	**1**	**Fixed factors**
Spindle speed; v (m/min)	150	200	250	Dry cutting
Feed rate; f (mm/rev)	0.10	0.15	0.20	Rake angle -6°
Depth of cut; d (mm)	1.0	1.2	1.4	
Nose radius; NR (mm)	0.4	0.8	1.2	

3. Results and discussions

The experiment results analyzed by statistics method that popularly used in many researches which is the analysis of variant (ANOVA) with 95% confidential and p-value 0.05 are considered as significant. The significant of cutting conditions can be ensured by p-value report from ANOVA. The effect of cutting condition on the response consists of tool flank wear (V_b) and the dynamic cutting force ratio (dFy/dFz) which are discussed through the main effect plot. The response of the Box-Benkhen experiment was provided by table 2.

Table 2. Experiment results of Box-Benkhen

Run order	Cutting conditions				Responses	
	Cutting speed (m/min)	Feed rate (mm/rev)	Depth of cut (mm)	Nose radius (mm)	V_b	dFy/dFz
1	150	0.15	1	0.8	0.1464	0.8787
2	150	0.15	1.4	0.8	0.1562	0.8573
3	200	0.15	1.2	0.8	0.1671	0.7435
4	200	0.1	1.2	0.4	0.1350	0.8290
5	250	0.15	1.4	0.8	0.1949	0.6577
6	200	0.2	1.2	1.2	0.1828	0.7250
7	200	0.15	1.4	0.4	0.1657	0.7664
8	150	0.15	1.2	0.4	0.1427	0.8771
9	200	0.1	1.4	0.8	0.1587	0.7910
10	200	0.15	1.2	0.8	0.1637	0.7436
11	200	0.2	1	0.8	0.1642	0.7346
12	200	0.2	1.4	0.8	0.1939	0.7080
13	250	0.2	1.2	0.8	0.1998	0.6576
14	200	0.1	1.2	1.2	0.1526	0.7161
15	200	0.15	1.4	1.2	0.1669	0.7400
16	250	0.15	1	0.8	0.1695	0.6809
17	200	0.2	1.2	0.4	0.1809	0.7257
18	200	0.15	1.2	0.8	0.1639	0.7499
19	150	0.1	1.2	0.8	0.1221	0.8892
20	250	0.1	1.2	0.8	0.1570	0.7032
21	200	0.15	1	0.4	0.1539	0.7888
22	200	0.1	1	0.8	0.1412	0.8506
23	200	0.15	1	1.2	0.1590	0.7724
24	150	0.15	1.2	1.2	0.1577	0.8606
25	250	0.15	1.2	1.2	0.1855	0.6747
26	250	0.15	1.2	0.4	0.1844	0.6780
27	150	0.2	1.2	0.8	0.1680	0.8539

3.1. Tool wear analysis

ANOVA analysis is as shown in table 3. It is indicated that all cutting conditions have statistically significant effect on the tool wear in order to p-valve less than 0.05. Thus, Cutting speed, feed rate, depth of cut and nose radius will be considered.

Table 3. ANOVA for response surface of too wear (V_b)

Source	DF	Adj SS	Adj MS	F-Value	P-Value	Remark
v (m/min)	1	0.008687	0.000620	22.14	0.000	Significant
f (mm/rev)	1	0.008423	0.002106	75.14	0.000	Significant
d (mm)	1	0.003263	0.003263	116.45	0.000	Significant
NR (mm)	1	0.004142	0.004142	147.79	0.000	Significant

The result of main effect plot reported that cutting speed is the most sensitive in term of direct variation on the tool wear followed by feed rate, depth of cut and nose radius respectively as shown in figure 2. The higher cutting speed accelerated the tool abrasion mechanical wearing due to frequently touching between work piece and tool on flank face. It causes a hard particle of work piece and removes a small portion of the tool.

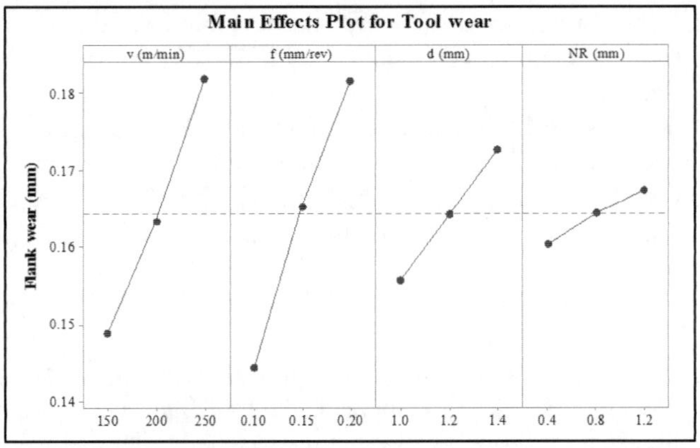

Figure 2. Main effect plot of tool wear (V_b).

3.2. Dynamic cutting force ratio analysis

After ANOVA analysis, table 4 indicates that all cutting conditions have statistically significant effect on the dynamic cutting force ratio when p-valve less than 0.05.

Table 4. ANOVA for response surface of dynamic cutting force ratio (dFy/dFz)

Source	DF	Adj SS	Adj MS	F-Value	P-Value	Remark
v (m/min)	1	0.134809	0.009629	23.75	0.000	Significant
f (mm/rev)	1	0.129119	0.032280	79.63	0.000	Significant
d (mm)	1	0.113052	0.113052	278.89	0.021	Significant
NR (mm)	1	0.010754	0.010754	26.53	0.030	Significant

The result of main effect plot reported that cutting speed is the most influential on the dynamic cutting force ratio (dFy/dFz) followed by feed rate, depth of cut and nose radius respectively as shown in figure 3, is an inverse variation on the tool wear. In the other words, the tool wear (V_b) is related to the dynamic cutting force directly and will be explained in the next. The higher cutting speed affects increasing of both dynamic feed force (dFy) and dynamic main force (dFz). Especially, dynamic main force (dFz) will increase more than dynamic feed force. Thus, cutting force ratio is rapidly decreased.

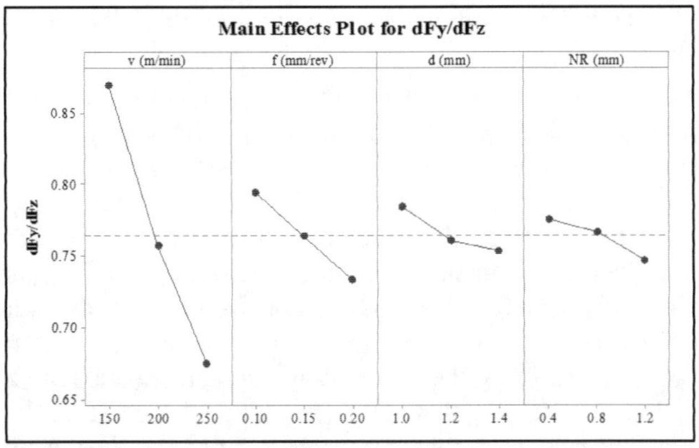

Figure 3. Main effect plot of dynamic cutting force (dFy/dFz).

Base on the main effect plot analysis, it is found that cutting speed, feed rate, depth of cut and nose radius have the influence on both the tool flank wear (V_b) and the dynamic cutting force (dFy/dFz). In addition, these responses have relation in each other.

3.3. Effect of dynamic cutting force on tool wear analysis

IOP Conf. Series: Materials Science and Engineering **389** (2018) 012008 doi:10.1088/1757-899X/389/1/012008

The result of experiment about relation of dynamic cutting force and tool wear will be provided for more understanding as shown in figure 4-7, under cutting conditions which varies from the lowest level (-1) to the highest level (1).

Figure 4 shows that the trend of both cutting speed 150 and 250 m/min are slightly decreasing of the dynamic cutting force ratio (dFy/dFz) when the tool flank wear grows up. In order to dynamic main force (dFz) is the same direction as the cutting speed. If the cutting speed is high, the tool flank wear will increase. It causes the nose radius to be bigger which make a large contacting area between work piece and cutting tool and also leads to a high vibration. For these reasons, the dynamic main force (dFz) is higher than the dynamic feed force (dFy) then produces decreasing of the dynamic cutting force ratio.

Figure 5 shows that the trend of both feed rate 0.1 and 0.2 mm/rev are slightly decreasing of the dynamic cutting force ratio (dFy/dFz) when amount of the tool flank wear is increased. According to faster feed rate leads to bigger tool wear. It causes cutting to be harder due to a large contacting area which directly affect to the dynamic main force (dFz). In addition, a faster feed rate leads to more surface roughness which affects the dynamic feed force (dFy) to be increased. However, the dynamic main force (dFz) is influenced by a faster feed rate much more than the dynamic feed force (dFy). Thus, the dynamic cutting force ratio (dFy/dFz) is decreased.

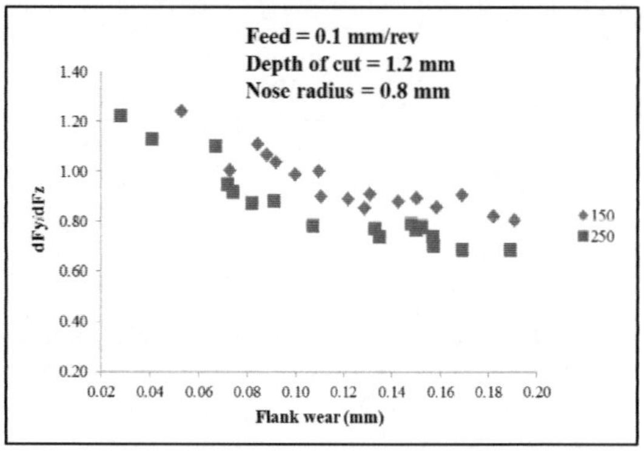

Figure 4. Example of dynamic cutting force ratio (dFy/dFz) versus tool flank wear under cutting speed 150 and 200 m/min, feed rate 0.1 mm/rev, depth of cut 1.2 mm and nose radius 0.8 mm.

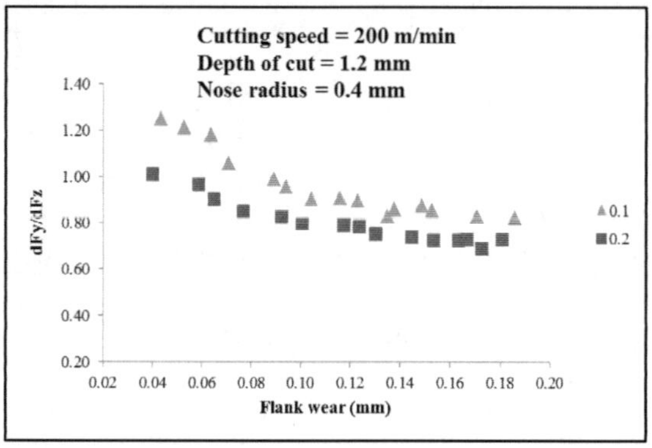

Figure 5. Example of dynamic cutting force ratio (dFy/dFz) versus tool flank wear under feed 0.1 and 0.2 mm/rev, cutting speed 200 m/min, depth of cut 1.2 mm and nose radius 0.4 mm.

Figure 6 shows that the trend of both depth of cut 1 and 1.4 mm are slightly decreasing of the dynamic cutting force ratio (dFy/dFz) when amount of the tool flank wear is bigger. The larger depth of cut means the larger contacting area between work piece and side cutting edge which leads to the cutting force is highly generated due to harder cutting and also there is risky vibration or chatter occurred which obtain more rough surface. Therefore, the dynamic main force (dFz) is higher than the dynamic feed force (dFy). Thus, the dynamic cutting force ratio (dFy/dFz) is decreased.

Figure 7 shows similar trend with other cutting conditions. The dynamic cutting force ratio (dFy/dFz) is decreased when the tool flank wear (V_b) is bigger for both nose radius 0.4 and 1.2mm. According to wearing not only occurs on the side cutting edge but also extend to the nose radius area that stimulated the vibration or chatter which affect roughing surface and a higher cutting force. In addition, the larger nose radius causes the contacting area to be larger, thus leading to the cutting force is highly generated and also less the dynamic cutting force ratio (dFy/dFz).

Base on analysis results, relation between the dynamic cutting force ratio (dFy/dFz) and the tool flank wear (V_b) were pointed out. The bigger tool flank wear means smaller dynamic cutting force. Even the cutting conditions is changed, the dynamic cutting force ratio still keeps the same trend.

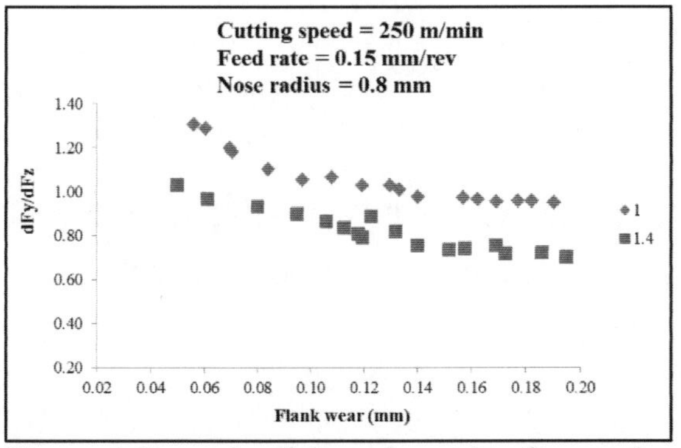

Figure 6. Example of dynamic cutting force ratio (dFy/dFz) versus tool flank wear under depth of cut 1 and 1.4 mm, cutting speed 250 m/min, feed rate 0.15 mm/rev and nose radius 0.8 mm.

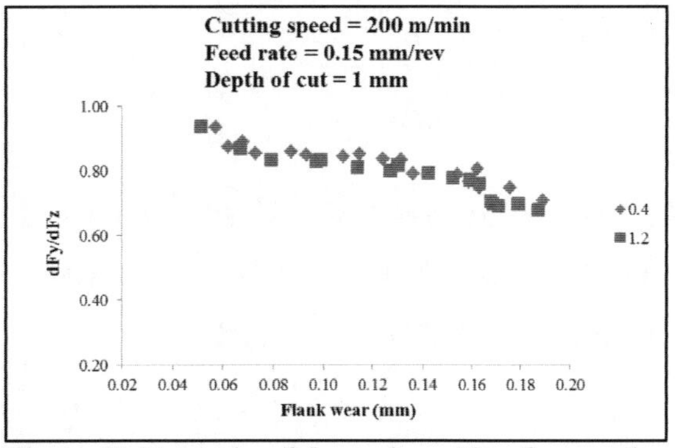

Figure 7. Example of dynamic cutting force ratio (dFy/dFz) versus tool flank wear under nose radius 0.4 and 1.2 mm, cutting speed 200 m/min, feed rate 0.15 mm/rev, depth of cut 1 mm.

4. Conclusions

According to experiment results, the conclusion can be explained as following.

- The most influential factors on the tool wear and the dynamic cutting force ratio are cutting speed followed by feed rate, depth of cut and nose radius respectively.
- The dynamic cutting force ratio directly associates to the tool wear. The dynamic cutting force ratio is slightly decreased when the tool wear is growth.
- The experiments demonstrate by varied cutting speed, feed rate, depth of cut or nose radius which was not differentiated the relation of dynamic cutting force ratio and the tool wear. The trend of dynamic cutting force ratio was slightly decreased even any cutting condition was changed.
- The most advantage of this study proved that it can predict the tool wear by utilizing of the dynamic cutting force ratio and also can be useful for the tool wear prediction mathematic model in turning process development in the future.

Acknowledgement

This work was performed by the partial funding of The Thailand Research Fund (TRF).

References

[1] Scheffer C, et al., *Development of tool wear-monitoring system for hard turning.* Int J Mach Tool Manuf, 2003. **43**: p. 973–985.

[2] Moriwaki, T., T. Shibasaka, and S. Tangjitsitchareon, *Development of in-process Tool Waer Monitoring System for CNC Turning.* JSME Internetional Journal, 2004. **47**.

[3] Jeyakumar, S., K. Marimuthu, and T. Ramachandran, *Prediction of cutting force, tool wear and surface roughness of Al6061/SiC composite for end milling operations using RSM.* Journal of Mechanical Science and Technology, 2013. **27**(9): p. 2813-2822.

[4] Tangjitsitcharoen, S., P. Thesniyom, and S. Ratanakuakangwan, *Prediction of surface roughness in ball-end milling process by utilizing dynamic cutting force ratio.* Journal of Intelligent Manufacturing, 2014.

[5] Tangjitsitcharoen, C.W.a.S., *A study of relation between tool wear and dynamic cutting force ratio in CNC turning process.* 2014.

[6] Somkait, T., *Development of Intelligent Monitoring and Identification of Cutting States by Spectrum Analysis for CNC Turning* Journal of Advance Mechanical Design, System, and Manufacturing 2008. **2(3)**: p. 366-377.

[7] J.K., *Application of wavelet transform technique to detect tool failure in turning operations.* The International Journal of Advanced Manufacturing Technology, 2005.

Synthesis and Evaluation of Titanium Oxide Modified Mussel Inspired Nanoparticles for Synergetic Photothermal and Photodynamic Effect

Nahla Rahoui[1a], Pei Zhou[1], Nadia Taloub[1], Mohammad Hegazy[1] and Yu Dong Huang[1b]

[1]MIIT Key Laboratory of Critical Materials Technology for New Energy Conversion and Storage, School of Chemistry and Chemical Engineering, Harbin Institute of Technology, P. O. Box: 1254, Harbin 150001, People's Republic of China.

E-mail: [a]nahlarahoui@hit.edu.cn; [b]huangyd@hit.edu.cn.

Abstract. Applying two therapeutical modality using one system is primordial for enhanced treatment efficiency and decreased side effects. The main objective of this work was to develop polydopamine (PDA) based hybrid nanoparticles with strong near infrared absorbance and good NIR to heat conversion efficiency. The results proved successful preparation of polydopamine nanoparticles (PDNS) decorated with metal oxide (TiO_2), spherical shaped of ~268 nm mean diameter and negative surface charge (ζ=−16.40 mV). The prepared PDNS@TiO_2 were characterized by FT-IR, UV-Vis, TEM and XPS. They exhibit a good thermal stability and significant photothermal conversion efficiency (η=34.9%). On other hand ESR spectra proves the ability of this kind of nanostructures to generate of ROS under UV irradiation due to the presence of TiO_2 on the surface. This paper describes the design and evaluation of novel nanostructures for combined photothermal and photodynamic effect. A remarkable synergistic therapeutic effect has been achieved compared with respective single treatments. This work suggests that the PDA-based nanoplatform can be a universal scaffold for single or combined therapy.

1. Introduction

The photothermal therapy (PTT) is considered as very promising tool to induce selective therapeutic effect by malignant tissues ablation. The main step in design of any PTT nanostructures is the selection of high absorptive agent in the NIR region. At the same time safe and efficient PTT agent [1]. The mussel inspired polydopamine polymer as NIR photothermal agent is known of its ease, durable deposition on all types of solid interface inorganic and organic substrates at different controllable thickness and remarkably high stability, thus PDA have been used as a versatile tool for surface modification and coating [2, 3].

The PDA was considered as a versatile material for surface modification due to its abundant amine cationic groups and also phenolic anionic groups known as indole, catechol, quinone and indolic/catecholic [3] that enhance metal ion chelation and the reducive effect of PDA on any metallic cations [4]. Besides the high photo-stability, PDA provides a high photo-protection effect to the coated material. Meanwhile PDA can absorb well in the NIR region and convert it into heat, especially PDA nanoparticles (PDNS) that have a high conversion efficiency of 40%. [5-7] To increase the therapeutic effect of the synthesised system and to get a combination of photothermal therapy and photodynamic therapy, PDNS could be decorated with TiO_2 nanoparticles using LPD method. The use of hybrid nanomaterials for combined therapeutic effect is more promising for NIR induced photothermal conversion efficiency through PDNS and UV induced generation of reactive oxygen species (ROS) to reduce the toxicity effect of metallic nanoparticles on normal tissues by the way limits the undesirable side effects [1,8]. Several studies investigated the titanium dioxide (TiO_2) as therapeutic agent, it has distinct properties such as: it low cost, structural stability, optical properties including high ultraviolet absorption, visible light induced photo-oxidation, remarkable photo-stability, good biocompatibility (widely used as food e.g., jelly, gum, food, excipient in medicine, toothpaste and cosmetics) [9, 10] besides the wide environmental application as photocatalysts for water, air, and soil decontamination [10]. Recently, the application of TiO_2 as photo absorbing agents that absorb well in the UV region ($\lambda = 275$–390 nm) at the same time can generate a cytotoxic ROS reactive oxygen species or free radicals (e.g. hydroxyl and superoxide radicals) [11].

In this work, in order to have an enhanced NIR photothermal therapeutic effect, PDNS and TiO_2 have been incorporated into one system (see figure 1) by design of PDNS interface with TiO_2 after chelation of Ti^{4+}, nucleation and growth through PDNS catechol groups [12]. The system have been characterised by several methods to prove successful synthesis. The photothermal measurements reveals the enhanced of photothermal conversion efficiency of TiO_2 designed PDNSs compared to unmodified PDNS besides the further generation of ROS after UV irradiation.

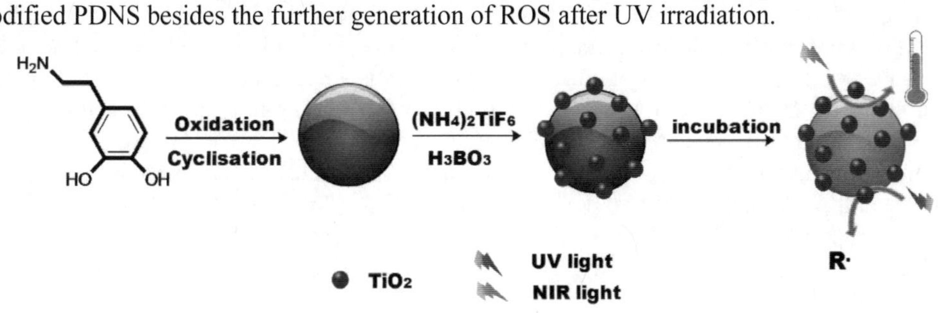

Figure 1. Experimental procedure for elaboration of PDNS@TiO_2 hybrid DDS

2. Experimental Materials and Methods

2.1. Materials and reagents

Ethanol absolute was obtained from Tianjin Hengxing Chemical preparation. The ultrapure water used for solution preparation was prepared using Millipak filter of 0.22 μm pore size Millipore in Milli-Q system. Dopamine hydrochloride DA-HCl (~35-37%) was purchased from Aladdin Industrial Corporation. Sodium hydroxide (NaOH) (≥96%) and Boric acid (H_3BO_3) (≥ 99.5%) were purchased from Sinopharm Chemical Reagent. All chemical reagents were used as received without any purification. Hexa-fluoro ammonium titatante $(NH4)_2TiF_6$ (Analytical reagent) was obtained from Shanghai Macklin Biochemical Co. Ltd.

2.2. Characterization methods

Fourier transform infrared (FTIR) spectra were recorded on a Nicolet FT-IR 5700 spectrometer (USA) the samples were prepared using potassium bromide (KBr) pastille (few mg in 100mg of KBr)
UV-Visible absorption spectra were collected on Lambda 750S spectrophotometer (PerkinElmer, USA) with a scanning range from 200 to 900 nm. X-ray photoelectron spectroscopy (XPS) measurements were carried on an ESCALAB 250 XPS system (Thermo Electron Corporation, USA) with an Al Kα X-ray source (1486.6 eV photons), the analysis deep was of 8~10nm.UV-Vis absorption spectra were performed on a Lambda 750 spectrophotometer. Transmission electron microscope (TEM; Hitachi H-7650 microscope, operating at an accelerating voltage of 100 Kv) was used to determine the size of synthesised and modified PDNS and to evaluate the morphology of the nanoparticles. Zeta potential

analysis were carried out on Malvern Instruments Zetasizer Nano series instrument of samples (2.5 mg/mL in deionized water). These Characterisation were conducted at room temperature. NIR photothermal effect were studied on NIR laser diode (STPC-510P) purchased from Xiamen Black and Decker Intelligent Technology (diameter of the laser spot: ~8.0 mm). The temperature of solution was monitored using a digital thermometer with a thermocouple. Electron spin resonance spectra was recorded on Bruker ESR (A200S-95/12) at 25° C with microwave power of 32 mW and a microwave frequency of 9.42 GHz.

2.3. Preparation of experimental materials

2.3.1. Preparation of PDNS: PDA nanoparticles (PDNS) were prepared by the oxidation and self-polymerization of dopamine though previously reported methods [3] The PDNS with ~ 250 nm average diameter were prepared as follow: Briefly, 180 mg of dopamine hydrochloride (DA-HCl) was dissolved in 90 mL of deionized water, at T=50°C and under vigorous stirring, 760 µL of NaOH (1N) solution was added. After 5h, the obtained dark black solution were centrifuged (12500 rpm) and washed and then all high molecular weight materials were removed by 4500 rpm centrifugation.

2.3.2. Insitu deposition of TiO_2 on PDNS: In this step, the PDNS surface has been decorated with TiO_2 through liquid phase deposition method (LPD). First, 5mg of PDNS were ultrasonically dispersed in LPD solution for about 30 minutes, LPD solution is a mixture of 5 mL $(NH_4)_2TiF_6$ (0.1 M) as precursor and 5mL H_3BO_3 (0.2 M) prepared in high purity water. The plentifully active hydroxyl groups on PDA coating can chelate the Ti^{4+} metallic cations in $(NH_4)_2TiF_6$ in order to boost the nucleation and the growth of TiO_2 on the PDNS surface. Subsequently, the mixture container was sealed and placed in water bath under mild agitation at 40 °C for 2h in dark to graft TiO_2 then cooled down to room temperature. Finally the reaction products were centrifuged at 12300 rpm for 15 min and washed several times with diH_2O until a constant pH the reaction products was freeze dried for 8h.

2.3.3. The study of photothermal effect. To study the photothermal effect amount of PDNS@TiO_2 were dispersed in high purity water (1mL) at different concentrations (from 25 to 200 µg/mL)and then put into quartz specimen holder and then irrdiated with 808 nm NIR laser lat 2.5 W cm^{-2} for 10 min.The temperature of the solutions was measured through submerged thermocouple microprobe in the solution, the increase of temperature during 10 minutes of NIR irradiation should be recorded every 30s in ordre to skectch the temperature profiles of the irradiated samples. The photothermal test was repeated for three times independently. The mean value was recorded as the final result.

2.3.4. The study of ROS generation. Two samples of PDNS@TiO_2 was prepared for ESR measurement, one left in dark for 15 min, the other one left in dark then irradiated with UV (320 nm) for 5min just before ESR measurement. The ESR measurements were done without spin trap because according to literature PDA have a scavenging effect what makes it suitable spin adduct [13].

3. Results and Discussion

3.1. Morphological analysis

Transmission electron microscope images of synthesised PDNSs (see figure 2) indicates well spherical shaped nanoparticles of uniform appearance. The average particle size was determined to be size ~ 238.32 nm, after TiO_2 deposition the particle's size enhanced slightly to ~ 269.55 nm.

3.2. Structural Analysis

The successful synthesis of polydopamine nanoparticles and it modification with TiO_2 layer was further confirmed by FTIR spectroscopy and UV-Visible spectrophotometer.

3.2.1. FTIR: The spectrum of PDNS exhibits peaks at about 3276 cm^{-1} that proves the abundant hydroxyl groups of the PDA catechol groups, this peak become more broadened and shifted to 3397.9 cm^{-1} after the incorporation of TiO_2 with an hydroxyl ends as shown in FTIR spectra (figure 3a), the peak at 1688 cm^{-1} indicates the existence of a large amount of carbonyl groups (C=O) on surface of PDA. On other hand the peaks at 1557 cm^{-1} and 1618 cm^{-1} corresponding to the N–H and C=C aromatic ring stretching, respectively [14], while the peak at 842 cm^{-1} corresponds to C–N stretching.

3.2.2. UV-Vis: UV-vis spectra depicted in figure 3(b) of PDNS and PDNS@TiO_2 at concentration of 0.5mg/mL were compared. As shown in figure 3b PDNS absorb in UV-vis and NIR range. Generally the TiO_2 nanostructures have bandgap of >3.0 eV, which increases their absorbance in UV range and limits their absorption in the NIR light [1], in our case the TiO_2 nanostructures labelled PDNS have a considerable absorption in UV-Vis and NIR region, this could be explained by tuning absorbance of TiO_2 and PDNS consequently, an enhanced absorbance was remarked for PDNS@TiO_2 compared to PDNS. The UV-vis-NIR spectra shows three absorption peaks in UV range at 201, 250, 320nm characteristics of TiO_2 particles [15] in Visible and NIR range the two spectra are similar. The absorption of PDNS@TiO_2 is slightly higher than as synthesised PDNS.

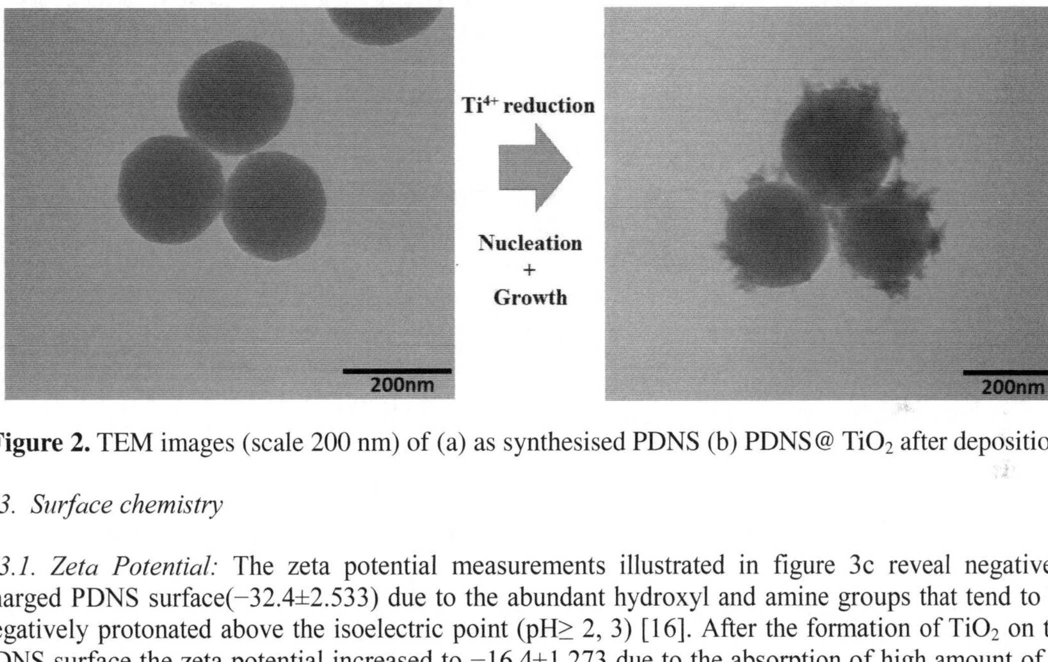

Figure 2. TEM images (scale 200 nm) of (a) as synthesised PDNS (b) PDNS@ TiO_2 after deposition

3.3. *Surface chemistry*

3.3.1. Zeta Potential: The zeta potential measurements illustrated in figure 3c reveal negatively charged PDNS surface(−32.4±2.533) due to the abundant hydroxyl and amine groups that tend to be negatively protonated above the isoelectric point (pH≥ 2, 3) [16]. After the formation of TiO_2 on the PDNS surface the zeta potential increased to −16.4±1.273 due to the absorption of high amount of Ti^{4+} cations, but it remains negative because a part of adsorbed Ti^{4+} have been reduced to TiO_2 with abundant hydroxyl groups on it.

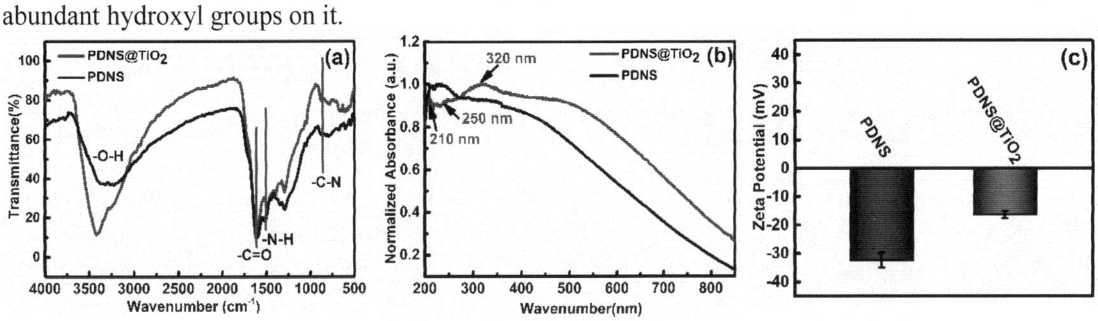

Figure 3. PDNS and PDNS@TiO_2 (a) FTIR spectra (b) UV-Vis spectra (c) Zeta potential

3.3.2. XPS: have been used for extraction of chemical surface information of unmodified and TiO_2 polydopamine, as shown in figure 4a the XPS survey spectra of three PDNS clearly show the peaks originated from C1s, O1s, and N1, while PDNS@TiO_2 shows four peaks originated from C1s, O1s, N1s and Ti2p. The high resolution C1s XPS spectra of PDNS in figure 4b was deconvoluted into four main peaks assigned to C-C/C-H (283.31 eV), C-N (284.36 eV), C-O (284.36 eV) and C=O (287.31 eV), respectively, the C=O appears due to oxidized form of PDA quinone group [2, 3, 17]. The XPS O1s and N1s spectra can further help to analyse the surface chemistry of the PDNS (figure 4c and 4g). N1s peaks in PDNS XPS spectra appeared at ~399.0 eV could be deconvoluted into three components; the peak at 398.0 eV assigned to -N= species of pyridine structure of PDA, the peak at 400 was attributed to the -N-H of the amine group in PDA cycle, last component at 402.61 eV corresponding to the primary amine-NH_2 and protonation primary amine –NH^{+3} [2, 14]. O1s peaks in PDNS XPS spectra appeared at ~500.0 eV. The O 1s deconvoluted spectra of PDNS could be deconvoluted into four components C–O (531.3 eV), O–C O (532.3 eV), and C=O (533.1 eV), attributed to catechol groups of PDA (figure 4c). The C1s and the N1s deconvoluted spectra (see figure 4e and figure 4h) of the PDNS@TiO_2 are similar to that of PDNS just some change in component amount. However the O1s (figure 4f). High resolution spectra of PDNS@TiO_2 can be deconvoluted into 5 components (Ti-O-Ti at 528.68 eV) (Ti-OH at 529.38 eV), (O-C at 530.88 eV), (O=C at 530.14 eV), (O-H at 531.71 eV), and adsorbed O_2 (at 532.59 eV). [15] Ti high resolution XPS spectra in figure 4d shows two peaks at 457.10 and 463.33 eV, corresponding to Ti $2p_{3/2}$ (Ti^{4+}) and Ti $2p_{1/2}$ (Ti^{3+}), respectively revealing the reduction of a part of the Ti^{4+} into Ti^{3+} [1, 15].

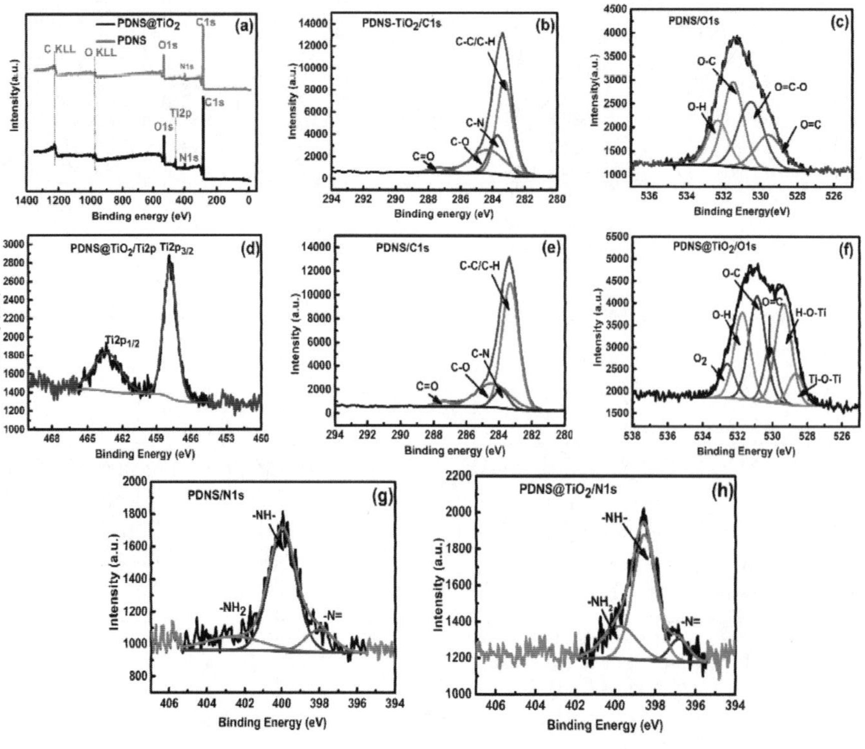

Figure 4. XPS data (a) XPS survey of PDNS and PDNS@TiO_2; (b) deconvoluted high resolution C1s spectra of PDNS powder (c) deconvoluted high resolution O1s spectra of PDNS powder; (d) deconvoluted high resolution Ti2p spectra of PDNS@TiO_2 powder; (e) deconvoluted high resolution C1s spectra of PDNS@TiO_2 powder (f) deconvoluted high resolution O1s spectra of PDNS@TiO_2powder; (g) deconvoluted high resolution C1s spectra of PDNS powder;(h) deconvoluted high resolution C1s spectra of PDNS powder

3.4. Photothermal measurements

In light of UV-Vis absorption spectra results, the PDNS and PDNS@TiO_2 can absorb in the NIR region (at 808 nm the absorbance of 200μg/mL was A_{808}=0.147). At different concentration PDNS@TiO_2, samples were irradiated with NIR laser (808nm) at power density of I=2.5 W/cm^2. The temperature profiles shows NIR induced temperature elevation dependence to nanostructure concentration as shown in figure 5a. The Temperature elevation records for PDNS@ TiO_2 were mainly attributed to the PDNS strong absorption and conversion of NIR light into heat. The photothermal conversion efficient (η) was calculated via equation (1), the heat capacity of solvent (Ultra-pure water) is ms Cs=4.2 J/°C. Maximum temperature elevation ΔT_{max}=29.8 was determined from heating curves (see figure 5b), whereas, the constant of time τ_s= 461.1s was determined from time plot (figure 5b) drawn using $t = -\tau_s \ln(\Delta T / \Delta T_{max})$ and cooling curve of 200 μg /mL PDNS@TiO_2 sample (figure5c). hS value (0.0091882 J/°C s) was extracted from $\tau_s = m_s C_s / hS$ and introduced in equation (1) along with other parameters, then η was calculated to be 34.9%, that slightly lower the PDA based nanostructures (40%) [18] due to the presence of TiO_2 that might limits the PDNS absorption of NIR light. However, our synthesised PDNS@TiO_2 nanostructures exhibits an efficient NIR to heat photothermal conversion enough for inducing sufficient hyperthermia.

$$\eta = \frac{hS\Delta T_{max} - Q_S}{I(1-10^{-A808})} \qquad (1)$$

3.5. Generation of Reactive oxygen species

ESR spectra (figure 5d) of PDNS@TiO_2 show several paramagnetic peaks but with major differences in the peaks intensities, this could be caused by the presence of various paramagnetic species, the synthesis process, and the light effect, washing and drying during samples preparation for analysis and also the thermal effect during the storage.

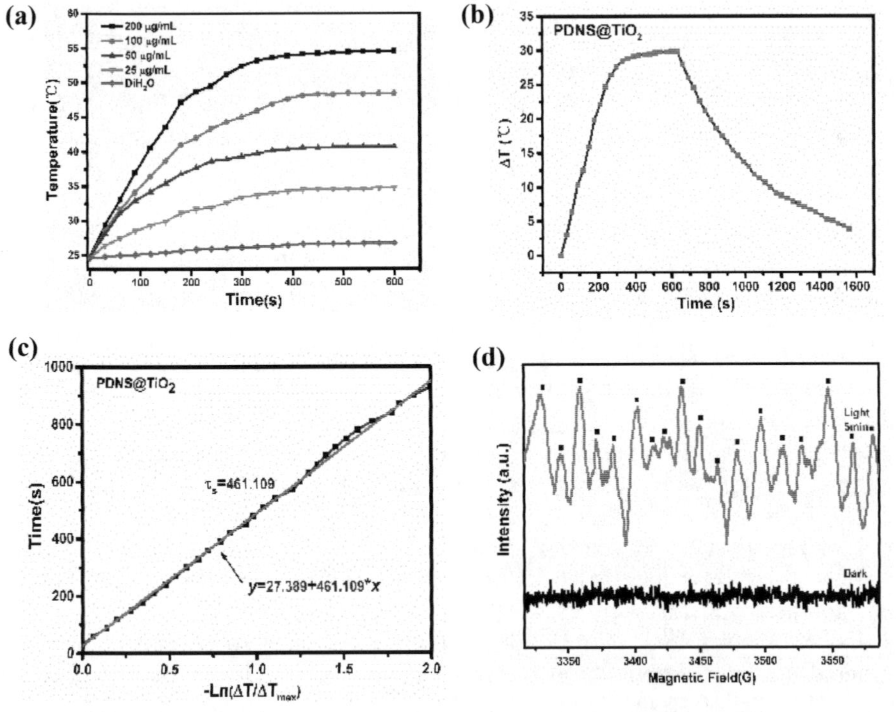

Figure 5. PDNS@TiO_2 (a) Temperature profiles at different concentration, (b) Heating cooling curve at 200 μg/mL, (c) Plot of cooling time versus negative natural logarithm of the temperature driving force, (d) ESR of generated ROS after PDNS@TiO_2 irradiation with UV.

On the other hand, the ESR spectra proves that the synthesised system generates ROS when it is photoxidized due to the presence of TiO_2 by using the polydopamine as spin trap agent that reacts with generated ROS and forms more stable radicals with half-life time longer enough for ESR analysis. The UV irradiated TiO_2 leads to PDA radical formation that has been detected by ESR [8]. The intensity of ESR signals depends to the concentration of free radicals. Our results proves the UV induced generation of toxic ROS using TiO_2 modified nanostructures.

4. Conclusion and future scope

Applying two therapeutically modality is primordial for enhanced treatment efficiency and decrease the side effects. The main objective of this work was to develop PDA based hybrid nanoparticles as UV and NIR photo absorbing agent. The results proved successful preparation of PDNS@TiO_2. The prepared PDNS@TiO_2 exhibit a good photothermal efficiency 34.9% due the PDNS substrate under NIR irradiation and also an effective generation of toxic ROS due to titanium metal oxide on the surface under UV irradiation. Our results further demonstrate that the as-obtained PDNS/TiO_2 nanostructures provides a synergetic NIR induced photothermal effect and UV induced ROS for photodynamic therapy application. A remarkable synergistic therapeutic effect has been achieved compared with respective single treatments. This work suggests that our synthesised nanoplatform could be applied in many filed such as biomedical and further photo-catalysis applications.

Acknowledgments

This work was financially supported by the National Natural Science Foundation of China (Grant No. 51673054), the Harbin City Science and Technology Innovation Talent Foundation (Grant No. 2017RAYXJ003), Shanghai Space Science and Technology Innovation Foundation (Grant No. SAST2017-114).

References

[1] Ou G, Li Z, Li D, Cheng L, Liu Z, Wu H.Photothermal therapy by using titanium oxide nanoparticles[J]. *Nano Research*, 2016, 29(5):1236-1243.
[2] Ye W, Chen Y, Zhou Y, Fu J, Wu W, Gao D, Zhou F, Wang C, Xue D.Enhancing the catalytic activity of flowerike pt nanocrystals using polydopamine functionalized graphene supports for methanol electrooxidation[J]. *Electrochimica Acta*, 2014, 142:18-24.
[3] Chen D, Zhao L, Hu W. Protein immobilization and fluorescence quenching on polydopamine thin films[J]. *Journal of Colloid & Interface Scienc*e, 2016, 477:123-130
[4] Ju KY, Lee Y, Lee S, Park SB, Lee JK. Bioinspired polymerization of dopamine to generate melanin-like nanoparticles having an excellent free-radical-scavenging property [J]. *Biomacromolecules,* 2011, *12*(3):625.
[5] Jiang Q, Luo Z, Men Y, Yang P, Peng H, Guo R, Tian Y, Pang Z, Yang W. Red blood cell membrane-camouflaged melanin nanoparticles for enhanced photothermal therapy[J]. *Biomaterials*, 2017,143:29-45.
[6] Rui Z, Sheng W, Ye T, Jiang X, Fu D, Shen S, Yang W, Polydopamine-coated magnetic composite particles with enhanced photothermal effect[J].*ACS Appl. Mater. Interfaces*, 2015,7:15876-84.
[7] Dong Z, Gong H, Gao M, Zhu W, Sun X, Feng L, Fu T, Li Y, Liu Z, Polydopamine nanoparticles as a versatile molecular loading platform to enable imaging-guided cancer combination therapy[J] *Theranostic*s, 2016, 6:1031-1042.
[8] He,W, Liu Y, Wamer WG, Yin JJ. Electron spin resonance spectroscopy for the study of nanomaterial-mediated generation of reactive oxygen species [J]. *Journal of Food and Drug Analysis*, 2014, 22(1):49-63
[9] Sun L, Li Z, Li Z, Hu Y, Chen C,Yang C, Du B, Sun Y, Besenbacher F, Yu M. Design and mechanism of core-shell TiO_2 nanoparticles as a high-performance photothermal agent[J], *Nanoscal*e, 2017, 9(42):16183-16192.

IOP Conf. Series: Materials Science and Engineering **389** (2018) 012009 doi:10.1088/1757-899X/389/1/012009

[10] Naserzadeh P, Ghanbary F, Ashtari P, Seydi E, Ashtari K, Akbari M. Biocompatibility assessment of titanium dioxide nanoparticles in mice feto-placental unit[J].*Journal of Biomedical Materials Research Part A*. 2017, 106(2):580–589.
[11] Kotagiri N, Sudlow GP, Akers WJ, Achilefu S.Breaking the depth dependency of phototherapy with cerenkov radiation and low-radiance-responsive nanophotosensitizers[J]. *Nature Nanotechnology*, 2015, 10(4):370-379.
[12] Zhang Y, Wang W, Ma X, Jia L. Polydopamine assisted fabrication of titanium oxide nanoparticles modified column for proteins separation by capillary electrochromatography[J]. *Analytical Biochemistry*, 2016, 512:103-109
[13] Ju, K. Y., Lee, Y., Lee, S., Park, S. B., & Lee, J. K. (2011). Bioinspired polymerization of dopamine to generate melanin-like nanoparticles having an excellent free-radical-scavenging property. *Biomacromolecules, 12*(3), 625-32.
[14] Han X, Zhang L, Li C. Preparation of polydopamine-functionalized graphene–Fe_3O_4 magnetic composites with high adsorption capacities[J]. *Rsc Advances*, 2014, 4(58):30536-30541.
[15] Su C, Liu L, Zhang M, Zhang Y, Shao C. Fabrication of Ag/TiO_2 nanoheterostructures with visible light photocatalytic function via a solvothermal approach[J]. *Crystengcomm*, 2012, 14(11):3989-3999.
[16] Su J, Sun H, Meng Q, Yin Q, Zhang P, Zhang Z, Yu H, Li Y. Long circulation red-blood-cell-mimetic nanoparticles with peptide-enhanced tumor penetration for simultaneously inhibiting growth and lung metastasis of breast cancer[J]. *Adv. Funct. Mater.* 2016, 26:1243-1252.
[17] Zangmeister RA, Morris TA, Tarlov MJ. Characterization of polydopamine thin films deposited at short times by autoxidation of dopamine[J]. *Langmuir,* 2013, 29:8619-8628.
[18] Jiang Q, Luo Z, Men Y, Yang P, Peng H, Guo R, Tian Y, Pang Z, Yang W, Red blood cell membrane-camouflaged melanin nanoparticles for enhanced photothermal therapy[J], *Biomaterials*, 2017, 143:29-45.

IOP Conf. Series: Materials Science and Engineering **389** (2018) 012010 doi:10.1088/1757-899X/389/1/012010

Study on Influence of Profile Modification on Transmission Performance of Spur Gear

Geng Hou[1a], Yinghui Zhang[1b], Weidong He[1] and Xiaochao Sun[2]

[1]School of Mechanical Engineering, Dalian Jiaotong University, Dalian 116028, Liaoning, China
[2]Yingkou Qingying Petrochemical Equipment Co., Ltd., Yingkou 115004, Liaoning, China

E-mail: [a]geng.hou@qq.com; [b]zyh@djtu.edu.cn

Abstract. Profile modification is an important method used for improving the spur gear transmission performance. By using the engineering software of Romax Designer, the transmission system model of single-stage cylindrical spur gear is established. And taking the elimination of impact loads in spur gear transmission process as the target, the full tooth profile modification of driving gear is completed with the modification curve of parabola. The flash temperature, oil film thickness and transmission error are taken as the indexes for measuring the transmission performance of spur gear. Through comparing the surface flash temperature, oil film thickness and transmission error before and after profile modification, the influence of profile modification on transmission performance of driving gear is obtained. The simulation and investigation results show that profile modification plays an important role in reducing the flash temperature, improving the tooth lubrication state and narrowing the fluctuation range of spur gear transmission error.

1. Introduction

During the actual operation of the involute spur gear pair, the tooth subjected to loading will occur elastic deformation, and normal pitch of the meshing teeth pair is not equal. Therefore, the interference and impact of mesh-in and mesh-out is generated [1-2]. This will cause the scraps on the tooth surface to accumulate near the pitch circle, and accelerate wear and agglutinate failure on tooth surface [3]. In addition, while the number of gear meshing teeth pairs is changing, the instantaneous impact load is also generated. In view of the above problems, the profile modification should be taken to alleviate the impact during gear meshing. Therefore, it is significance to study the influence of profile modification on the transmission performance of spur gear. In this paper, based on the numerical calculation of Romax Designer software, eliminating impact load is taken as the target of profile modification, the profile modification value of driving gear is obtained. Furthermore, through comparing the flash temperature, the oil film thickness and the transmission error before and after profile modification, the influence of profile modification on these three transmission performance indexes is analyzed. In order to provide reference for the studying on profile modification of spur gear.

2. Basic parameters of model used for simulation

The simulation model used for study is a single-stage cylindrical spur gear transmission system. The gear transmission system model established by Romax Designer is shown in Figure 1. (The box is not considered in this model.) In Figure 1, n is the rev of the input shaft, which equal to 970 rpm; M is the load torque of output shaft, which equal to 886 N▪m. In this simulation model, the material of gears is 40CrNiMo (steel brand of China), and the basic parameters of gear pair are shown in Table 1; The material of shafts is 45 carbon steel (steel brand of China); The bearings selected are all conical roller bearings of the type of SKF 31313; The lubricating mode of the gear transmission system is oil lubrication, and the oil brand is ISO VG 100 PAG Synthetic.

In addition, in order to make the simulation results easy to study, and avoid the influence of unbalanced load on the study of profile modification, the driving gear of the model has been axial modified by the method of helix angle modification. Therefore, the gear pair in this model has no unbalanced load [4].

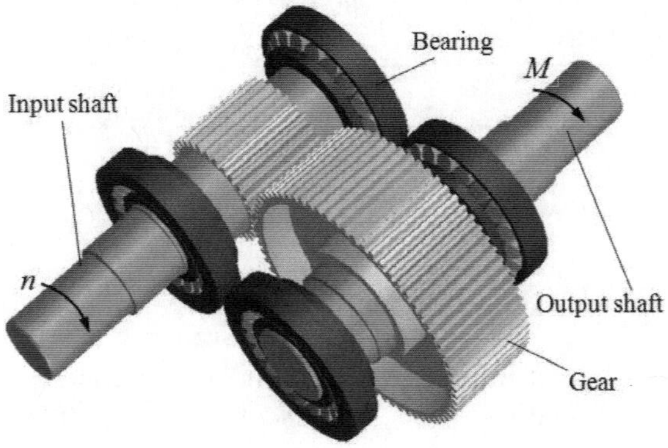

Figure 1. Model of spur gear transmission system

Table 1. Basic parameters of spur gear pair

	The number of teeth	Modulus /mm	Tooth width /mm	Pressure angle /°
Driving gear	30	3.0	78	20
Driven gear	71	3.0	78	20

3. Determination of profile modification value

3.1 The theory and method of profile modification

As shown in Figure 2 is the load distribution diagram of one tooth pair of spur gear in one meshing cycles. Before profile modification, the variation trend of the tooth load is polyline AMNHIOPD in whole meshing period. From the figure, Δq_1, Δq_2 and Δq_3 is the impact load of mesh in, the impact load of mesh out and the impact load during the number of gear meshing tooth pairs is changing, respectively. Sudden change of load can result in shock excitation. In order to avoid the adverse effect caused by the shocking load, the profile modification of the spur gear is required. And through profile modification, the variation trend of the tooth load can change from polyline AMNHIOPD to polyline ABNHIDP.

At present, there are many kinds of methods used for calculating profile modification value in the world. And different countries and enterprises generally adopt different methods of profile modification according to their actual experience. In order to have better manufacturability in the process of gear machining [5-6], the profile modification method of pinion (driving gear) is full tooth profile modification in this paper. The diagram of profile modification is shown in Figure 3, the C_α is the value of profile modification by the modification curve of parabola [7].

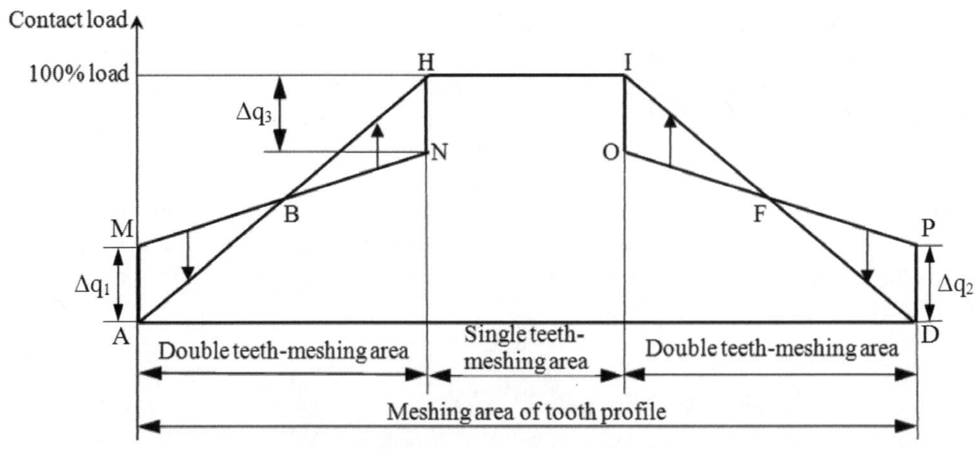

Figure 2. The diagram of the load distribution before and after the profile modification

3.2 Calculating the profile modification value

Follow the procedure shown in Figure 4, and input different values of full tooth profile modification of driving gear manually to software of Romax Designer, the target value profile modification value C_α^* can be obtained. Finally, it is calculated that the target profile modification value is 14μm.

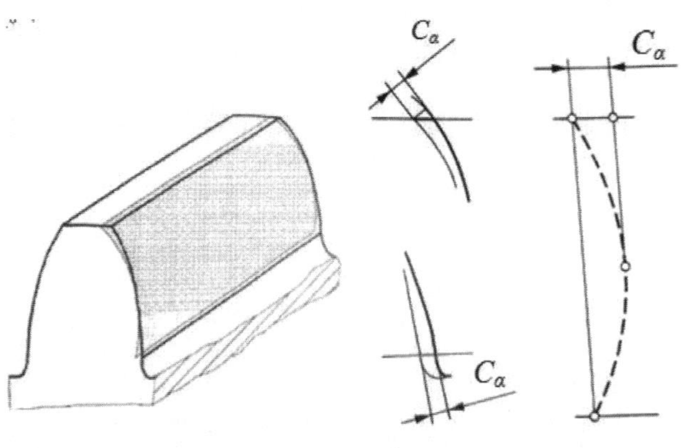

Figure 3. The diagram of full profile modification

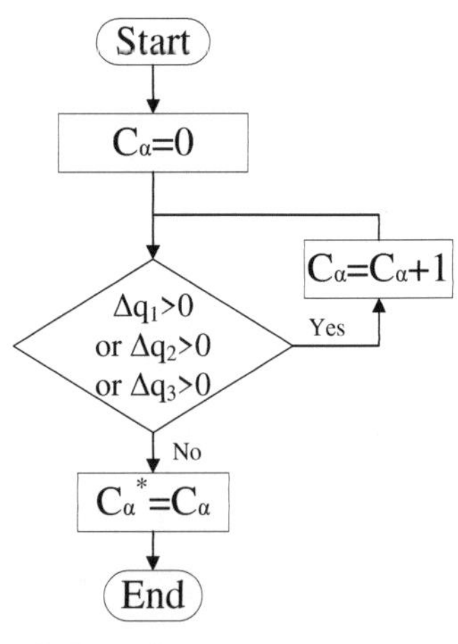

Figure 4. The flow diagram of profile modification

Per unit length loads of tooth before and after profile modification are shown in Figure 5(a) and Figure 5(b). (In both Figure 5(a) and Figure 5(b), the meaning of abscissa are roll angle, which is the expansion angle of the tooth profile along the base circle, and the same below.) The comparison shows, the load acting on the top and root of the tooth of driving gear decreases significantly after profile modification. Moreover, it eliminates the instantaneous impact load during the changing of the number of meshing tooth pair.

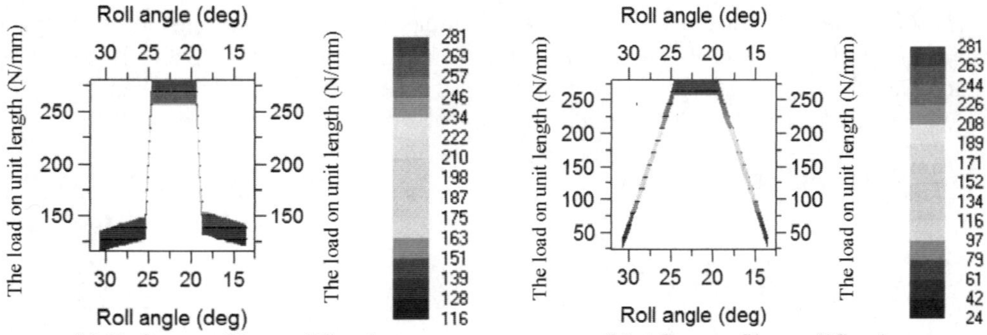

Figure 5. Per unit length load of tooth before and after profile modification

4. The influence of profile modification on transmission performance of spur gear

4.1 The influence of profile modification on flash temperature of tooth surface

4.1.1 The theory of calculating the flash temperature of tooth surface. Using the software of Romax Designer, based on the flash temperature method, the flash temperature of gear tooth surface before and after profile modification are calculated [8]. And the formula of this theory is

$$\theta_{\max} = 0.62\mu\omega_b^{3/4}\left[E_{red}\left(\frac{1}{\rho_1}+\frac{1}{\rho_2}\right)\right]^{1/4}\frac{\left|\upsilon_{\rho1}-\upsilon_{\rho2}\right|}{B_1\sqrt{\upsilon_{\rho1}}+B_2\sqrt{\upsilon_{\rho2}}} \tag{1}$$

In formula (1), $\theta_{\max}$ is flash temperature, which the unit is ℃; μ is friction coefficient; ω_b is normal force on unit tooth width, which the unit is N/mm; E_{red} is the synthetical elastic modulus of gear pair materials, which the unit is MPa; ρ_1 and ρ_2 are respectively the radius of curvature of driving gear and driven gear at the meshing point, which the unit are mm; $\upsilon_{\rho1}$ and $\upsilon_{\rho2}$ are the tangential velocities of meshing gears along the contact surface at the meshing point, which the unit are mm/s; B_1 and B_2 are the thermal contact coefficients of meshing gears.

4.1.2 Simulation and analysis of flash temperature of tooth surface. Through numerical calculation of Romax Designer, the flash temperature the driving gear tooth surface before and after tooth profile modification are obtained, as shown in Figure 6(a) and Figure 6(b). Comparing the flash temperature before and after the profile modification can be known, profile modification makes the maximum temperature location of the tooth surface move from the top and root to the location near the pitch line. Moreover, the maximum flash temperature of the tooth surface has been decreased after profile modification. Therefore, profile modification is conductive to avoid or alleviate the agglutination of tooth surfaces.

4.2 The influence of profile modification on oil film thickness of tooth surface

4.2.1 The theory of calculating the oil film thickness of tooth surface. Using the software of Romax Designer, based on the film thickness formula of Dowson-Higginson line contact elastohydrodynamic lubrication, the minimum oil film thicknesses of gear tooth surface before and after profile modification are calculated [9-10]. And the formula of this theory is

$$h_{\min} = \frac{2.65\alpha^{0.54}\left(\eta_0 U\right)^{0.7}R^{0.43}}{E'^{0.03}W^{0.13}} \tag{2}$$

In formula (2), $h_{\min}$ is the minimum oil film thickness, which the unit is μm; α is viscosity coefficient of lubricating oil, which the unit is mm^2/N; η_0 is dynamic viscosity of lubricating oil, which the unit is

(N▪s)/mm^2; E' is the synthetical elastic modulus of gear pair materials, which the unit is MPa; R is the equivalent radius of curvature of a gear pair at the meshing point, which the unit is mm; U is the entrainment velocity of lubricating oil of a gear pair at the meshing point, which the unit is m/s; W is the load on unit contact length, which the unit is N/mm.

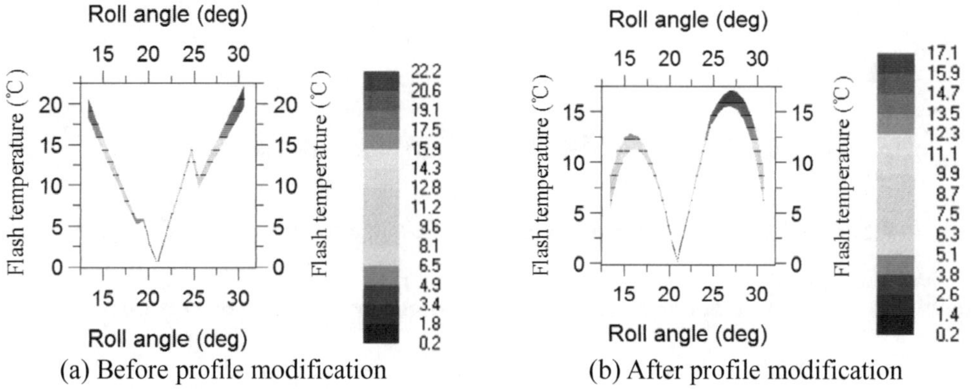

Figure 6. The flash temperature of driving gear before and after profile modification

4.2.2 Simulation and analysis of oil film thickness of tooth surface. Through numerical calculation of Romax Designer, the minimum oil film thicknesses of the driving gear tooth surface before and after tooth profile modification are obtained, as shown in Figure 7(a) and Figure 7(b). Comparing the Figure 7(a) and Figure 7(b) can be known, because the load of the single teeth-meshing area is larger, the oil film thickness in this area has a downward trend compared with the double teeth-meshing area. In addition, compared with before profile modification, the overall trend of oil film thickness after modification is increasing. And during the transition from double teeth-meshing area to single teeth-meshing area, the instantaneous decrease of oil film thickness is not appeared. Therefore, profile modification is conductive to improve the lubrication state of tooth surface.

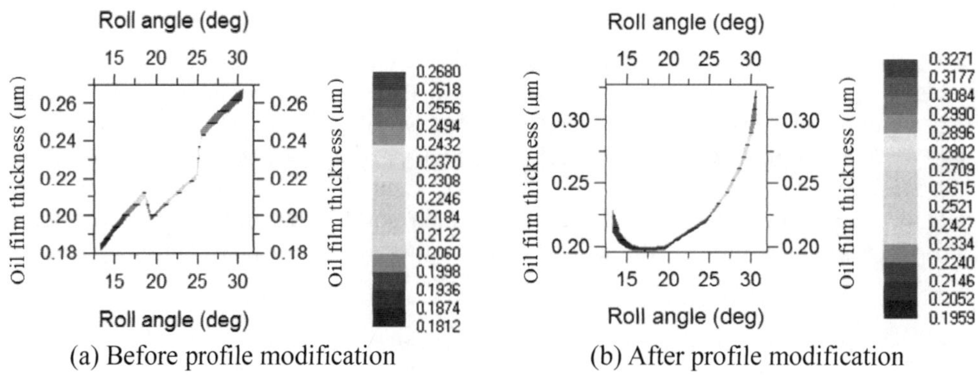

Figure 7. The oil film thickness of driving gear before and after profile modification

4.3 The influence of profile modification on transmission error of gear

4.3.1 The theory of calculating the transmission error of gear. In the actual working process of gear mechanism, the actual transmission ratio of the gear pair fluctuates in the vicinity of the theoretical value, due to the factors such as manufacturing error and elastic deformation. In this paper, the transmission error of gear is measured by the displacement error on meshing line [11], and its value is

$$TE = r_{b2}\theta_2' - r_{b1}\theta_1 \tag{3}$$

In formula (2), *TE* is the displacement error on meshing line, which the unit is μm; r_{b1} and r_{b2} are respectively the base circle radius of driving gear and driven gear, which the unit is mm; θ_1 is the theoretical rotation angle of driving gear, which the unit is rad; θ_2' is the actual rotation angle of driven gear, which the unit is rad.

4.3.2 Simulation and analysis of transmission error of gear. Through numerical calculation of Romax Designer, the transmission error of the driving gear before and after tooth profile modification are obtained, as shown in Figure 8(a) and Figure 8(b). Comparing with before profile modification, the increase of maximum transmission error is 6.52%, and increase of minimum transmission error is 59.3% after profile modification. However, fluctuation range of gear transmission error is reduced by 73%. Thus, the profile modification is conductive to reduce the fluctuation range gear transmission error, vibration and noise.

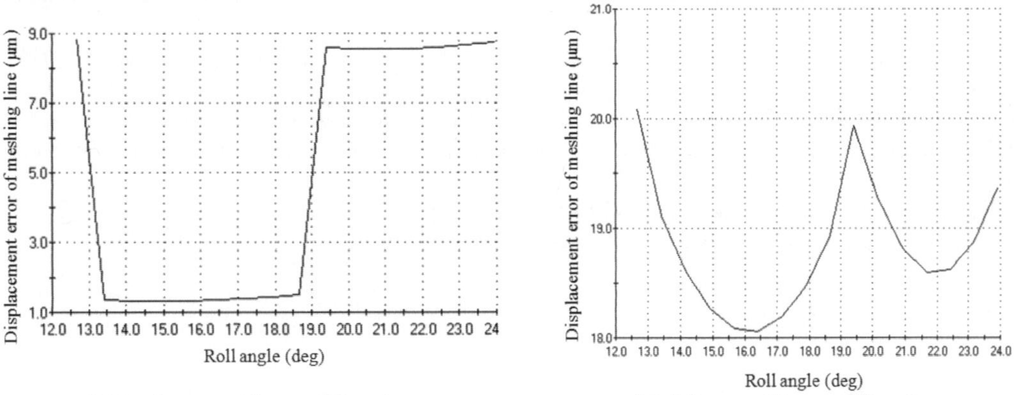

(a) Before profile modification (b) After profile modification

Figure 8. The transmission error of driving gear before and after profile modification

5. Conclusions

In this paper, using the engineering software of Romax Designer as the numerical computing tool, with the modification curve of parabola, the full tooth profile modification of driving gear of single-stage cylindrical spur gear transmission system is completed. Based on that, the influence of profile modification on the transmission performance of spur gear is simulated and analyzed. Comparing the flash temperature, the thickness of oil film and the transmission error of driving gear before and after profile modification, the results show that gear profile modification plays a positive role in reducing the flash temperature, improving the lubrication state and narrowing the fluctuation range of gear transmission error. That is, the profile modification of gear is conductive to improve gear transmission performance.

Acknowledgments

This work is supported by the 2011 Collaborative Innovation Center Project of Ministry of Education of China, and the Ph.D. Initial Scientific Research Fund of Liaoning Province, China (NO. 201501183).

References

[1] Li R M, Liu L P and Tan Z. Research on tooth profile modification related problems of machine tool main transmission system involute cylindrical gear *J. Manufacturing Technology & Machine Tool* No.5, (2015) pp 72-75.

[2] Ke H Y and Liu H L. Dynamic Characteristic Analysis on Spur Gear with Tooth Profile Modification Based on Dynamic Transmission Error *J. Mechanical Engineering & Automation* No.4, (2018) pp 14-17.

[3] Li R F. 1998 *Stiffness analysis and modification method of gear transmission* (Chongqing, China: Chongqing University Press) pp 1-3.

[4] S R Hipsley, R J Davey and R T Wheway. Optimization of Gear Tooth Contact by Helix Angle Modification *J. Gear Solutions* No.1, (2015) pp 42-48.

[5] Editorial Committee of Gear Manual. 2000 *Gear Manual* (*2nd volume*).2nd Ed. (Beijing, China: China Machine Press) pp 143-146.

[6] Cui W B. Study on lightweight technology of gear transmission system of reducer *J. Low Carbon World* No.4, (2017) pp 77-78.

[7] Romax. 2013 *Wind Turbine Drivetrains-Design and Analysis* (Shanghai, China: Shanghai Scientific & Technical Publishers) pp 81-87.

[8] Zhu X L and E Z K. 1992 *Analysis of Load Capacity of Gears* (Beijing, China: Higher Education Press) pp 261-262.

[9] Wen S Z and Huang P. 2002 *Principles of Tribology* .2nd Ed. (Beijing, China: Tsinghua University Press) pp 73-74.

[10] Ye Y D and Zhou Z B. Research on influence of gear parameters on lubrication oil film thickness under condition of elasto bydrodynamic lubrication *J. Journal of Machine Design* Vol.29, No.9, (2012) pp 73-76.

[11] Xiong H G. 2015 *Modeling, analysis and application of mechanical transmission system Romax Designer* (Wuhan, China: Huazhong University of Science and Technology Press) p 144.

IOP Conf. Series: Materials Science and Engineering **389** (2018) 012011 doi:10.1088/1757-899X/389/1/012011

Simulation of Part Surface Residual Stress Based on Grind-hardening Process

M H Liu[1], Z Yuan[1], Z Wang[1] and J Sun[1]

[1] School of Mechanical Engineering, Shenyang Jianzhu University, Liaoning Shenyang 110168, China

The corresponding e-mail address: rty26845@163.com

Abstract. Grind-hardening is a new technology for steel parts surface enhancing which uses thermal and mechanical composite effects on the parts in the grinding process. During the process, the workpiece surface generates hardened layer which improves the surface quality. As a key performance indicator of measuring the part surface quality, the residual stress of the hardened layer affects the strength, fatigue life and corrosion performance of the part directly. Therefore, the study of residual stress in hardened layer plays an important role in the effective control of hardened layer quality. First, grinding temperature field is analysed for studying the distribution characteristics of surface residual stress after grind-hardening process. Then, different grinding parameters an effect on the temperature field is studied by establishing the mathematical model and the heat transfer model of the grinding zone temperature field. Furthermore, combined with the grinding force generated in the grinding process, the thermo-mechanical coupling analysis with finite element method is adopted to obtain the residual stress and its distribution of the part. Also, the influence mechanism of the part residual stress formation is revealed.

1. Introduction

Grind-hardening technology is a newly built composite processing technology. It implements hardening using grinding heat on the surface, which makes the surface temperature rising to austenitizing temperature quickly, and the then rapidly cooling to the temperature of martensitic transformation. The process improves the surface hardness and wear resistance, which achieves same performance as surface hardening treatment [1-3]. The new technology combines the grinding technology with heat treatment well, which comply the increasingly advocated concept of process integration and green manufacturing in the engineering. Grind-hardening effect is performed by hardened layer. The performance of hardened layer varies from different microstructure [4]. Brinksmeier and Brockhoff from Gernany [5-6] proposed the technology that grinding heat can be used to quench steel surface first in 1994. Then grind-hardening technology attracted the attention of scholars from different countries. Up to now, most researches focus on local microstructure and microhardness and hardness penetration depth of the workpieces [7-9].

During the grind-hardening process, the combining effect of grinding force and grinding heat produce a lot of grinding force on the part surface. After grinding process, part of the grinding force remains in the surface, which formating the grinding residual stress. It has a decisive impact on part performance, dimensional stability, fatigue strength and wear resistance. For this consideration, it has some practical significance to carry out the residual stress study on part surface.

IOP Conf. Series: Materials Science and Engineering **389** (2018) 012011 doi:10.1088/1757-899X/389/1/012011

2. Grinding temperature field model

The grinding process is a random cutting process of many abrasive. A collection of heat source at many random grinding points can be approximately considered constant plane heat source. The total heat flux intensity of contact area is the average heat per unit area at per unit time, and is expressed as [10]

$$q = \frac{F_t v_s}{bl} \tag{1}$$

The heat flux intensity of the workpiece is [11]

$$q_w = R_w \frac{F_t v_s}{bl} \tag{2}$$

where, F_t is tangential grinding force, and v_s is the grinding wheel speed. b is the width of the grinding wheel or the workpiece (Whichever is the smaller value). l is contact arc length, and is expressed as

$$l = \sqrt{a_p d_s} \tag{3}$$

The width of the heat source is equal to the width of the wheel or workpiece (Whichever is the smaller value). The heat ratio of the incoming workpiece is expressed as [12]

$$R_w = \frac{1}{\left[1 + \frac{(k\rho c)_s}{(k\rho c)_w} \frac{v_s}{v_w} \left(\frac{A_R}{A}\right)_s\right]^{\frac{1}{2}}} \tag{4}$$

where, v_w is the workpiece feed rate. $(k\rho c)_s$ and $(k\rho c)_w$ are the thermal conductivity parameter of the grinding wheel and the workpiece, respectively, and is relation to the material properties mainly. Since plane grinding, $(v_s/v_w)(A_R/A)_s \approx 1$, therefore

$$R_w = \frac{1}{\left[1 + \frac{(k\rho c)_s}{(k\rho c)_w}\right]^{\frac{1}{2}}} \tag{5}$$

3. Temperature field simulation results and discussion

3.1. Finite element modeling and load

The workpiece is modeled by Finite element software ANSYS. In the grinding process, the body structure has not changed. So hardening effect only generated on the surface. Therefore, the refined mesh on the surface of the workpiece makes the simulation results more accurate. 45 steel as the workpiece material is chosen in this study. And the model size is 90mm×9mm×14mm. Parameters for grinding process are shown in table 1.

Table 1. Grinding process parameters

No.	1	2	3	4	5	6
a_p/μm	350	400	450		400	
v_w/(m/min)		9		0.54	0.6	0.66

Triangle heat source model is used and optimized for simulation. In temperature field study, a moving heat source can be loaded directly on the workpiece surface [13]. To solve this problem, we suggest that heat source is moving forward continuously. Each step the heat source moving forward, load the heat flow density load on the corresponding unit of the current step instead of the former load. Besides, the last calculation result is used as the initial condition in the new step. Repeat this process step by step until realizing the moving heat source load in the finite element model. The movement of heat source is implemented by *DO, *ENDDO. Due to the temperature rising of workpiece produces temperature difference from the air outside, convective heat transfer coefficient is loaded on the workpiece surface. The ambient temperature is T_{∞}=20 ℃.

3.2. Temperature field simulation results and discussion

Figure 1 shows the temperature nephogram of the 218th load step (t=2.75s) for grind-hardening simulation. From Figure 1, it can be seen that the highest temperature is 1240℃, and the lowest temperature is 20℃ in the 218th step.

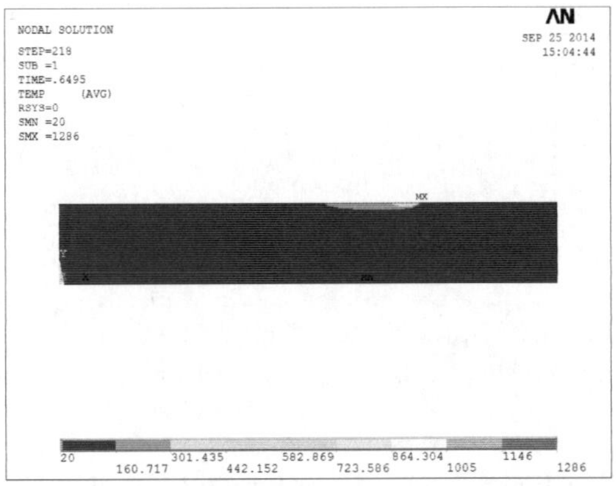

Figure 1. Temperature field simulation results for the 218th step load when a_p=400μm,v_w=0.6m/min

From Figure 2, when moving heat source arrives at a certain position of the middle zone, a lot of grinding heat will generated. In this case the temperature of the workpiece may rise quickly to the maximum value 1240℃ which is much more than Ac3 phase change threshold of 45 steel(720℃). When the heat source moves away from this node, temperature decreases rapidly lower than martensitic transformation temperature which is affected by the thermal conductivity of the workpiece itself and convective heat with the surrounding air. After cooling, martensite is produced so as to achieving the surface hardening on the workpiece surface.

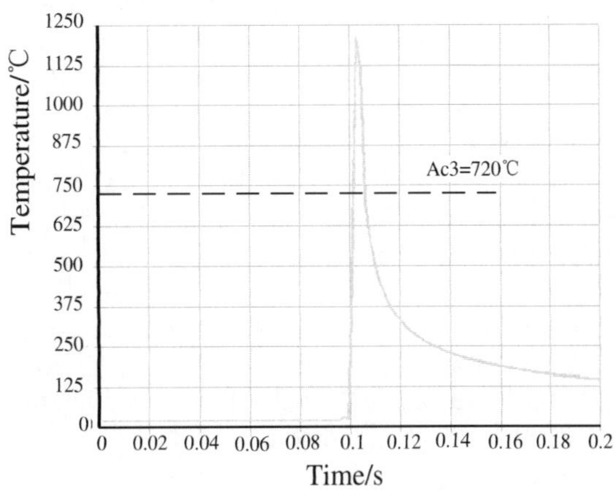

Figure 2. Temperature variation from time when a_p=400μm, v_w=0.6m/min

When the heating temperature of 45 steel is above Ac3, the body structure begins to change for the austenite. With the temperature dropped below Ms(transition temperature of martensite) rapidly, austenite changes into martensite. Therefore, the zone that heating temperature above Ac3 is approximate seen as grind-hardening area. Hardness penetration depth can be predicted as well. As shown in Figure 3, the surface temperature is much higher than austenite phase transition temperature. At 0.1mm from the surface, the maximum temperature is higher than the Ac3 temperature. In this consideration, the region from the surface to 0.1mm away from the surface is recognized as the hardened layer.

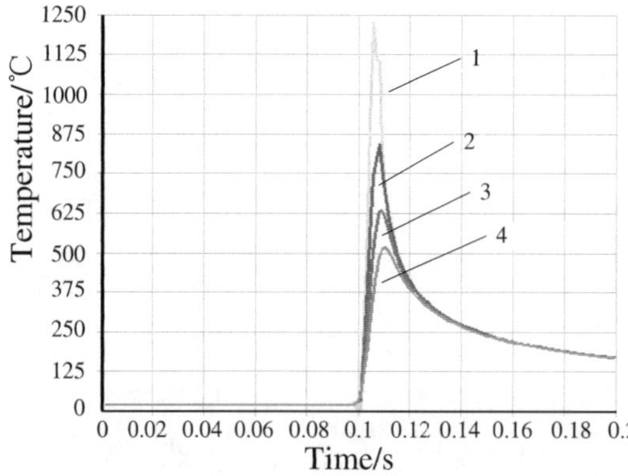

Figure 3. The temperature change with different depth of nodes when a_p=400μm, v_w=0.6m/min (Distance from the surface 1—0mm 2—0.1mm 3—0.2mm 4—0.3mm)

4. Residual stress simulation

During the grinding process, grinding force produces from the wheel acts on the part. Meanwhile, the temperature change caused by grinding heat affects the structural deformation and material phase transformation of part, which causing changes in load and thermal boundary conditions. Therefore, the thermo-mechanical coupling analysis can reflect the residual stress distribution accurately for thc part. Due to the high concentration of instantaneous heat input, greater stress will produced in the grinding process.The grinding stress field shows the material nonlinearity and geometric nonlinearity. Considering the complexity of the grinding process, the stress field is regarded as the material nonlinear transient problem for simplicity. Meanwhile, the elastoplastic model is selected for modeling. On the basis of thermoelastic analysis, the following assumptions are made:

(1) The part has no original tensile stress before the grinding process.

(2) Part material is an ideal material for isotropic, whose mechanical properties, such as yield strength, elastic modulus, linear expansion coefficient and tangent modulus, varies with temperature.

(3) Material deformation of viscoelastic and viscoplastic nature is ignored during the analysis. We suppose that plastic deformation satisfies the Von-Mises yield criterion.

(4) The plastic zone satisfies Prandtl-Reuss plastic flow criterion and reinforcement criterion.

Figure 4 shows the residual stress distribution on part surface. Residual tensile stress exists on part surface, whose maximum value appears in the intersection of the cutting-in zone and the middle zone. Since the increase of grinding depth results in large temperature rise rate and temperature gradient on part surface when part goes from the cutting-in zone to the middle zone, the residual tensile stress reaches maximum value 222.2MPa(as shown in figure 4(a)), 239.5MPa(as shown in figure 4(b)) and 272.6MPa (as shown in figure 4(c)) respectively with different grinding depth. The residual stress decreases first and then increases along the depth direction of the part.

IOP Conf. Series: Materials Science and Engineering **389** (2018) 012011 doi:10.1088/1757-899X/389/1/012011

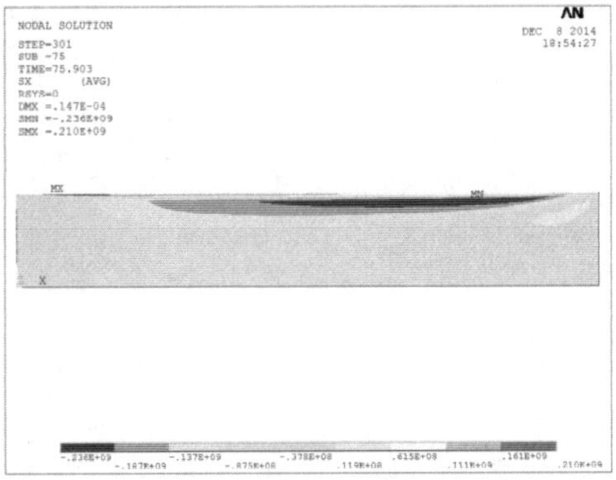

(a) a_p=350μm

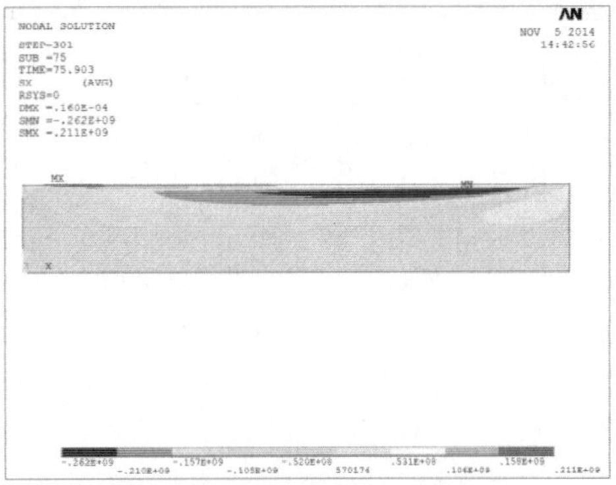

(b) a_p=400μm

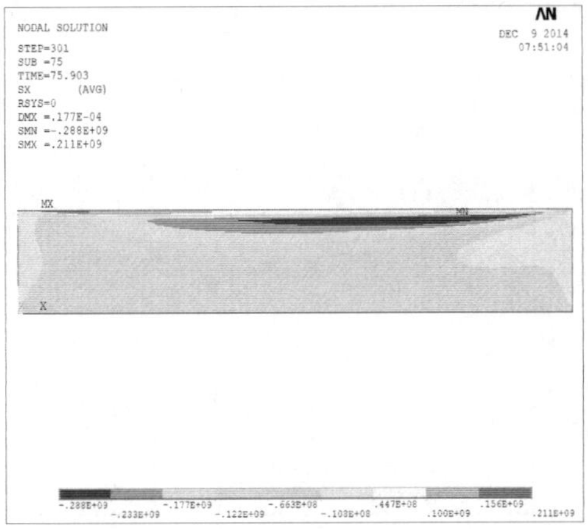

(c) a_p=450μm

Figure 4. Residual stress under the condition of different grinding depth when v_w=9m/min

Figure 5 depicts the residual stress distributions in the depth direction of the part at ap=350μm(as shown in figure 5(a)),ap=400μm(as shown in figure 5(b)),ap=450μm(as shown in figure 5(c)). Residual tensile stress is manifested on the part surface. Tensile stress shows a gradually decrease trend with the depth increases along the depth direction until transferring into compressive stress at a certain depth. Finally, the residual stress tends to zero. Also, at the same distance from the surface, the residual compressive stress increases with the grinding depth ap increases. This is due to increase of grinding depth and temperature enhances thermal stress significantly. Generally speaking, thermal stress always produces residual tensile stresses on the surface, which increases the residual tensile stress on part surface. The greater grinding depth makes the greater hardness penetration depth, which achieve more obvious hardened effect. As the residual tensile stress on part surface increases, more internal residual compressive stress is needed to balance the residual tensile stress. From this point of view, the maximum value of residual compressive stress increases as well.

As the distance from the surface is far, the internal residual stress will tend to zero, since it is too far to effect on the part for the grinding thermal stress, phase change force or mechanical force on the surface.

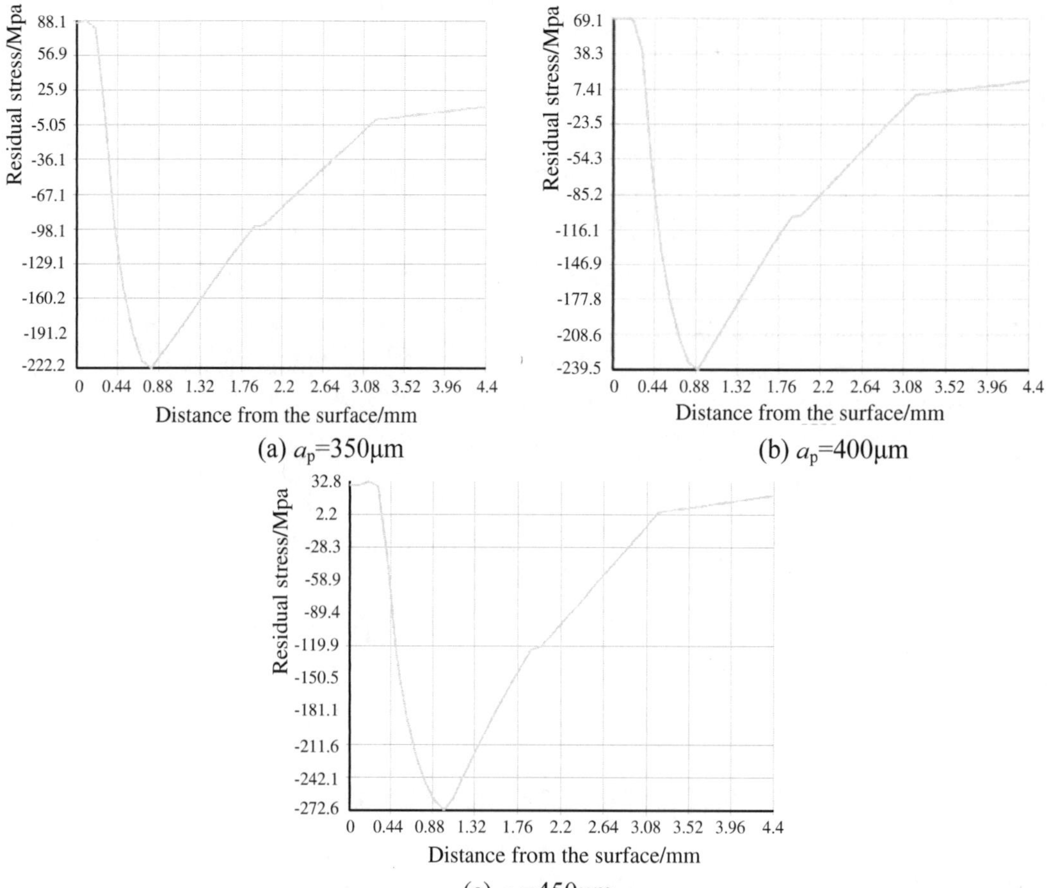

(c) a_p=450μm

Figure 5. Residual stress distribution with the condition of different grinding depth

5. Conclusions

(1) From temperature field simulation, it can be seen that the hardened layer is formed on the part surface when grinding temperature is higher than Ac3. Therefore, the zone whose temperature is higher than Ac3 is approximately regards as hardened zone. Further, hardness penetration depth can be predict as well.

(2) The simulation study on the surface residual stress of the hardened part shows that residual compressive stress exists on the part due to material expansion after grinding. While residual tensile stress is manifested on the part surface. Tensile stress shows a gradually decrease trend with the depth increases along the depth direction until transferring into compressive stress at a certain depth. Finally, the residual stress tends to zero.

Acknowledgments

This project is supported by the National Natural Science Foundation of China (No. 51705342), Education Research Project of Liaoning Provincial Department (No. LJZ2016007), National Natural Science Foundation of Liaoning Provincial (No. 201602638) and Research Innovation Cultivation Project of Shenyang Jianzhu University (No. CXPY2017021).

References

[1] Nguyen T, Zhang L C 2011 *International Journal of Machine Tools & Manufacture* 51(4) 309-

IOP Conf. Series: Materials Science and Engineering **389** (2018) 012011 doi:10.1088/1757-899X/389/1/012011

319
[2] Stohr R, Heinzel C 2002 *Abrasives magazine* 6 22-30
[3] Li C H, Hou Y L, Liu Z R, Ding Y C 2011 *International Journal of Abrasive Technology* 4(1) 77-89
[4] Foeckerer T, Zaeh M F, Zhang O B 2013 *International Journal of Heat and Mass Transfer* 56 224-237
[5] Brinksmeier E, Brockhoff T 1999 *La Metallurgia Italiana* 91(4) 19-23
[6] Brockhoff T, Brinksmeier E 1999 *Annals-Manufacturing Technology* 48(1) 255-260
[7] Fricker D C, Pearce T R C, Harrison A J L 2004 *Journal of Engineering Manufacture* 218(10) 1339-1356
[8] Zaeh M F, Brinksmeier E, Heinzel C, Huntemann J W and Foeckerer T 2009 *WGP Ann Production Eng* 3 271-279
[9] Zarudi I, Zhang L C 2002 *Journal of Materials Science* 37(18) 3935-3943
[10] Guo C, Wu Y, Varghese V and S. Malkin 1999 *Annals of the CIRP* 48(1) 247-250
[11] Salonitis K, Kolios A 2015 *Int J Adv Manuf Technol* 79 1443-1452
[12] Lavine A S 1988 *ASME Journal of Engineering for Industry* 110(1) 1-6
[13] Rowe W B, Black S C E, Mills B 1997 *Proc.R. Soc. Lond.A. London* 1083-1104

IOP Conf. Series: Materials Science and Engineering **389** (2018) 012012 doi:10.1088/1757-899X/389/1/012012

Effect of Microstructure on the Instantaneous Springback and Time-dependent Springback of DP600 Dual Phase Steel

Rongting Li [a,b] *, Ruimin Mai [a,b*], Yongxi Jin [a,b], Ruiguo Han [a,b] and Shu Wang [a,b]

[a] Science and Technology on Transient Impact Laboratory, Beijing 102202, PR China
[b] No.208 Research Institute of China Ordnance Industries, Beijing 102202, PR China

* Corresponding author: Email: lirongting818@163.com

Abstract: Springback subject to tensile test of dual phase steel DP600 with different phase volume fractions has been observed and investigated in this study. Finite element simulations using representative volume element (RVE) generated from the real microstructure are carried out. It's revealed after tensile test unloading that the instantaneous springback values (l_{is}) of DP600 increases with the martensite volume fraction increasing, and the proportion ψ of the time-dependent springback value (l_{tds}) in total springback ($l_{tds}+l_{is}$) increases with martensite volume fraction decreasing. In FE simulation of the RVEs, the effective plastic stress (EPS) σ_e is extracted to represent the residual stress and the higher l_{is} with more martensite volume fraction corresponds to larger variation $\Delta\sigma_{eM}$ of martensite phase in RVE during instantaneous springback period, which agrees with the present accepted law that the instantaneous springback is the residual stress-driven elastic recovery. After unloading finishing, the time-dependent springback proportion ψ increases with the proportion v of EPS of ferrite phase (σ_{eF}) in total of RVE (σ_{eRVE}) increasing, which verifies the influence of the residual stress in "softer phase" on the time-dependent springback.

1. Introduction

Attributed to the good formability, metal sheets are widely used in automotive forming. Various quality defects (like springback, wrinkling and crack) in metal sheet are induced after plastic forming processes including tensile, bending, drawing and so on. Assembly precision is affected and the development of related forming technology is restricted due to the above defects. Present day with the amount applications of high strength steel, the springback has been attached great attention [1-3]. Springback is the elastic recovery after the deforming force exerted on a material is removed, leading to a dimensional change. The precise measurement, prediction and compensation for springback are quite important in industry production and the related research [4-6] have been widely carried out. The most studied springback features were under the assumption that the springback is instantaneous, time-independent and finished at the exact moment the forming load is removed [7-9].

Most of the experimental results about the springback were gained right after unloading and some were perhaps obtained after a period. However, it has been found in large amount of experiments that the springback after plastic forming consists of the instantaneous springback induced by the releasing of macroscopic elastic stress at the unloading moment and the time-dependent springback driven by residual stress, occurring with the time going [10-12]. In most experiments, the time intervals between the two kinds of recovery were ignored. It has also been investigated that the releasing of residual stress after plastic forming results in the time-dependent springback of metal component without the external force [13, 14].

The time-dependent springback is the potential risk in the forced assembly of auto body, the geometric tolerance of gyro assembly and the high-pressure fluid tube distortion which may induce explosion in the conveying [15]. Despite the importance of time-dependent springback, only a few relevant studies have been carried out until now. The first systematic complication to time-dependent springback prediction and control was reported by Wagoner et al. in 1997 [10] and further completed by Wang et al. [14]. The continuous research of Wagoner [16] revealed that time-dependent springback took up a significant portion (e.g. 18% for 6022-T4) of the total springback. Both novel elasticity tests and finite element simulations on time-dependent springback were performed and compared with that of draw-bend tests. In the finite element simulation, the elastic-plastic material model was replaced by the residual stress-driven creep model after unloading and the simulation value of time-dependent springback was shown nearly twice as much as the experiment value. Time-dependent springback is more observed for aluminum alloys while present days the time-dependent springback at room temperature in some high strength autobody-used steels like stainless steel 1Cr18Ni9Ti [4] and TRIP steel [11] et al. are also revealed. Lim et al. [12] carried out series study of the time-dependent springback of several auto-used steels. It was revealed that the no time-dependent spingback phenomenon were discovered on HSLA, DQSK and AKDQ, while the behavior was obvious on the DP600, DP80, DP980 and TRIP780 steels.

Dual phase is characterized by a composite microstructure like combination of soft ferrite matrix and hard dispersed martensite and ensures good formability and high strength, being widely used in the automotive forming. Based on the special microstructure and the time-related springback of dual phase steel, the dual phase steel DP600 with different martensite volume fractions were loaded and then unloaded in the tensile test to record the instantaneous and time-dependent springback values. Finite element simulation of the loading tensile test and the springback from unloading moment were carried out with representative volume elements (RVEs) generated from the real microstructure to discuss the influence of the two phases of DP600 on springback.

2. Experimental details

The commercial dual phase steel produced by Baosteel is studied in this paper. The chemical composition of the steel is given in table 1. The specimens with different phase volume fractions are produced by different hot rolling process. The calculation of the phase volume fraction of DP600 is shown in figure 1. The quantitative image analysis is performed by using the image-Pro Plus to calculate the volume fraction of ferrite and martensite. The volume fraction of the phases were approximately represented by the corresponding pixel numbers identified and counted by the software. As shown in figure 1(a)(b)(c), martensite is colored in green, and the volume fraction can then be calculated. The metallography of three specimens with different martensite volume fractions is shown in figure 2(a)(b)(c).

Table 1. Chemical composition (Wt %) of the DP600 steel

Element	C	Si	Mn	P	S	Al	Nb
Wt (%)	≤0.12	≤1.5	≤1.5	≤0.03	≤0.91	≤0.06	≤0.1

IOP Conf. Series: Materials Science and Engineering **389** (2018) 012012 doi:10.1088/1757-899X/389/1/012012

Figure 1. Calculation of phase volume fraction (a) SEM of DP600 (b) ferrite (c) pixel of martensite

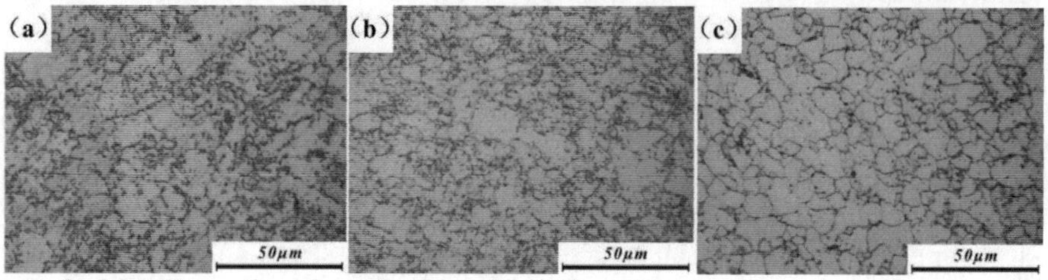

Figure 2. Metallography of specimens with different martensite volume fractions (a)13.8% (b)10.6% (c)7.6%

2.1. Uniaxial tensile test
Uniaxial tensile tests were performed with samples of different martensite volume fractions and the loading direction was parallel to the rolling direction of the sheet. The longitudinal strain was measured with a mechanical extensometer with a gauge length of 50mm and measurement precision of 0.001m. The tensile tests were carried out on standard qualified sample with the thickness of 0.9mm (figure 3). The true tensile stress-strain curves were calculated from engineering stress-strain switched from the original loading versus longitudinal displacement measured by a mechanical extensometer.

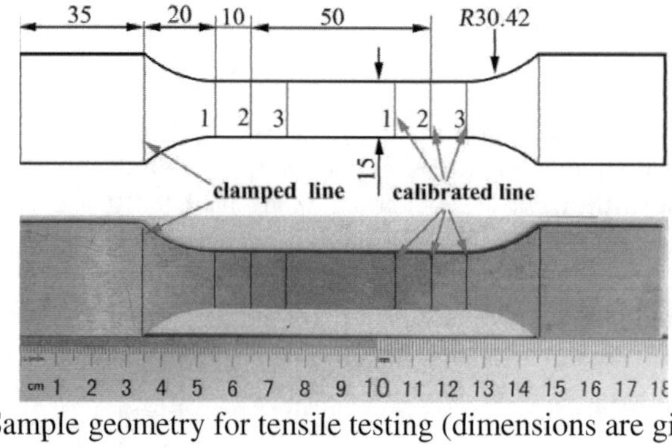

Figure 3. Sample geometry for tensile testing (dimensions are given in mm)

2.2. Springback experiment
The uniaxial tensile test is the basis to investigate the time-dependent springback. When the specimens were loaded to gain the pre-strain of 10%, they were instantaneously unloaded and the deformation of the calibrated part were continually recorded by the mechanical extensometer in the following 32 hours. The whole process cost about 5s. The time-dependent springback period was defined from the 30th second after unloading moment. The time-dependent springback curve was obtained by minus the instantaneous springback data from the data recorded by the extensometer.

3. Finite element modeling of springback after tensile test
The evaluating of the 2D representative volume elements (RVEs) model deformation, converted from the real metallography and SEM images, were performed in the finite element simulation using the FE

software ANSYS. The mechanical properties of the two phases in DP600 steel were considered independently based on the accepted information in the literatures [17], and were investigated to reveal the influence of the micro deformation of ferrite and martensite on the springback process of the heterogeneous material.

3.1. Model and mesh

The image is a two-dimensional continuous function essentially and the amplitude of the image is a continuous function of its location. The digital image constituted of discrete pixel units is obtained by uniformly collecting of the pixels of the two-dimensional image. As is known that the gray value of each pixel in the digital image is presented by a corresponding integer value and the digital image is constituted of pixels with different gray value. Then a discrete function $f(x,y)(x=1\sim M, y=1\sim N)$ is constituted by all pixels in a digital image as equation (1):

$$f(x,y)=\begin{bmatrix} f(1,1) & f(1,2) & \cdots & f(1,N) \\ f(2,1) & f(2,2) & \cdots & f(2,N) \\ \vdots & \vdots & \vdots & \vdots \\ f(N,1) & f(N,1) & \cdots & f(N,N) \end{bmatrix} \tag{1}$$

where (x, y) is the coordinate position of pixel, $f(x, y)$ is the gray value of the pixel (x, y).

Every gray value function f(x, y) has a corresponding threshold T. The threshold segmentation is the most common image segmentation method. The pixels with different gray value could be distinguished by appropriately selected threshold. As is shown in figure 4(a), the selected RVE of DP600 steel SEM image consists of the ferrite and martensite islands and the RVE is divided into two parts according to different gray value, the binary function $f(x, y)$ is expressed as:

$$f(x,y)=\begin{cases} 1 & f(x,y) \in [0,T] \\ 2 & f(x,y) \in (T,255] \end{cases} \tag{2}$$

where 1 is the martensite component , 2 is the ferrite component. The threshold $T$is generally expressed as following:

$$T = T[(x,y), f(x,y), p(x,y)] \tag{3}$$

where $p(x, y)$ is the partial property of the adjacency of (x, y). Then the phases with different gray value in the RVE is binarized by adaptive threshold binarization method [18], as shown in figure 4(b), white and black colors indicating the ferrite and martensite islands respectively. After being binarized into vectorized image, the RVE image is imported into the finite element analysis software ANSYS and finite element meshed model is generated in figure 4(c), taking the heterogeneity of the material into consideration. Three different RVEs (figure 5(a)(b)(c)) with different martensite volume fractions are prepared to conduct the finite element simulation.

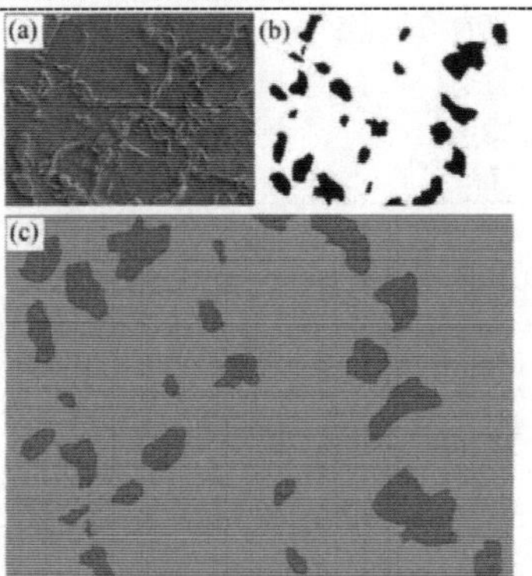

Figure 4. Selected RVE from real microstructure (a) SEM image (b) binarized microstructure (c) meshed finite element model

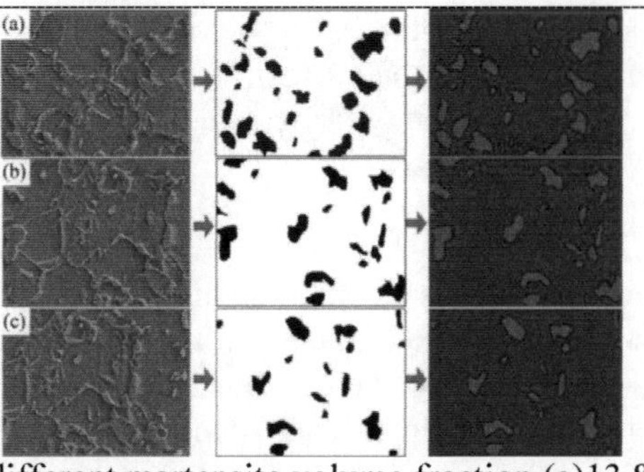

Figure 5. RVEs with different martensite volume fraction (a)13.8% (b)10.6% (c)7.6%

3.2. Material property

In the current finite element analyses of DP steels, a dislocation density-based strain hardening model [19] is used to describe the flow curve of each phase at room temperature. The model constants are quantified by [20] and are expressed in equation (4) and (5).

$$\sigma = \sigma_0 + \Delta\sigma + \alpha M\mu\sqrt{b}\sqrt{\frac{1-\exp(M_T k_r \varepsilon)}{k_r L}} \quad (4)$$

$$\sigma_0 = 77 + 80\%\mathrm{Mn} + 75\%\mathrm{P} + 60\%\mathrm{Si} + 80\%\mathrm{Cu} + 45\%\mathrm{Ni} + 60\%\mathrm{Cr} + 11\%\mathrm{Mo} + 5000\mathrm{N_{ss}} \quad (5)$$

Where σ and ε are respectively the flow stress and true strain. The α is a constant and a value of 0.33, M_T is the Taylor factor and have the value of 3. The μ is the value of shear modulus with a value of 80GPa and b is the Burgers vector (b=2.5×10^{-10}m). L and k_r are respectively the dislocation mean free path and the recovery rate, listed in table 2. The value of d_α is the ferrite grain size (10^{-5}). The $\Delta\sigma$ provides strengthening by precipitation or the carbon in solution.

In ferrite it is:

$$\Delta\sigma \text{ (in MPa)} = 50*(\%\ \mathrm{C_{ss}^{f}}) \quad (6)$$

In martensite it is:

$$\Delta\sigma \text{ (in MPa)} = 30.65*(\%\ \mathrm{C_{ss}^{m}}) - 161 \quad (7)$$

Where $\%\ \mathrm{C_{ss}^{f}}$ and $\%\ \mathrm{C_{ss}^{m}}$ respectively represent the wt% carbon in solid solution in ferrite and martensite.

Table 2. Parameter in equation 1

	$\Delta\sigma$	L(m)	k_r
Martensite	$3065\mathrm{C_{ss}^{m}} - 161$	(i) 3.5×10^{-8}	(ii) 41
(iii) **Ferrite**	(iv) $5000\mathrm{C_{ss}^{f}}$	(v) d_α	(vi) $10^{-5}/d_\alpha$

3.3. Boundary and loading conditions

The boundary conditions applied to the RVE model are shown in figure 6. In the uniaxial tensile simulation, the strain ε_x is applied to the specimen edge as displacements, $u_x = \varepsilon_x \times l_m$. The RVE was loaded to gain the plastic pre-strain of 10% and then unloaded to springback driven by residual stress releasing. The behavior of the instantaneous springback is nonlinear elastic-plastic deformation. The simulation is based on the assumption that the two phases is homogeneous and isotropic and behave separately in the deformation.

While the material constitutive model of the time-dependent springback has not been uniformly confirmed and the models used in simulations of the time-dependent springback mostly are based on some approximation and hypothesis. As it has been accepted that the time-dependent springback is driven by internal residual stress [12, 14]. The average effective stress σ_e in both the RVE and the two phases are calculated to correspond to the experiment law that the time-dependent springback is influenced by different phase volume fractions.

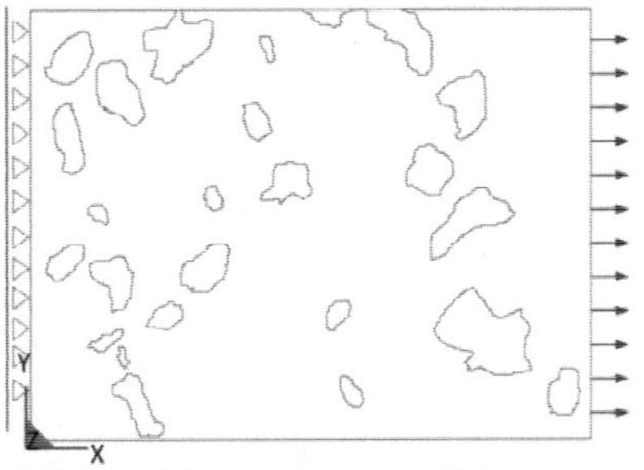

Figure 6. Boundary condition

4. Results and discussion

4.1. Tensile test results

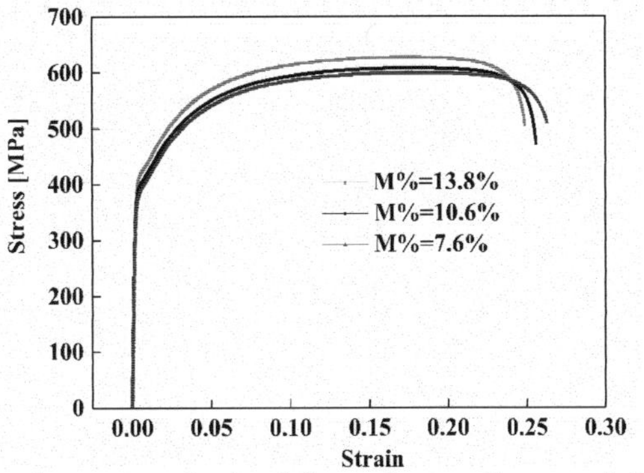

Figure 7. Stress-strain curves obtained from the tensile test of all conditions

Figure 7 shows the tensile test stress-strain curves of the DP600 steel with three different martensite volume fractions. It has been shown that the sample with martensite volume fraction of 13.8% has the yield strength of 410 MPa and fracture strain of 0.20. The sample with martensite volume fraction of 10.6% and 7.6% respectively have the yield strength of 390 MPa, 380 MPa and fracture strain of 0.23, 0.26. It has been widely investigated that the strength of dual phase steel is only influenced by the volume fraction of martensite [21]. The yield strength is found to increase with increasing of the martensite volume fraction and the fracture strain is shown to decrease with the increasing of the martensite volume fraction.

4.2. Springback after tensile test unloading

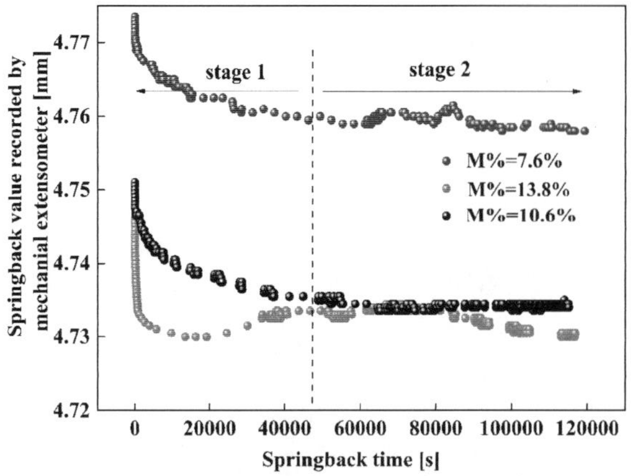

Figure 8. Springback values after unloading

The springback value measured by mechanical extensometer 32 hours (~115200s) following unloading are shown in figure 8. There are local fluctuations on curves, the curves can be divided into a fast stage and a slow stage and are shown to enter the stable period almost at the same time. The springback datas, namely instantaneous springback l_{is}, time-dependent springback l_{tds} and the proportion ψ of l_{tds} in total springback value $l_{tds}+l_{is}$ are summarized in table 3.

Table 3. Springback values of DP samples with different martensite volume fractions

M%	l_{is} (mm)	l_{tds} (mm)	$\psi=l_{tds}/\ l_{tds}+l_{is}$
7.6%	0.227	0.0160	6.6%
10.6%	0.250	0.0168	6.5%
13.8%	0.255	0.0160	5.9%

As shown in table 3, the instantaneous springback values l_{is} are respectively 0.255mm, 0.250mm and 0.227mm for samples with martensite volume fraction of 13.8%, 10.6% and 7.6%. The time-dependent springback l_{tds} is 0.0160mm for the DP600 sample with martensite volume fraction of 7.6%, with the proportion ψ=6.6% of the total springback value. The l_{tds} of samples with martensite volume fraction of 10.6% and 13.8% are respectively 0.0168mm and 0.0160mm, accounting for 6.5% and 5.9% of the total springback.

Combining figure 8 with table 3, the instantaneous springback value l_{is} increases with increasing of the martensite volume fraction in DP600 samples. The time-dependent springback l_{tds} shows no regular change with the martensite volume fraction ranging, while the proportion ψ of the time-dependent springback in the total springback value decreases with the martensite volume fraction increasing. It's accepted that [17] the martensite with a higher Young's modulus responses more obviously to the releasing of the elastic deformation energy in the instantaneous springback period, which may contribute to the regular change of the instantaneous springback value with the martensite volume fraction increasing. In the studies of the time-dependent springback, because of the similar time-related macro deformation characteristics, the time-dependent springback behavior are mostly approximated like the room temperature creep or viscoplastic behavior [22], in which the stored energy of the "softer phase" ferrite is higher than it in the "harder phase" martensite, deriving the time-dependent springback more obviously. Therefore, the proportion ψ of the time-dependent springback in the total springback value increases with increasing of the "softer phase" ferrite volume fraction.

4.3. Simulation result discussion

It's widely accepted that the time-dependent springback is attributed to the residual stress releasing. In order to study the micro mechanism of springback and the effect of martensite volume fraction on the springback behavior of the RVE, three different models are created and the "effective plastic stress" distribution of the models in unloading process are studied. In the simulation, the residual "effective plastic stress" after unloading is considered as the deriving force of the springback. The effective plastic stress distribution in RVEs and single martensite phase at the unloading beginning moment with pre-strain of 10% are shown in figure 9(a)(b)(c)(d)(e)(f). It's shown that the effective plastic stress in the RVE area is distinctly heterogeneous and mainly distributes to the ferrite adjacent to martensite.

The instantaneous springback deformation during the unloading period is elastic recovery and not time-related. The material model used in the finite element simulation of this instantaneous period is still the model mentioned in Section 3.2. The average effective plastic stress variation Δσe from the unloading beginning moment to the ending moment of elastic recovery in RVES (ΔσeRVE) and in single martensite phase (ΔσeM), and the proportion u=ΔσeM /ΔσeRVE are calculated as is shown in figure 10(a) . The ΔσeM of the three models are respectively 131.15 MPa (M%=7.6%), 133.62 MPa (M%=10.6%) and 137.95 MPa (M%=13.8%), accounting in ΔσeRVE with proportion u of 87.8% （M%=7.6%） , 92.2% (M%=10.6%) and 94% (M%=7.6%). As is revealed in figure 10(b), the instantaneous springback values l_{is} increases with both the martensite volume fraction and the ΔσeM increasing, which agrees with the present accepted law that the more releasing of plastic residual energy, the more instantaneous springback value.

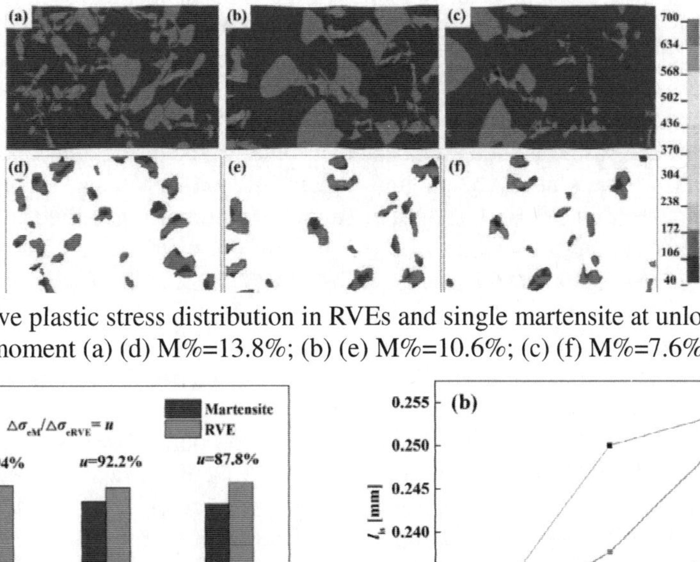

Figure 9. Effective plastic stress distribution in RVEs and single martensite at unloading beginning moment (a) (d) M%=13.8%; (b) (e) M%=10.6%; (c) (f) M%=7.6%

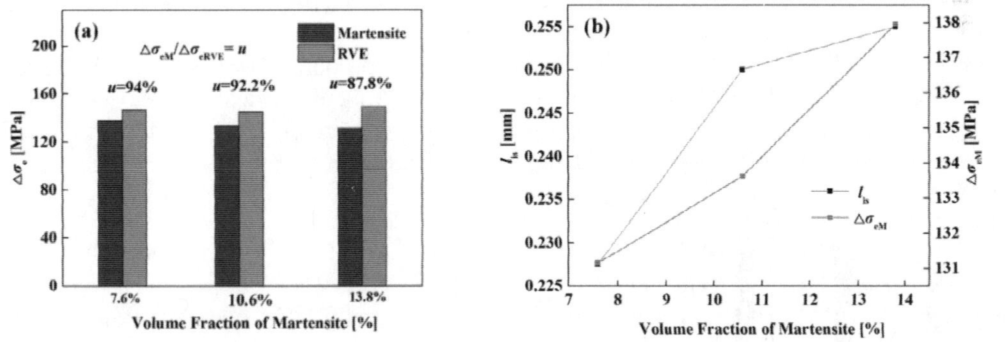

Figure10. Effect of martensite volume fraction on the instantaneous springback (a) releasing of effective plastic stress in RVE and single martensite (b) relation of l_{is} , ΔσeM and martensite volume fraction

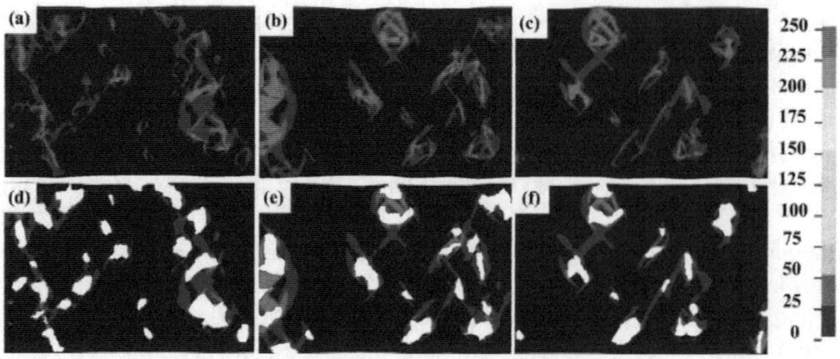

Figure 11. Effective plastic stress distribution in RVEs and single Ferrite at elastic recovery ending moment (a) (d) M%=13.8%; (b) (e) M%=10.6%; (c) (f) M%=7.6%

Driven by the releasing of residual stress, the time-dependent springback is time-related and the material model is different from that of instantaneous springback. However, at present, the material model of time-dependent springback has not been identified yet. It's not available to accurately simulate the special time-related deformation with the current elastic-plastic material models. According to the macroscopic deformation law and the influence of microstructure, the average effective plastic stress σ_{eF} at the beginning moment of time-dependent springback (the ending moment of elastic recovery) is extracted to approximately predict the driving force, which directly influences the following time-dependent springback tendency. The effective plastic stress distribution in RVEs and single ferrite phase at the ending moment of elastic recovery (unloading ending moment) are shown in figure 11(a)(b)(c)(d)(e)(f).. In comparison with the unloading beginning moment, the effective plastic stress decreases distinctly and also distributes to the ferrite adjacent to martensite. The average effective plastic stress at the beginning moment of time-dependent springback (the ending moment of elastic recovery) in RVES (σ_{eRVE}) and single ferrite phase (σ_{eF}), and the proportion $v=\sigma_{eF}/\Delta\sigma_{eRVE}$ are calculated as is shown in figure 12(a). The v of the three models are respectively 83.5% （M%=7.6%）, 78.6% (M%=10.6%) and 76.3% (M%=7.6%). In figure 12(b), no obvious corresponding relationship is observed between the martensite volume fraction and the time-dependent springback value l_{tds}, while the proportion ψ of the time-dependent springback in the total springback value decreases with the ferrite volume fraction decreasing (the martensite volume fraction increasing). As mentioned in Section 4.2, the stored energy of the "softer phase" has effect on the time-dependent springback. The relation of phase volume fraction, and the proportion $v=\sigma_{eF}/\Delta\sigma_{eRVE}$ is discussed in figure 12(c). It's revealed that the proportion ψ of the time-dependent springback in the total springback value decreases with the proportion $v=\sigma_{eF}/\Delta\sigma_{eRVE}$ decreasing, which verifies the influence of the stored energy in "softer phase" on the time-dependent springback.

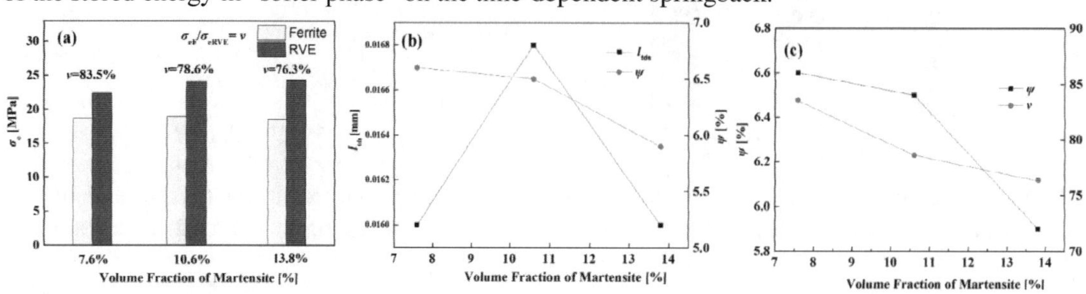

Figure12. (a) Effective plastic stress in RVE and single ferrite phase at the beginning moment of time-dependent springback; (b) relation of ψ, v and martensite volume fraction

5. Conclusions

The effect of phase volume fractions on the instantaneous springback and time-dependent springback of dual phase steel DP600 is studied in the paper. DP600 steel specimens with different martensite volume fractions were loaded in tensile test to gain the pre-strain of 10% and then unloaded to record the instantaneous springback values and time-dependent springback values. In the finite element simulation, representative volume elements (RVEs) generated from the real microstructure were also loaded and unloaded to discuss the influence of the two phases of DP600 and the residual stress in the two phases on springback.

The instantaneous springback values (l_{is}) of DP600 specimens are revealed to show corresponding increasing with the increasing of the martensite volume fraction, while the time-dependent springback values (l_{tds}) present no obvious relation with variation of the phase volume fraction. But the proportion ψ of time-dependent springback in the total springback value is shown to increase with the ferrite volume fraction increasing (namely the martensite volume fraction decreasing).

In the simulation of the RVEs deforming, the higher l_{is} with more martensite volume fraction corresponds to larger variation $\Delta\sigma_{eM}$ of martensite, which agrees with the present accepted law that the instantaneous springback is the elastic recovery and the more releasing of plastic residual stress, the more instantaneous springback value. The proportion ψ of the time-dependent springback of the total springback value increases with both the ferrite volume fraction increasing (the martensite volume fraction decreasing) and the proportion $\nu=\sigma_{eF}/\sigma_{eRVE}$ increasing, which verifies the influence of the stored energy in "softer phase" on the time-dependent springback.

Acknowledgements
We gratefully acknowledge the financial support of the program for the National Natural Science Fund item (11772303).

References

[1] L. M. Geng, R. H. Wagoner, Role of plastic anisotropy and its evolution on Springback, Int J Mech Sci. 44 (2002) 123-48.
[2] M. Bakhshi-Jooybari, B. Rahmani, V. Daeezadeh, A. Gorji, The study of springback of CK67 steel sheet in V-die and U-die bending processes, Mater Des. 30 (2009) 2410-9.
[3] W. Gan, R. H. Wagoner, Die design method for sheet springback, Int J Mech Sci. 46 (2004) 1097-13.
[4] D. X. E, R. X. Ning, C. T. Tang, Y. Yan, T. Gu, Experiments and analysis on the springback in rotary-drawing tube bending, Trans. Beijing Ins Technol. 26 (2006) 410-32 [In Chinese].
[5] R. Kazan, M. Fırat, A. E. Tiryaki, Prediction of springback in wipe-bending process of sheet metal using neural network, Mater Des. 30 (2009) 418-23.
[6] N. Asnafi, Springback and fracture in v-die air bending of thick stainless steel Sheets, Mater Des. 21 (2000) 217-36.
[7] H. A. Al-Qureshi, A. Russo, Spring-back and residual stresses in bending of thin-walled aluminium tubes, Mater Des. 23 (1999) 217-22.
[8] A. Megharbel, G. A. Nasser, A. Domiaty, Bending of tube and section made of strain-hardening materials, J Mater Process Technol. 203 (2008) 372-80.
[9] H. A. Al-Qureshi HA, Elastic-plastic analysis of tube bending, Int J Mach Tools Manuf. 39 (1999) 87-104.
[10] R. H. Wagoner, W. Carden, D. Matlock, Springback after drawing and bending of metal sheets, In: T. Chandra T, S. Leclair, J. Meech, B. Verma, M. Smith, B. Balachandran, editors. IPMM097-Intelligent Processing and Manufacturing of Materials, University of Wollongong, Intelligent Systems Applications, 1997, p. 1-10.
[11] H. Lim, M. Lee, J. Sung, R. H. Wagoner, Time-dependent springback, Int J Mater Form. 1 (2008) 157-60.
[12] H. Lim H, M. Lee, J. Sung, J. Kim, R. H. Wagoner, Time-dependent springback of advanced high strength steels, Int J Plast. 29 (2012) 42-59.

[13] M. Zhan, H. Yang, L. Huang, R. J. Gu, Springback analysis of numerical control bending of thin-walled tube using numerical-analytic method, J Mater Process Technol. 177 (2006) 197-201.
[14] J. Wang, R. H. Wagoner, W. Carden, D. Matlock, F. Barlat, Creep and anelasticity in the springback of aluminum, Int J Plast. 20 (2004) 2209-32.
[15] D. X. E and Y. F. Liu, Springback and time-dependent springback of 1Cr18Ni9Ti stainless steel tubes under bending, Mater Des. 31 (2010) 1256-61.
[16] R. H. Wagoner, M. Li, Simulation of springback: through-thickness integration, Int J Plast. 23 (2007) 345-60.
[17] H. Hosseini-Toudeshky, B. Anbarlooie, J. Kadkhodapour, G. Shadalooyi, Microstructural deformation pattern and mechanical behavior analyses of DP600 dual phase steel, Mater Sci Eng A. 600 (2014) 108-21.
[18] W. Shen, Q. B. Fan, F. C. Wang, Z. Ma, Y. W. Wang, Finite Element Study of Influence Factors on Thermal Conductivity of Plasma Sprayed Zirconia Coatings Based on Digital Image Processing, J Inorg Mater. 25 (2010) 1217-20.
[19] R. Rodriguez, I. Gutierrez, Unified formulation to predict the tensile curves of steels with different microstructures, Mater Sci Forum Vpls. 426-432 (2003) 4525-30.
[20] R. Rodriguez, I. Gutierrez, in: M. Lamberigts (Ed.), Proceedings of the TMP 2004, Verlag Stahleisen GMBH, Düsseldorf, 2004, p.356.
[21] S. Yu, R. Y. Fu, L. Li, Effect of martensite content on the mechanical properties of dual phase steels of St12Q1 and St37-2G, Shanghai Metal. 26 (2004) 8-12 [In Chinese].
[22] S. Sun, D. X. E, A novel model based on viscoelastic theory to predict the time-dependent springback for DP600 steel sheet, Act Metall Sin. 51 (2015) 1356-64.

IOP Conf. Series: Materials Science and Engineering **389** (2018) 012013 doi:10.1088/1757-899X/389/1/012013

Analysis of Steady State Characteristics for Single-axial Groove Sleeve Bearing

Li-li Wang *, Yu-liang Wei, Guo-teng Yuan and Shao-hui Guo

College of Mechanical and Electronic Engineering, Shandong University of Science and Technology, Qingdao, 266590, China

*E-mail: wang-ly-ly@163.com

Abstract: The finite difference method is used to solve the Reynolds equation and the steady state characteristics of single-axial groove bearings are obtained theoretically. It is assumed that the groove is located in four different positions on the top, the bottom, the left and the right of the bearings. The pressure distribution curves and bearing capacity of groove at four different positions are analyzed under the predefined boundary conditions. It has been shown that the oil film pressure distribution and bearing capacity remain unchanged with the common sleeve bearing when the groove is located in the divergence area of oil film; when the groove is located in the convergence area of oil film, the oil film pressure distribution will change greatly. Compared to the sleeve bearing with groove located in the divergence area, the bearing capacity of the sleeve bearing with groove located in the convergence area will decrease to a certain extent, the amplitude of the decrease will change with the different positions of groove in the convergence area of oil film. When oil groove is in the convergence area, though the carrying capacity and pressure decreases, the pressure forms the two pressure strips and the stable performance of sleeve bearing is improved.

1. Introduction

Sleeve bearing is widely used in various fields, due to its higher bearing capacity, higher turning accuracy, higher impact resistance, smaller noise and vibration. For larger size and higher bearing capacity requirements, the sleeve bearings with oil groove have better lubrication performance and heat dissipation effect.

Axially grooved journal bearing is widely used in industry with its better stability and better lubrication performance. Wang [1] analyzed the influence of groove depth on oil film pressure and bearing capacity of single-axial journal bearing. Roy et al. [2-3] studied the steady state and dynamic characteristics of single-axial journal bearing. Brito et al. [4] compared the performance of a journal bearing with a single and a twin axial groove configuration. Through the experimental study of two-axial oil groove journal bearing, the influence of rotated speed and load on the oil film pressure of two-axial oil groove journal bearing is analyzed by Chatterton et al. [5]. Yu et al. [6] analyzed double oil grooved round sliding bearing performance with disturbance. The linear and nonlinear transient motion of a 2-lobe geometrically imperfect hybrid journal bearing system compensated with constant flow valve restrictor is analyzed by Jain and Sharma [7]. Majumdar et al. [8] studied the state and dynamic characteristics of journal bearing with three axial grooves. Mallya et al. [9] analyzed the static characteristics of misaligned three-axial water-lubricated journal bearing in the turbulent regime. On the basis of previous studies, the influence of eccentricity and groove location on the oil film pressure and bearing capacity of single-axial groove sleeve bearing is studied

2. Theory Model

2.1. Reynolds equation

The two-dimensional Reynolds equation of incompressible fluid lubrication and finite long bearing can be expressed as:

$$\frac{\partial}{\partial x}\left(h^3\frac{\partial p}{\partial x}\right)+\frac{\partial}{\partial y}\left(h^3\frac{\partial p}{\partial y}\right)=6U\eta\frac{\partial h}{\partial x} \tag{1}$$

where h is the oil film thickness, p is the oil film pressure, U is the rotation speed of the sleeve, η is lubricating oil dynamic viscosity, x is the circumferential coordinate, y is the axial coordinate.

Geometry of single-axial groove bearing is shown in Figure 1, bearing has one groove along the axial direction, and the axial groove angle of the sleeve bearing is 36°. As shown in Figure 1, it is assumed that the groove is located in the position of A, B, C and D of the bearing. The oil film thickness without oil groove: $h=c[1+\varepsilon\cos(\varphi-\theta)]$, and the oil film thickness with oil groove: $h=c[1+\varepsilon\cos(\varphi-\theta)]+h_g$. c is the radius clearance, ε is the eccentricity, θ is the attitude angle, h_g is the depth of the oil groove, φ is circumferential coordinate .

Under steady state condition, equation (1) can be reduced to

$$\frac{\partial}{\partial\varphi}\left(H^3\frac{\partial P}{\partial\varphi}\right)+\left(\frac{d}{l}\right)^2\frac{\partial}{\partial\lambda}\left(H^3\frac{\partial P}{\partial\lambda}\right)=3\frac{\partial P}{\partial\varphi} \tag{2}$$

where H is the non-dimensional oil film thickness: $H=\frac{h}{c}$, P is the non-dimensional oil film pressure, $P=\frac{pc^2}{2U\eta R}$, d is the width of bearing, l is the length of bearing, λ is axial coordinate, $\lambda=\frac{y}{l/2}$.

The dimensionless oil film thickness without oil groove: $H=1+\varepsilon\cos(\varphi-\theta)$, the dimensionless oil film thickness with oil groove: $H=1+\varepsilon\cos(\varphi-\theta)+H_g$, $H_g=h_g/c$.

2.2. Boundary conditions

For axially grooved sleeve bearings, the Reynolds boundary condition is used.

Pressure at both edges of the bearing is zero.

$$P(\varphi,\pm1)=0 \tag{3}$$

Pressure distribution is symmetric about the mid plane along axial direction.

$$\left(\frac{\partial P}{\partial\lambda}\right)_{\lambda=0}=0 \tag{4}$$

Pressure is set equal to 0 when the pressure falls below zero.

2.3. Loading capacity

For single-axial groove sleeve bearing, the non-dimensional loading capacity components along the line of centers and its perpendicular direction are as following:

$$\bar{W}_\xi=-\int_0^{2\pi}\left(\int_0^1 P\,d\lambda\right)\sin(\varphi-\theta)\,d\varphi \tag{5}$$

$$\bar{W}_\eta=-\int_0^{2\pi}\left(\int_0^1 P\,d\lambda\right)\cos(\varphi-\theta)\,d\varphi \tag{6}$$

IOP Conf. Series: Materials Science and Engineering **389** (2018) 012013 doi:10.1088/1757-899X/389/1/012013

where $\bar{W}_\xi$ is the non-dimensional loading capacity components along the line of centers, $\bar{W}_\eta$ is the non-dimensional loading capacity which is perpendicular to the line of centers.
The non-dimensional loading capacity and attitude angle are as following:

$$\bar{W} = (\bar{W}_\xi + \bar{W}_\eta)^{1/2} \tag{7}$$

$$\theta = tan^{-1}(\frac{\bar{W}_\xi}{\bar{W}_\eta}) \tag{8}$$

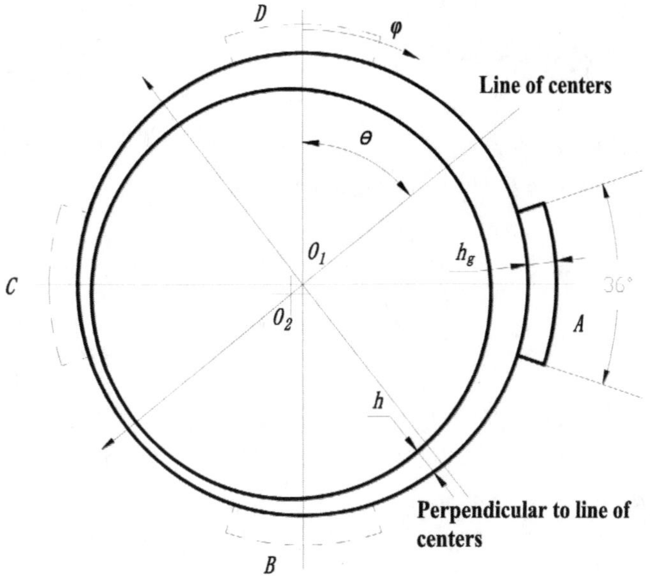

Figure 1. Geometry of single-axial groove bearing

3. Results and discussion

The Reynolds equation with different locations of groove A, B, C and D is solved, and the non-dimensional oil film pressure and bearing capacity are obtained in Figure 2(a), (b), (c), (d), Figure 3(a), (b) and Figure 4.

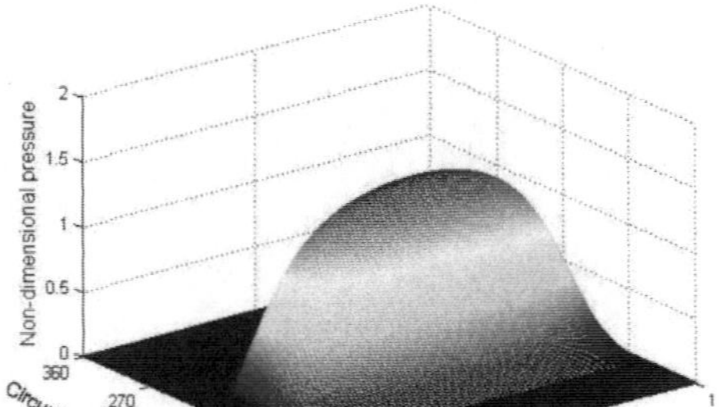

(a) Position A

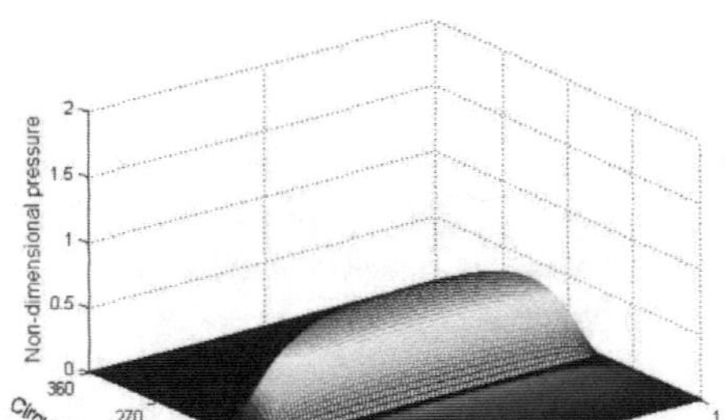

(b) Position B

IOP Conf. Series: Materials Science and Engineering **389** (2018) 012013 doi:10.1088/1757-899X/389/1/012013

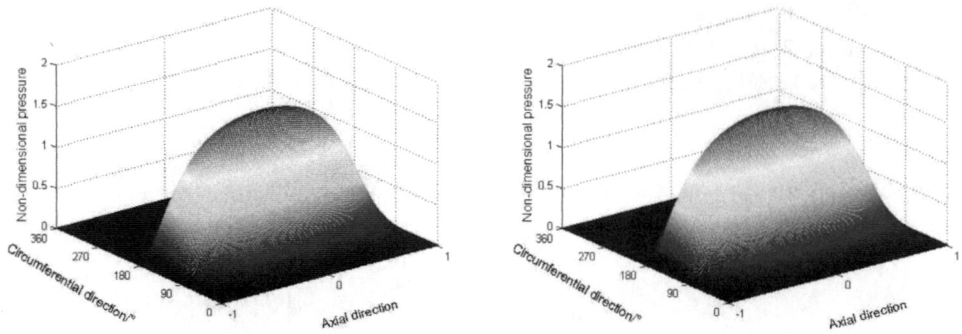

(c) Position C (d) Position D

Figure 2. Three dimensional pressure distribution

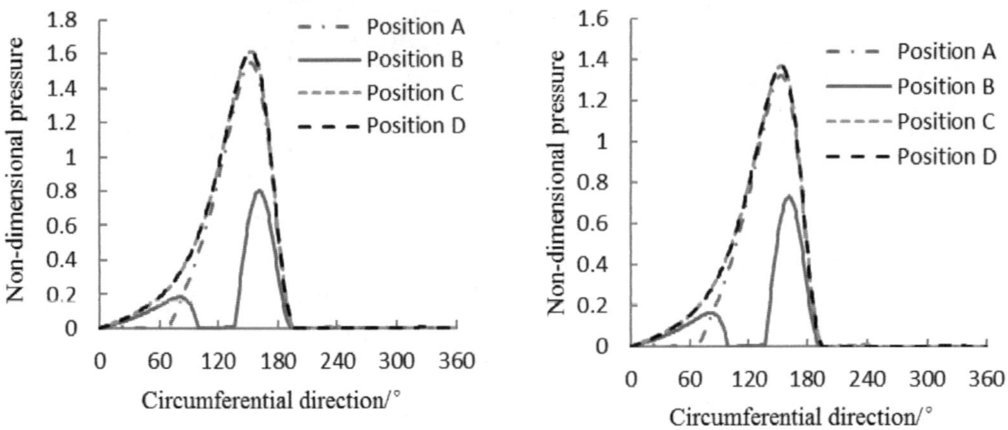

(a) $\lambda = 0$ (b) $\lambda = \pm 0.5$

Figure 3. Circumferential pressure distribution

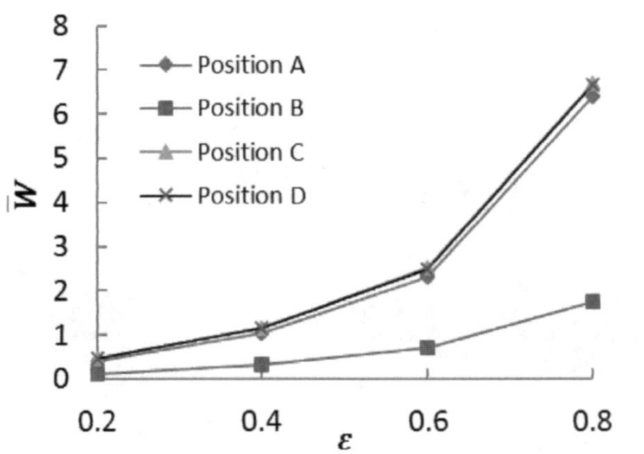

Figure 4. Non-dimensional bearing capacity

It can be seen from Figure 2(a), (b), (c), (d), Figure 3(a), (b) and Figure 4 that the non-dimensional oil film pressure distribution and non-dimensional bearing capacity of position A and B are obviously different with common sleeve bearing, and the non-dimensional oil film pressure distribution and non-dimensional bearing capacity of position C and D are basically the same with the common sleeve bearing. According to the location of each groove in Figure 1, the non-dimensional oil film pressure distribution and non-dimensional bearing capacity remain unchanged when the groove C, D are located in the divergence area of oil film; when the groove A , B are located in the convergence area of oil film, the non-dimensional oil film pressure distribution will change greatly. Compared to the sleeve bearing with groove located in the divergence area, the non-dimensional bearing capacity of the sleeve bearing with groove located in the convergence area will decrease to a certain extent, the amplitude of the decrease will change with the different position of groove in the convergence area of oil film. As shown in Position A of Figure 3(a) and (b), because the location of groove is near to the oil film start position, the non-dimensional oil film pressure is close to zero. However, when the location of groove has a certain distance from the oil film start position, as shown in Position B of Figure 3(a) and (b), the oil film forms two distinct pressure regions. So when oil groove is in the convergence area, though the carrying capacity and pressure decreases, the pressure forms the two pressure strips and the stable performance of sleeve bearing is improved. In the future study, the structure of single-axial groove bearing should be optimized to get the optimal groove position.

4. Conclusions

The steady state characteristics of bearing are affected by the location of groove when the oil groove is located in the convergence area of oil film. The bearing capacity of the sleeve bearing with groove located in the convergence area will decrease to a certain extent, the amplitude of the decrease will change with the different position of groove in the convergence area of oil film. When the location of groove is a certain distance from the oil film start position, the oil film forms two distinct pressure regions, which is helpful to improve the dynamic characteristics of sleeve bearing.

Acknowledge

This work was supported by the grant from China Postdoctoral Science Foundation funded project (No. 2017M612304), Shandong Provincial Postdoctoral Innovation Foundation (No. 201701016), supported by SDUST Research Fund (NO. 2015JQJH104), Qingdao Postdoctoral Research funded project, and supported by Key Laboratory of High-efficiency and Clean Mechanical Manufacture at Shandong University, Ministry of Education.

References

[1] Wang Longgang. The design of journal groove sliding bearing test stand's simulation software. Beijing Jiaotong University, 2008.
[2] L. Roy, B. C. Majumdar. Effect of the location of axial groove on the steady state and dynamic characteristics of oil journal bearings. In: 12th national conference of machines and mechanisms, 2005, 261–268.
[3] L. Roy, S. K. Laha. Steady state and dynamic characteristics of axial grooved journal bearings. Tribology International, 2009, 42(5): 754-761.
[4] F.P. Brito, A. S. Miranda, J.C.P. Claro, M. Fillon. Experimental comparison of the performance of a journal bearing with a single and a twin axial groove configuration. Tribology International, 2012, 54:1-8
[5] Steven Chatterton, Phuoc Vinh Dang, Paolo Pennacchi, et al. Experimental evidence of a two-axial groove hydrodynamic journal bearing under severe operation conditions. Tribology International, 2017, 109: 416-427.
[6] Yu Guichang, Liu Shulian, Zheng Shuiying. Double oil grooved round sliding bearing performance analysis with disturbance. Journal of Vibration and Shock, 2012, 31(05):46-49+54.
[7] Dharmendra Jain, Satish C Sharma. Dynamic analysis of a 2-lobe geometrically imperfect journal bearing system. Proceedings of the Institution of Mechanical Engineers, Part J: Journal of Engineering Tribology, 2017, 231(7):934-950.
[8] B C Majumdar, R Pai, D J Hargreaves. Analysis of water-lubricated journal bearings with multiple axial grooves. Proceedings of the Institution of Mechanical Engineers, Part J: Journal of Engineering Tribology, 2004, 218(2):135-146
[9] Ravindra Mallya, Satish B Shenoy, Raghuvir Pai. Static characteristics of misaligned multiple axial groove water-lubricated bearing in the turbulent regime. Proceedings of the Institution of Mechanical Engineers, Part J: Journal of Engineering Tribology, 2017, 231(3):385-398

IOP Conf. Series: Materials Science and Engineering **389** (2018) 012014 doi:10.1088/1757-899X/389/1/012014

Effect of Oxide Film Crack Length on Film-induced Stress in Stress Corrosion Cracking Tip

YANG Hongliang[1], XUE He[2,*], CUI Yinghao[2] and NI Chenqiang[2]

[1.] Center of Engineering Training, Xi'an University of Science and Technology, Xi'an 710054, China

[2.]School of Mechanical Engineering, Xi'an University of Science and Technology, Xi'an 710054, China

*Corresponding author: Tel: 0086-29-83856250, E-mail: xue_he@hotmail.com

Abstract. Oxide film rupture theory has become one of the most popular models to quantitatively predict the stress corrosion cracking rate at crack tip of Ni-based alloys in high temperature water nuclear environment. To understand the micro-mechanical state at the tip of environment assisted cracking (EAC) without considering the external load, the stress-strain induced by oxide film in the base metal at the EAC tip was simulated and discussed using a commercial finite element analysis code. Results show the stress-strain state induced by oxide film is different because of crack length, and film-induced stress is one of the main factors that can't be ignored in the stress corrosion crack growth driving force. The study also provides a foundation to improve the quantitative predication accuracy of EAC growth rate of nickel-based alloys and austenitic stainless steels in the important structures of nuclear power plants.

1. Introduction

There have been reports on stress corrosion cracking (SCC) in austenitic alloys used in light water reactor (LWR) components over the last thirty years [1,2]. The pressure vessel and the steam generator used in the nuclear power tend to occur stress corrosion in the process of its service, and the surface form a layer of oxide film[3].Usually, the oxide film plays a protective role in stress corrosion. However, under the complex environmental factors, the oxide film can produce film-induced stress, which can promote the dislocation emission and motion, and make the crack tip produce local plastic deformation[4]. The stress can make the intergranular film fracture, leading to the formation of deep matrix crack[5,6]. In the quantitative prediction model of crack growth rate based on slip dissolution theory, crack length is an important mechanical parameter. The crack length will influence the mechanical field of the crack tip induced by film-induced stress[7].In order to understand the effect of crack length on stress and strain field caused by film-induced stress at crack tip, the stress and strain fields in the SCC crack tip were analyzed by using the sub model technology of ABAQUS and the nickel base alloy as the research object. Attempt to obtain effect of crack length on stress and strain field caused by film-induced stress.

2. Theory model

The oxide film rupture theory considers that the EAC process in nickel-based alloys and austenitic stainless steels in high temperature water environment can be divided into two periods: I-oxide film forming at the crack tip, II-oxide film aging and rupture at the crack tip with a high stress and strain [8,9].

Because the aging and rupture of the oxide film takes most of the time in EAC period, according to the Faraday's law, Ford and Andresen in GE deduced a formula to estimate the EAC growth rate of nickel-based alloys and austenitic stainless steels in high temperature water environment of nuclear power plants[8-10]:

$$\frac{da}{dt}=\frac{M}{Z\rho F}\cdot\frac{i_0}{1-m}\left[\left(\frac{t_0\cdot\dot{\varepsilon}_{ct}}{\varepsilon_f}\right)^m-m\cdot\frac{t_0\cdot\dot{\varepsilon}_{ct}}{\varepsilon_f}\right] \quad (1)$$

Where, da/dt is the EAC growth rate, M and ρ are the atomic mass and the density of metals, respectively, F is Faraday's constant, i_0 is the bare surface oxidation current density, z is the change in charge due to the oxidation process, m is the exponent in the current decay curve, ε_f is the degradation strain of the oxide film, t_0 is the time before onset of the current decay, and $d\varepsilon_{ct}/dt$ is the strain rate at the EAC tip.

Eq.(1) is the most widely used in prediction equation of the EAC growth rate in nickel-based alloys and austenitic stainless steels in high temperature water environment of nuclear power plants. Fracture mechanism of film-induced stress in stress corrosion cracking can be show as: Without considering the external load, there is an additional tensile stress at one side of the interface matrix. It makes the local plastic deformation at crack tip, and the stress can causes the dislocation emission, which can cause the oxide film intergranular cracking, and then leading to the occurrence and extension of stress corrosion cracking. Thus it can form deep matrix crack, and stress corrosion cracking can be produced[5,6].

3. Finite element modeling

3.1 Specimen model

Half inch compact tension specimen (0.5T-CT) was used in this numerical calculation with the virtual experiment process according to the American Society for Testing and Materials Standard[11]. The geometric shape and size of 0.5T-CT specimen are shown in Figure1. The thickness of the oxide film formed by Ni-based alloys in high temperature and high pressure water is about 1μm to 2 μm [12], and 2 μm is adopted in this research.

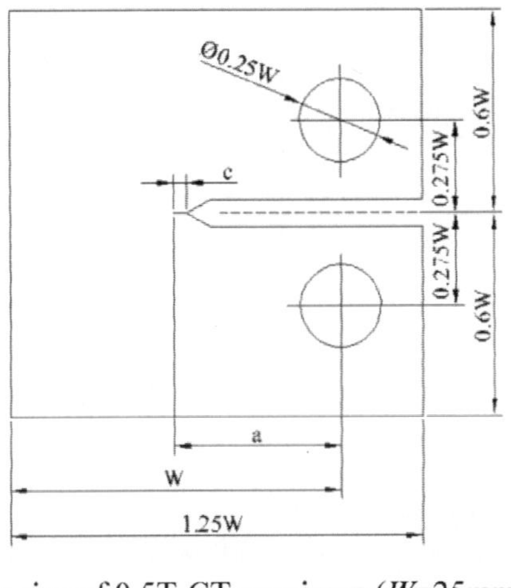

Figure1 Geometric size of 0.5T-CT specimen (W=25mm, a=0.5W, c=2mm)

3.2 Material model

IOP Conf. Series: Materials Science and Engineering **389** (2018) 012014 doi:10.1088/1757-899X/389/1/012014

The non linear relationship between stress and strain beyond yield at crack tip of nickel-based alloys is described by Ramberg-Osgood equation in this numerical simulation[13]:

$$\frac{\varepsilon}{\varepsilon_0} = \frac{\sigma}{\sigma_0} + \alpha\left(\frac{\sigma}{\sigma_0}\right)^n \tag{2}$$

Where ε is strain; σ is stress; E is Young's modulus of the material; σ_0 is the yield strength of the material; α is the yield offset and n is the hardening exponent for the plastic.

The material parameters of Ni-based alloys 600 and oxide film are shown in Table 1.

Table 1. Mechanical properties of alloy 600 and the oxide film formed in PWR primary water at 288 °C [14,15]

Material	Alloy 600	Oxide film
Young's modulus, E /GPa	189.5	140
Poisson's ratio, ν	0.286	0.31
Yield strength, σ_0 /MPa	436	--
Yield offset, α	3.075	--
Hardening exponent, n	6.495	--

3.3. FE Model

Using ABAQUS sub model to calculate the stress and strain distribution in the SCC crack tip , the base metal and the oxide film are set up as a whole, and the different material properties are given in the modle. Using the plane deformation model and adopting 8 node and the 2 plane strain element, the finite element model is shown in Figure 2.In order to improve the accuracy of the calculation, the oxide film and the base metal at the crack tip are refined to obtain a more detailed and accurate data[16].

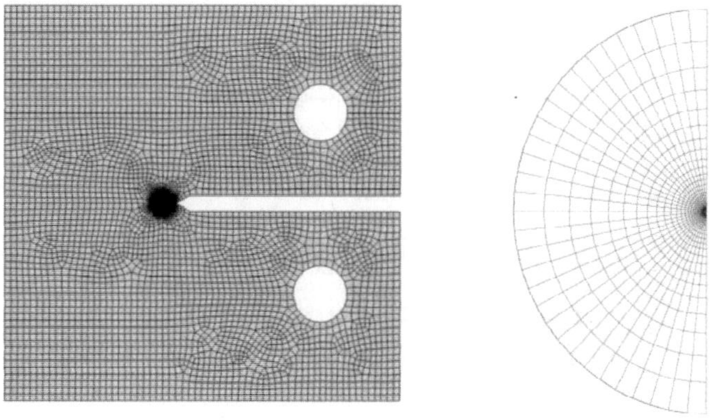

(a) Integral model (b) Crack tip model

Figure 2 Finite element mesh

4. Results and Discussion

4.1. Effect of crack length on Mises stress

Figure 3 shows the influence of crack length on the Mises stress induced by film-induced stress around the crack tip. When a=0, the Mises stress have the maximum values at θ=0° and decrease gradually as θ increase, and the smallest value appears at θ=90°. When a>0, Mises stress changes most dramatically in the range of about -30°～30°. Stress concentration phenomenon occurs because of the existence of the crack, and the maximum value of Mises stress occurs at θ=0°. When the crack is initially formed in the oxide film, the Mises stress is gradually increased with the increasing of the crack length. When the crack is extended to the Ni-based alloys, the Mises stress decreases slightly with the increasing of the crack length. This is due to the existence of the crack, the film-induced stress is released, and the stress on the crack tip is reduced.

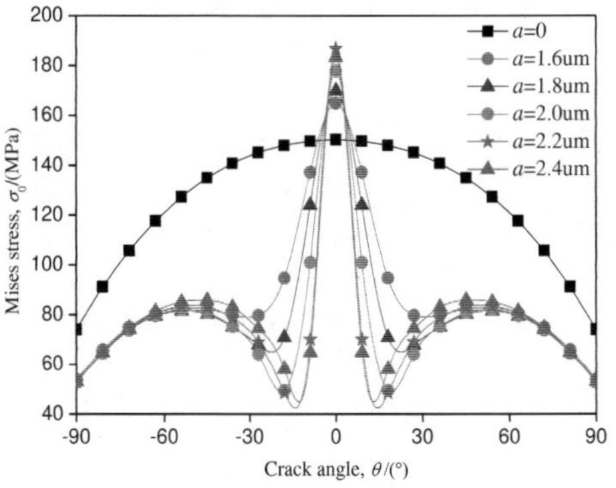

Figure 3 Effect of crack length on Mises stress

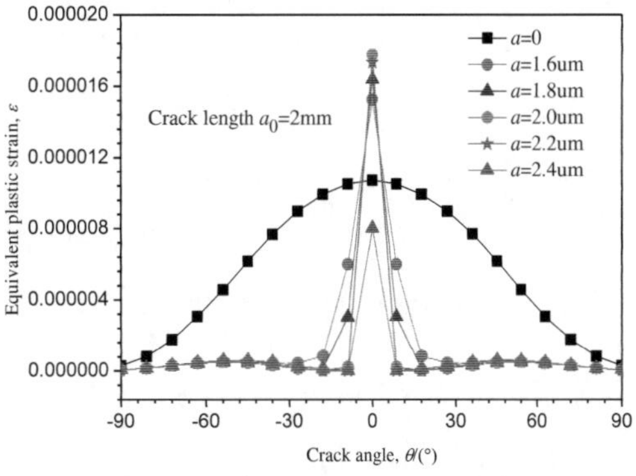

Figure 4 Effect of crack length on equivalent plastic strain

4.2. Effect of crack length on equivalent plastic strain

Figure 4 shows the effect of crack length on equivalent plastic strain induced by film-induced stress around the crack tip. When a=0, the equivalent plastic strain have the maximum values at θ=0° and decrease gradually as θ increase, and the smallest value appears at θ=90°. When a>0, equivalent plastic strain changes most dramatically in the range of about -30°~30°, and the maximum value of equivalent plastic strain occurs at θ=0° because of stress concentration phenomenon. When the crack is initially formed in the oxide film, equivalent plastic strain is gradually increased with the increasing of the crack length. When the crack is extended to the Ni-based alloys, the equivalent plastic strain decreases sharply with the increasing of the crack length because of the film-induced stress is released.

4.3 Effect of crack length on tensile stress

The effect of crack length on tensile stress induced by film-induced stress around the crack tip is shown in Figure 5. When a=0, the tensile stress have the maximum values at θ=0° and decrease to negative gradually as θ increase, and the tensile stress is changed from tensile stress to compressive stress. When a>0, tensile stress changes sharply in the range of about -30°~30°, and the maximum value of tensile stress occurs at θ=0°. The tensile stress is changed into compressive stress with the increase of crack length when the crack is extended to the Ni-based alloys.

4.4. Effect of crack length on tensile plastic strain

The effect of crack length on tensile plastic strain induced by film-induced stress around the crack tip is shown in Figure 6. When a=0, the tensile plastic strain have the maximum values at θ=0° and decrease to negative gradually as θ increase. When a>0, tensile plastic strain changes sharply in the range of about -30°~30°, and the maximum value of tensile stress occurs at θ=0°. When the crack is initially formed in the oxide film, tensile plastic strain is slightly increased with the increasing of the crack length. When the crack is extended to the Ni-based alloys, the tensile plastic strain decreases clearly with the increasing of the crack length.

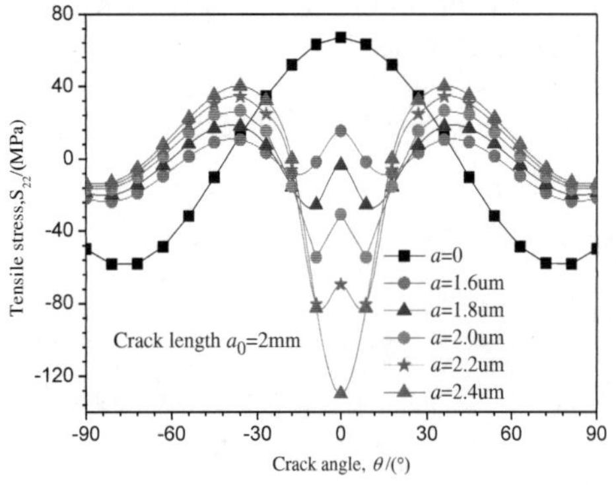

Figure 5 Effect of crack length on tensile stress

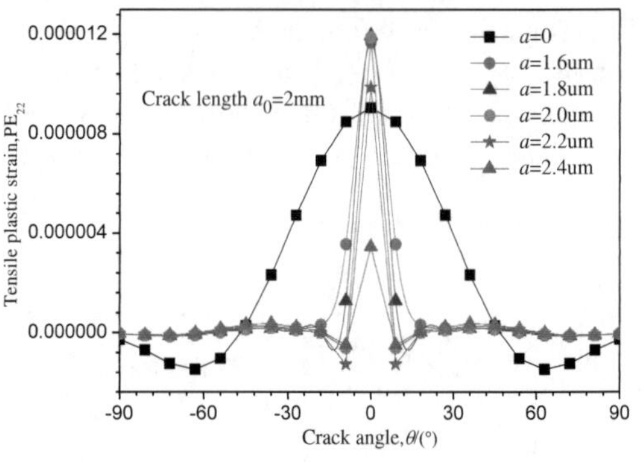

Figure 6 Effect of crack length on tensile plastic strain

4.5 Effect of crack length on tensile plastic strain gradient

Xue et al[17] proposed a method to replace the crack tip strain rate by using the crack tip strain gradient. In the strain gradient model, it is assumed that the crack tip strain is equal to the plastic strain at the front of the crack tip. In order to calculate conveniently, the crack tip strain gradient can be calculated by simplified formula (3).

$$\frac{d\varepsilon_p}{da} \approx \frac{\Delta\varepsilon_p}{\Delta a} = \frac{\varepsilon_{p2} - \varepsilon_{p1}}{a_{i+1} - a_i} \tag{3}$$

Figure 7 shows the effect of crack length on tensile plastic strain gradient at crack tip. When the crack grows inside the oxide film, tensile strain rate is negative. the tensile plastic strain gradient is gradually increased with the increasing of the crack length. When the crack is extended to the Ni-based alloys, the tensile plastic strain gradient increases with the increasing of the crack length at $r <$ 0.4μm, and the tensile plastic strain gradient gradually approaches to 0 with the increase of r.

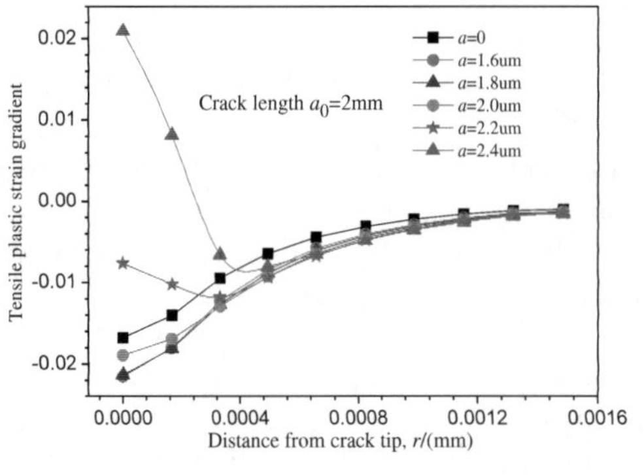

Figure 7 Effect of crack length on tensile plastic strain gradient at crack tip

5. Conclusion

(1) The change of stress and strain is obvious in the range of about -30°~30°, and the maximum value of stress and strain occurs at θ=0°. When the crack is initially formed in the oxide film, stress and strain gradually increase with the increasing of the crack length. When the crack is extended to the Ni-based alloys, the stress and strain decreases with the increasing of the crack length.

(2) Without considering the external load, the effective stress of the crack increases with the increasing of film-induced, and promotes the effective of plastic strain. When the critical condition is reached, the nucleation and expansion of micro cracks are caused, which leads to stress corrosion cracking. Film-induced stress is one of the main factors that can't be ignored in the driving force of stress corrosion crack growth.

Acknowledgments

Foundation item: National Natural Science Foundation of China (51475362), Scientific Research Program Funded by Shaanxi Provincial Education Department (16JK1493)

References

[1] Seiya Yamazaki, Zhanpeng Lu, Yuzuru Ito et al. The effect of prior deformation on stress corrosion cracking growth rates of Alloy 600 materials in a simulated pressurized water reactor primary water[J], Corrosion Science, 2008, 50(3) 835–846.
[2] R.W. Staehle. 1-Historical views on stress corrosion cracking of nickel-based alloys : The Coriou effect[J]. Stress Corrosion Cracking of Nickel Based Alloys in Water-cooled Nuclear Reactors, 2016:3-131.
[3] WANG Wenwen, LUO Ji, GUO Leichen, et al. Finite element analysis of stress corrosion cracking for copper in an ammoniacal solution [J]. Rare Metals, 2015, 34(6): 426-430.
[4] Chu Wuyang, Qiao Lijie. Hydrogen embrittlement and stress corrosion -- basic part [M]. 2013.
[5] Hong Lu, Kewei Gao, Wuyang Chu. Determination of tensile stress induced by dezincification layer during corrosion for brass[J]. Corrosion Science, 1998, 40(10):1663-1670.
[6] Guo X J, Gao K W, Qiao L J, et al. The correspondence between susceptibility to SCC of brass and corrosion-induced tensile stress with various pH values[J]. Acta Metallrugica Sinica, 2002, 44(10):2367-2378.
[7] Han Enhou, Wang Jianqiu, Wu Xinqiang. Corrosion Mechanisms of Stainless Steel and Nickel Base Alloys in High Temprature High Pressure Water[J]. Acta Metallurgica Sinica, 2010, 46(11): 1379-1390.
[8] Tetsuo Shoji, Zhanpeng Lu, Hiroyoshi Murakami. Formulating stress corrosion cracking growth rates by combination of crack tip mechanics and crack tip oxidation kinetics[J]. Corrosion Science, 2010, 52(3): 769-779.
[9] Peter L Andresen, F. Peter Ford. Fundamental modeling of environmental cracking for improved design and lifetime evaluation in BWRs[J]. International Journal of Pressure Vessel Technology, 1994, 59(1--3): 61-70.
[10] Peter L. Andresen, F. Peter Ford. Response to " On the modeling of stress corrosion cracking of iron and nickel base alloys in high temperature aqueous environments" [J]. Corrosion Science, 1996, 38(6): 1011-1016.
[11] ASTM standard E399-90, 2002. Annual Book of ASTM Standards [S].
[12] YANG Hongliang, XUE He, ZHAO Lingyan et al. Effect of Oxide Film Shape on Stress-Strain at Stress Corrosion Cracking Tip of Ni-Based Alloy[J]. Hot Working Technology. 2016, 45(20): 58-60.
[13] Ramberg W, Osgood W R. Description of stress-strain curves by three parameters. NACA Tech Note[R], No. 902, 1943.
[14] P.Goudeau, P.O.Renault, P.Villain, C.Coupeau. Thin Solid Films[J], 2001, 496-500.
[15] Yang Hongliang, Xue He, Yang Fuqiang, Zhao Lingyan. Effect of Film-Induced Stress on Mechanical Properties at Stress Corrosion Cracking Tip[J]. Rare Metal Materials and Engineering, 46(2017): 3595-3600.
[16] YANG Hongliang, XUE He, YANG Fan. Study on Oxide Film Rupture Behavior Caused by Film-induced Stress[J]. FOUNDRY TECHNOLOGY, 38(6): 1274-1277.
[17] Xue H, Sato Y, Shoji T. Quantitative estimation of the growth of environmentally assisted cracks at flaws in light water reactor components[J]. Journal of Pressure Vessel and Technology: Transactions of the ASME, 2009, 131(1):61-70.

IOP Conf. Series: Materials Science and Engineering **389** (2018) 012015 doi:10.1088/1757-899X/389/1/012015

Effect of Crack Length on Mechanical Field of Stress Corrosion Crack Tip of Cold Working 316L Stainless Steel

YANG Hongliang[1], XUE He[2,*] and ZHANG Jianlong[2]

[1.] Center of Engineering Training, Xi'an University of Science and Technology, Xi'an 710054, China
[2.] School of Mechanical Engineering, Xi'an University of Science and Technology, Xi'an 710054, China

*Corresponding author: Tel: 0086-29-83856250, E-mail: xue_he@hotmail.com

Abstract. The effect of crack length at different cold working level on stress and strain state and fracture parameters of stress corrosion cracking tip of 316L stainless steel in high temperature water environment of the nuclear pressure vessel was studied by sub-model technology of elastic plastic finite element method. It combined with the mechanical properties of 316L stainless steel under different cold working level, and the effect of crack length on the Mises stress, equivalent plastic strain, tensile stress, tensile strain and *J* integral of 316L stainless steel under different cold working level was compared. The results indicate that crack length of stress corrosion cracking has certain effect on stress and strain state at crack tip. With the increase of crack length, Mises stress and equivalent plastic strain, tensile stress and tensile strain increase at crack tip, at the same time, *J* integral increases with increase of crack length. Cold working level has changed the mechanical properties of 316L stainless steel. When the stress intensity factor *K* is constant, with the increase of cold working level, Mises stress increases and equivalent plastic strain decreases at crack tip, while tensile stress increases and tensile strain decreases at crack tip. The rate of change of stress strain curve from 0° to 90° is different because of cold working level. The higher the level of cold working, the faster the Mises stress and the tensile stress increase at 90° in front of the crack tip, and equivalent plastic strain and tensile strain decrease most slowly at 90° in front of the crack tip. As the degree of cold working increases, the *J* integral of crack tip also increases. The increase of crack length and the level of cold working aggravate the stress corrosion cracking of 316L stainless steel under high temperature and high pressure in a certain range.

1. Introduction

Tensile stress, sensitivity of material and corrosion environment are three conditions for the ccurrence of stress corrosion cracking[1,2]. Cold working in the process of pressure vessel manufacturing often leads to large residual plastic deformation, resulting in a larger residual tensile stress, which tends to make the material reach yield strength[3-5], such as cold rolling, cold bending, straightening, welding, assembly and so on. In the process of crack propagation, the length of the crack changes continuously, and the stress and strain of the crack tip varies with the length of the crack. Therefore, the effective prediction and control of residual stress in the manufacturing process of pressure vessels and the study of the influence of crack tip stress on material sensitivity of different length cracks help to further understand the mechanism of stress corrosion cracking (SCC), so as to effectively predict and control nuclear structural materials damage the purpose of failure. Andresen et al. studied the stress corrosion

IOP Conf. Series: Materials Science and Engineering **389** (2018) 012015 doi:10.1088/1757-899X/389/1/012015

cracking behavior of stainless steel and nickel-base alloy after cold working. The results show that the cold working deformation introduced during the processing of the structural material is the main factor to promote and accelerate the stress corrosion cracking of the material. The crack growth rate of the cold rolled nickel base alloy and stainless steel is more than 100 times higher than that of the cast alloy[6-8]. France EDF's pressurized water reactor (PWR) nuclear power operation experience shows that there is no stress corrosion cracking phenomenon in the part without obvious cold processing. Cold working is the main factor to produce SCC[9]. The residual tensile stress on the austenitic stainless steel surface caused by cold working is now considered to be one of the major causes of cracking[10-13]. Ilevbare[14] et al. reported that 53% of PWR assembly failures were related to cold working in material processing. Different cold working methods and degrees will affect the mechanical properties of the material, and the crack tip stress-strain state and propagation law of different crack lengths under different mechanical properties also become more complex.

Therefore, the effect of crack length on SCC with different degrees of cold work has become a hot issue. It is an important practical significance to analyze the effect of crack length on the mechanical state of crack tip of SCC under different degrees of cold working. In order to understand the effect of crack length on mechanical state of cold working 316L stainless steel at crack tip, the stress-strain field at crack tip of the stress corrosion crack was analyzed by using the ABAQUS simulation software in the hope of obtaining the influence of cold working on the stress and strain field at crack tip of 316L stainless steel in stress corrosion cracking.

2.Finite element modeling

2.1.Geometric model

The use of finite element analysis of crack tip stress and strain state selection of compact tensile specimen 1T-CT sample, the sample geometry shown in Fig.1.

Using finite element analysis of crack tip stress and strain state, the of compact tensile specimen 1T-CT sample is selected, and the sample geometry is shown in Fig.1. The geometric size and experimental process of the sample according to American Society for Testing and Materials Standard. Initial prefabricated crack length c=2mm, It changes the crack length a, the initial crack length a=0.5W=25mm, and takes Δa=1mm, 2mm, 3mm respectively.

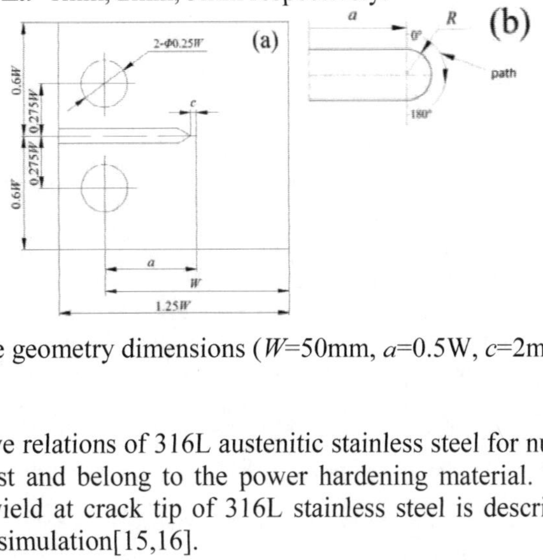

Figure 1. 1T-CT sample geometry dimensions (W=50mm, a=0.5W, c=2mm) (a) and crack tip (b)

2.2.Material model

The mechanical constitutive relations of 316L austenitic stainless steel for nuclear power materials are obtained by the tensile test and belong to the power hardening material. The relationship between stress and strain beyond yield at crack tip of 316L stainless steel is described by Ramberg-Osgood equation in this numerical simulation[15,16].

$$\varepsilon = \frac{\sigma}{E} + \alpha \frac{\sigma_0}{E}\left(\frac{\sigma}{\sigma_0}\right)^n \tag{1}$$

Where ε is strain; σ is stress; E is Young's modulus of the material; σ_0 is the yield strength of the material; α is the yield offset and n is the hardening exponent for the plastic.

The mechanical properties of 316L austenitic stainless steel for nuclear power in high temperature and high pressure water environment are shown in table 1.

Table 1 mechanical parameters of 316L stainless steel[17]

Cold work level	5%	10%	20%
Yield strength, MPa	388	478	643
Young's modulus, GPa	197	193	191
Hardening exponent	3.19	3.54	4.95
Yield offset	8.06	7.87	4.47
Poisson's ratio	0.3	0.3	0.3

2.3. Finite element model

Using ABAQUS to calculate stress and strain distribution in crack tip , it is a commonly software used to precision calculation. Using the plane deformation model and adopting 8-node biquadratic plane strain quadrilateral reduced integration[18].

3. Results and Discussion

3.1. Effect of crack length on Mises stress of 316L stainless steel with different cold working level

The cold working level was 5%, 10%, 20% respectively. According to Fig.1, the initial crack length is a=0.5W=25mm. Change the crack length a, and respectively take Δa = 1mm, 2mm, 3mm. Effect of crack length on the Mises stress of 316L stainless steel with different cold working is shown in Fig.2. The cold working degree of the graph (a) is 5%, (b) is 10%, and (c) is 20%.

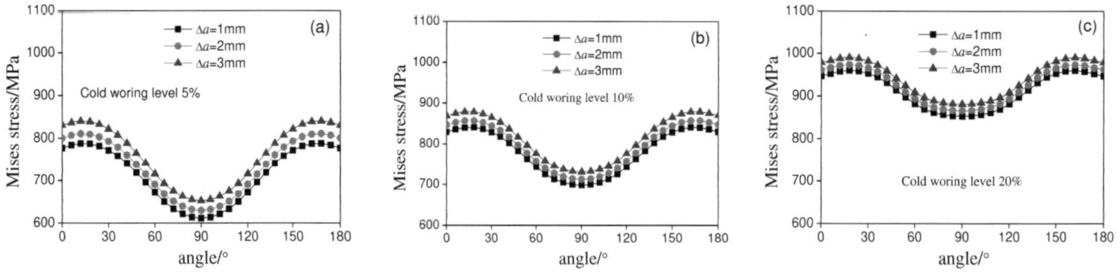

Figure 2. Mises stress distribution at crack tip

From Fig. 2 (a), (b) and (c), it can be seen that the Mises stress at the crack tip increases with the increase of the crack length. From 0° to 90°, the Mises stress first increases slightly and then decreases, and the minimum Mises stress occurs at 90° just in front of the crack tip. The level of cold working is different, and the Mises stress at the crack tip is also different. The Mises stress increases with the increase of cold working level. However, with the increase of cold working level, the Mises stress curve gradually flattened and the minimum Mises stress at the crack tip approaches to the maximum value. That is, the increasing rate of Mises stress is maximum at 90°, and the cold working level could change the Mises stress distribution . This is due to the mechanical properties of the materials changed by cold working. The 316L stainless steel produced cold working plastic deformation. The yield strength $\sigma_{0.2}$ of 316L stainless steel is increased by plastic deformation, and the hardening exponent n is increased to a certain degree. Under the condition of certain stress intensity factor K, it is known from the formula (3) that the Mises stress at crack tip will also increase, therefore, it will aggravate the stress corrosion cracking to a certain extent.

3.2. Effect of crack length on equivalent plastic strain of 316L stainless steel with different cold working level

The effect of crack length on the equivalent plastic strain of 316L stainless steel at different cold working level is shown in Fig.3.

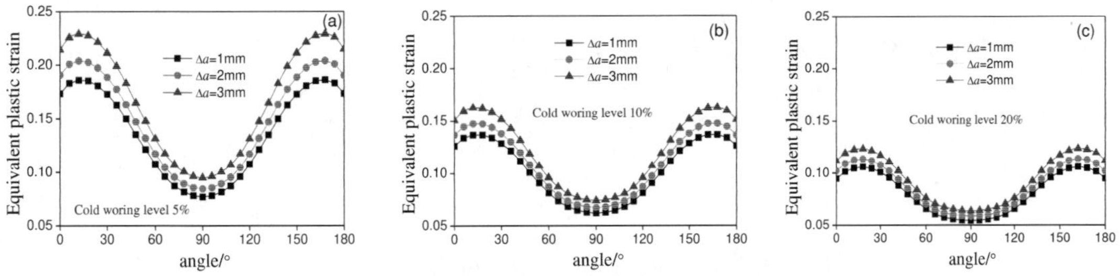

Figure 3. Equivalent plastic strain distribution at crack tip

It can be seen from Fig.3 (a), (b) and (c), the equivalent plastic strain at crack tip increases with the increase of the crack length. From 0° to 90°, the equivalent plastic strain at crack tip first increases and then decreases, and the minimum value of equivalent plastic strain appears at 90° in front of the crack tip. The cold working level is different, the equivalent plastic strain at the crack tip is also different, and the equivalent plastic strain decreases with the increase of cold working level. However, with the increase of cold working level, the equivalent plastic strain curve gradually flattened and the minimum at the crack tip approaches to the maximum value. The decreasing rate of equivalent plastic strain is maximum at 90°. Cold hardening occurs after cold working of 316L stainless steel. hardening exponent n is improved, and yield strength $\sigma_{0.2}$ is also increased. Therefore, when the stress intensity factor K is constant, the equivalent plastic strain at the crack tip decreases with the increase of the cold working degree.

3.3 Effect of crack length on tensile stress of 316L stainless steel with different cold working level

Effect of crack length on tensile stress of 316L stainless steel with different cold working level is shown in Fig.4. It can be seen from Fig.4 (a), (b) and (c), the tensile stress at the crack tip increases with the increase of the crack length. From 0° to 90°, the increase rate of tensile stress increases, and the maximum tensile stress occurs at 90° in front of the crack tip. The tensile stress at the crack tip is different because of cold working, and the tensile stress increases with the increase of cold working level. Tensile stress plays an important role in stress corrosion cracking, so the increase of tensile stress will aggravate the stress corrosion cracking to a certain exten.

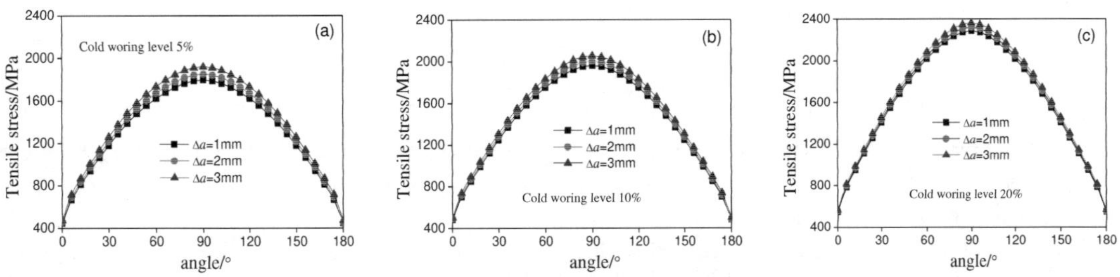

Figure 4. Tensile stress distribution at crack tip

3.4. Effect of crack length on tensile strain of 316L stainless steel with different cold working level

Effect of crack length on tensile strain of 316L stainless steel with different cold working level is shown in Fig.5.

It can be seen from Fig.5 (a), (b) and (c), the tensile strain of the crack tip increases with the increase of the crack length. From 0° to 90°, the tensile strain at crack tip first increases and then decreases, and the minimum value of tensile strain appears at 90° in front of the crack tip. The tensile strain at the crack tip decreases with the increase of cold working level. The decreasing rate of tensile strain is minimum at 90°.

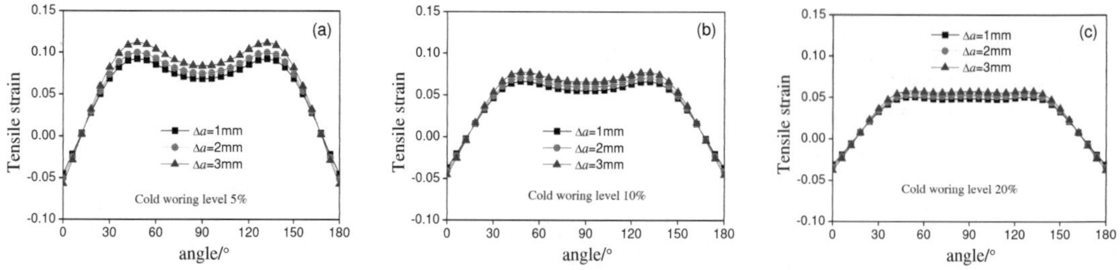

Figure 5. Tensile strain distribution at crack tip

3.5. Analysis of J integral at crack tip

In fracture mechanics, J is a parameter that can quantitatively characterize the stress and strain field strength at the crack tip, and can also be used as a parameter to characterize elastic-plastic fracture. Crack length also affect the fracture parameter. Ignoring the influence of the elastic part on J, only taking the J integral of the plastic part into considered, and the calculation formula of J can be simplified to the formula (8).

$$J = \frac{K^2(1-\nu^2)}{E} \tag{2}$$

The change of J integral with crack length under different cold working is shown in Fig.6. It can be seen from Fig.6 that the J integral at crack tip increases with the increase of crack length. Cold working level also has an impact on the distribution of J integral. At the same time, J integral also increases with the increase of cold working level.

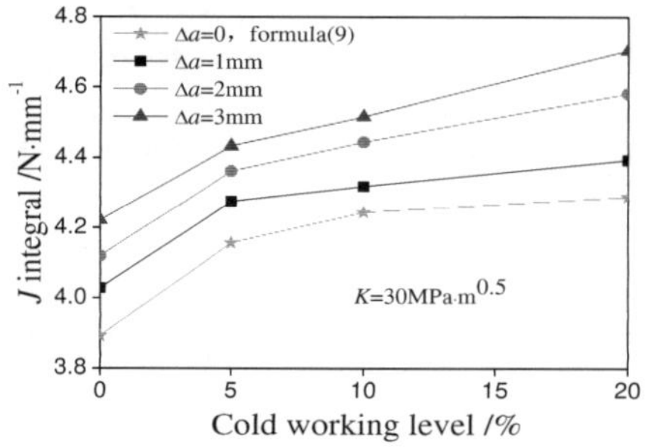

Figure 6. J integral changes at crack tip

4. Conclusion

(1) The crack length of (SCC) has a certain influence on the stress and strain at the crack tip. Mises stress, equivalent plastic strain, tensile stress and tensile strain of stress corrosion crack tip increase with the increase of crack length.

(2) The J integral at crack tip increases with the increase of crack length. The cold working level also affect the distribution of the J integral at crack tip. J integral increase with the increase of cold working level. The increase of crack length and the level of cold working aggravate the stress corrosion cracking of 316L stainless steel under high temperature and high pressure in a certain range.

(3) Cold working level changes the mechanical properties of 316L stainless steel, and also affects the stress corrosion cracking stress- strain state. When the stress intensity factor K is constant, with the increase of cold working level, Mises stress increases and equivalent plastic strain decreases at crack tip, while tensile stress increases and tensile strain decreases at crack tip.

(4) The rate of change of stress strain curve from 0° to 90° is different because of cold working level. The higher the level of cold working, the faster the Mises stress and the tensile stress increase at 90° in front of the crack tip, and equivalent plastic strain and tensile strain decrease most slowly at 90° in front of the crack tip.

Acknowledgments
Foundation item: National Natural Science Foundation of China (51475362), Scientific Research Program Funded by Shaanxi Provincial Education Department (16JK1493).

References

[1] Yang Hongliang, Xue He, Yang Fuqiang, Zhao Lingyan. Effect of Film-Induced Stress on Mechanical Properties at Stress Corrosion Cracking Tip. Rare Metal Materials and Engineering, 46(2017): 3595-3600.
[2] I.I. Ahmed , B. Grant, A.H. Sherry et al. Deformation path effects on the internal stress development in cold worked austenitic steel deformed in tension. Materials Science & Engineering A 614 (2014) 326–337.
[3] HOU Juan, PENG Qunjia, SHOJI Testuo et al. Study of microstructure and stress corrosion cracking behavior in welding transition zone of Ni-based alloys[J]. ACTA METALLURGICA SINICA, 2010, 46(10): 1258-1266.
[4] Donghai Du, Kai Chen, Hui Lu et al. Effects of chloride and oxygen on stress corrosion cracking of cold worked 316/316L austenitic stainless steel in high temperature water, Corrosion Science 110 (2016) 134–142.
[5] J. Hou, T. Shoji, Z. P. Lu et al. Residual Strain Measurement and Grain Boundary Characterization in the Heat Affected Zone of a Weld Joint between Alloy 690TT and Alloy 52[J]. Journal of Nuclear Materials, 397(2010)109-115.
[6] P. L. Andresen, M. M. Morra. IGSCC of non-sensitized stainless steels in high temperature water. Journal of Nuclear Materials 2008, 383:97-111.
[7] S. Yamazaki, Z. P. Lu, Y. Ito, Y. Takeda, T. Shoji. The effect of prior deformation on stress corrosion cracking growth rates of Alloy 600 materials in a simulated pressurized water reactor primary water[J]. Corrosion Science, 2008, 50: 835.
[8] Q.J. Peng, J. Hou, T. Yonezawa, T. Shoji, Z.M. Zhang, F. Huang, E.-H. Han, W. Ke, Environmentally assisted crack growth in one-dimensionally cold worked Alloy 690TT in primary water[J]. Corros. Sci. 57 (2012) 81-88.
[9] T. Couvant, L. Legras, C. Pokor, F. Vaillant, Y. Brechet, J-M. Boursier, P. Moulart. Investigations on the mechanisms of PWSCC of strain hardened austenitic stainless steels. Proc. of the 13th International Conference on Environmental Degradation of Materials in Nuclear Power Systems Whistler, British Columbia, 2007. p.1.
[10] A. Sáez-Maderuelo, D. Gómez-Briceño. Stress corrosion cracking behavior of annealed and cold worked 316L stainless steel in supercritical water, Nuclear Engineering and Design 307 (2016) 30-38.
[11] Johsei Nagakawa , K. Ueno, Y. Murase , N. Yamamoto. Effect of cold-work on the radiation-induced deformation of austenitic stainless steels. Journal of Nuclear Materials 367-370 (2007) 910-914.
[12] Litao Zhang, Jianqiu Wang. Effect of dissolved oxygen content on stress corrosion cracking of a cold worked 316L stainless steel in simulated pressurized water reactor primary water environment. Journal of Nuclear Materials 446 (2014) 15-26.
[13] T. Terachi , T. Yamada , T. Miyamoto , K. Arioka. SCC growth behaviors of austenitic stainless steels in simulated PWR primary water. Journal of Nuclear Materials 426 (2012) 59-70.
[14] G.O. Ilevbare, F. Cattant, N.K. Peat, SCC of Stainless Steels under PWR Service Conditions. 42, INIS, 2011, RN:42088756.
[15] Xue He, Li Yongqiang. Micro-mechanical State at Tip of Environmentally Assisted Cracking in Nickel-based Alloy[J]. Rare Metal Materials and Engineering, 2016, 45(3): 0537-0541.
[16] YANG Hongliang, XUE He, CUI Yinghao. Effect of material mechanical properties on stress distribution at crack tip of Ni-based alloys in high temperature water environment [J]. Hot Working Technology. 2016, 45(24): 83-85.
[17] Masayuki Kamaya. Elastic-plastic failure assessment of cold worked stainless steel pipes[J]. International Journal of Pressure Vessels and Piping, 131 (2015) 45-51.
[18] YANG Hongliang, XUE He, ZHAO Lingyan et al. Effect of Oxide Film Shape on Stress-Strain at Stress Corrosion Cracking Tip of Ni-Based Alloy[J]. Hot Working Technology. 2016, 45(20): 58-60.

Roughness Effects on the Fracture Energy between FRP and Concrete

YIN Yu-shi[12,*] and FAN Ying-fang[1]

[1] Department of Civil Engineering, Dalian Maritime University, Dalian 116026;
[2] Liaoning Provincial College of Communications, Shenyang110122）

*Corresponding author:yys00080568@163.com

Abstract: The single test is carried out on totally 54 concrete samples with 6 different interface-roughness conditions, as to delve into the influence of concrete surface roughness on the concrete-FRP interface. Accordingly the closed-form solution based on the compressive strength and interface roughness parameters of concrete is acquired. As the research result indicates, the extremum and extreme point exist in terms of influence of concrete surface roughness on fracture energy, the model acquired in line with the test data is appropriately applicable under the condition $0.34 \le f_i \le 0.56$, and such model is overall applicable under the concrete strength of $30\text{MPa} \le f_{c0} \le 50\text{MPa}$. The model proposed hereof in this paper is superior to the existing models, as not only the significant contribution made by concrete strength to fracture energy is considered, but the influence of concrete surface roughness on interface is highlighted.

1 Introduction

The research on the interface between FRP and concrete (after referred to as interface) is always the emphases and hotspot at home and abroad. Because the action of shear stress, the interface failure pattern mainly divides into four modes: (1)Debonding from the interface between FRP and epoxide resin (2) Failure from the epoxide resin inside (3) Debonding from the interface between concrete and epoxide resin (4) Cracking and debonding form the concrete surface layer; No matter whatever the failure mode is, it is a complicated mechanics process, considering the failure process contains energy transfer, Hillerborg[1]proposed fracture energy in the first place on the ground of the virtual crack model,due to the fracture can not rely on failure pattern, so using G_f to interpret interface degradation rule is more appropriate and convenient.

There are many interface fracture energy models .According to the experiment phenomenon, in the 1990 s Holzenkämpfer [2] found many damage occurred in the concrete surface, so he thought tensile strength of concrete took great impact on interface, interface fracture energy calculation model is put forward for the first time, only one variable named a tensile strength of concrete in the model;In 1997 Neubauer and Rostasy [3] found important role for the first time that the relationship between FRP sheet and concrete bonding surface width ratio $\frac{b_f}{b_c}$ on the interface, thus proposedthemodel using twovariable parameters named concrete tensile strength f_t, width ratio $\frac{b_f}{b_c}$;In 2000, Niedermeier[4] revised Neubauer and Rostasy model, the new stripping bearing capacity model was put forward, but the model of the fracture energy was nottimely update; Dai andUeda[5] in 2003 noticed epoxide resin failure mode, therefore presented new model based on the concrete compressive strength f_c and shear stiffness $\frac{G_a}{t_a}$ of epoxide resin; Dai[6] proposed fracture energy model twice in 2005, first of all, he emphasized the axial compressive strength of concrete contribution to the interfacial energy, in addition in the new model, he considered the stiffness $E_f t_f$ of the FRP plate, epoxide resin shear stiffness $\frac{G_a}{t_a}$, and the compressive strength of concrete f_c that above three parameters supported the new fracture energy model; In 2005, a new interfacial fracture model based on a new fine unit was proposed by Lu et al. [7], the model weakened the stiffness of CFRP sheet, emphasized the great impact such as the concrete tensile strength f_t, shear stiffness $\frac{G_a}{t_a}$ the width ratio $\frac{b_f}{b_c}$ on the fracture energy. In 2013 YuFei Wu and ChenJiang[8], put forward interface fracture model the based on concrete compressive strength f_c, width ratio $\frac{b_f}{b_c}$, and took the comparison with other existing models.

There are many fracture models nowdays, and more and more scholars reached a unified viewpoint that the concrete tensile (pressure)strength, width ratio contributed a lot to interface , but how much contribute to fracture energy based on the morphology between FRP and concrete and roughness of concrete surfacehad not report. From experimental research domestic and abroad in recent 20 years found [9-12], concrete surface roughness, have important influence on bearing capacity of interface. So far, the countries about the reinforcement project guide has a clear pretreatment of interface, it can take a positive influence on reinforce the performance [13, 14], thus if we ignore the research on the effects of surface roughness of the RC beam concrete, so this paper through the single test with totally 54 concrete samples and 6 different interface-roughness conditions to examine the effects of concrete roughness on fracture energy.

2 Experimental Program

2.1 Test Specimens

Test sets concrete specimen size is 80×80×200mm^3, as shown in figure 1. Considering surface roughness and concrete strength would affect the interfacial bonding in local area, there are fifty-fourspecimens all of which are C30, C40, C50, each strength degree assumes six kinds of interface roughness, and three specimens are in a group unit. HICOMA - HITEX series carbon fiber cloth made in Nanjing Hitech Composites Co. Ltd was used in this experiment, epoxy resin AB glue, by 2:1 compound the same corporation made, all details are shown in table 1.

Table 1. Material physical parameter

		f_c (MPa)	f_t (MPa)	E_f (MPa)	t_f (mm)	m_f (g/m^2)
	C30	35.0	5.3			
Specimen	C40	46.0	6.4			
	C50	57.5	7.4			
CFRP			3400	2.3×10^5	0.167	300
Epoxide resin			38	2.4×10^3		

Note: $f_t = \frac{(f_c)^{2/3}}{2}$

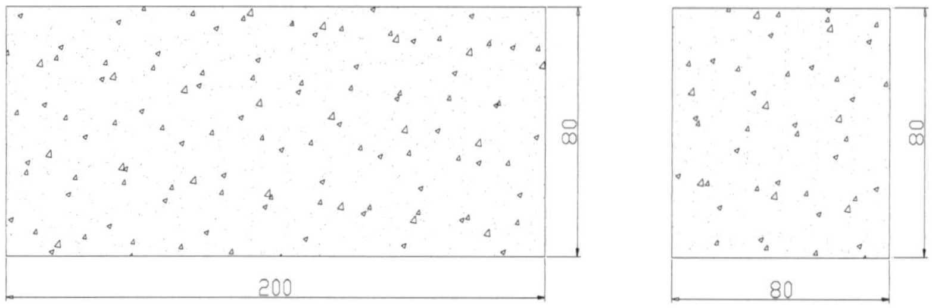

Figure 1 Specimen size and physical figure (unit:mm)

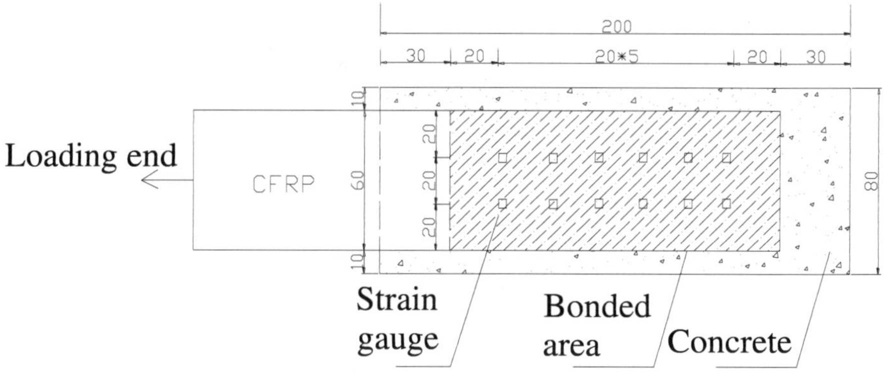

Figure 2. Area of test concrete to be tested

2.2 Test preparation

Make fifty-four FRP-concrete pieces of single shear specimens, paste a group strain gages on each of the pieces' scope of specimen surface ($60\times 140mm^2$), and detect the change in strain in the direction of bonded length in the process of the test. The way to paste way is shown in figure 2. For the purpose of homogenizing influence on the interface due to the interfacial random roughness, two columns strain gauges were made along the interface bonding length, where spacing is 20 mm and the one gauge bonding area is 5mmx3mm. Strain calculation used in this work is average of the two columns strain gauges of each section.

2.3Roughness quantitative definition

At present Chinese commonly used sand filling method [15]to detect the roughness, its measurement method is as follows: the concrete specimen is surrounded with plastic plank, the highest point of the plastic plank and the highest point of concrete surface convex is level, the standard sand is poured into plastic plank, removed all the sand and weigh its volume, the average depth h of the sand filling can be represented by sand volume V divided by the bond area ab . Detailed expressed as:

$$h=\frac{V}{ab} \tag{1}$$

Where: a, bis the length and width of the concrete specimen respectively.

Roughness can be calculated as follows[16] (2):

$$f_i=\frac{h}{\delta} \tag{2}$$

Where: f_i is the interfacial roughness, it is dimensionless value from zero to one, h is the average depth, mm, δ is biggest value of the concrete surface, mm. The calculation result is as follows from table 2.

Table 2 Quantitative interfacial roughness values

Interfacial specimens	*f0*	*f1*	*f2*	*f3*	*f4*	*f5*
Roughness	0.25	0.34	0.44	0.56	0.68	0.88

2.4 Test procedure

All trials are loading with 20 t electro-hydraulic materials testing machine, as shown in figure 3.Loading rate is 5mm/min. During the experiment process, IMC dynamic acquisition system is connected to the strain gauge, the status of strain is observed with the change of the loading force in real time. At the same time, collect ultimate load, the limit displacement and ultimate displacement in the process of CFRP debonding, and observe the specimen failure pattern.

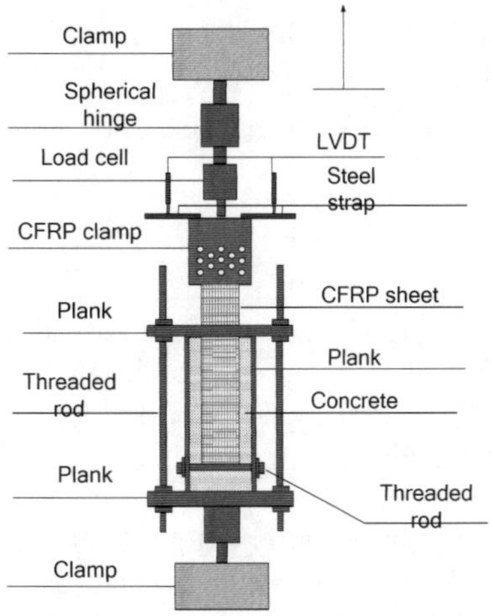

Figure 3 Single shear experimental setup

3. Experiment results and analysis

3.1 Important parameters in the bonding experiment

For single shear test of concrete specimens whose roughness is f0-f5, the important parameters related to interfacial load slip curves are as shown in table 3.

Table 3 Experimental results of single shear test

Specimens	Roughness	*Fu* (kN)	τ_m (MPa)	s_τ (mm)	s_f (mm)
	f0	12.500	3.280	0.003	0.348
	f1	15.030	5.790	0.010	0.288
	f2	17.770	7.350	0.018	0.312
C30	f3	12.230	3.170	0.013	0.345
	f4	6.770	1.650	0.010	0.202
	f5	2.560	0.970	0.008	0.049
	f0	13.200	3.530	0.003	0.358
	f1	15.780	5.920	0.010	0.305
	f2	19.240	8.550	0.021	0.315
C40	f3	13.550	4.080	0.017	0.328
	f4	5.760	1.680	0.010	0.143
	f5	2.440	1.020	0.009	0.044
	f0	14.000	3.760	0.003	0.379
	f1	19.770	9.410	0.016	0.301
	f2	21.230	11.900	0.030	0.275
C50	f3	17.560	8.130	0.034	0.276
	f4	5.770	2.490	0.015	0.099
	f5	2.560	1.070	0.009	0.046

IOP Conf. Series: Materials Science and Engineering **389** (2018) 012016 doi:10.1088/1757-899X/389/1/012016

3.2 The influence of roughness on interface behavior

G_f is the fracture energy of the interface, is the area contained in the bond stress-slip curve of the interface. Such energy determines the bearing capacity of the interface. Additionally this data can be perceived as the important parameter of the bond stress-slip curve. The research indicates that the bearing capacity of the CFRP- concrete interface in in direct proportion to the interface energy square [11][17]. Here are the following expressions:

$$G_f = \frac{P_u^2}{2b_f^{\,2}E_f t_f} \tag{3}$$

Where b_f refers to the width of CFRP, P_u refers to the failure load, E_f and t_f respectively refer to the CFRP elasticity module and the calculated thickness. The interface fracture energies under various types of roughness are calculated as in Table 4.

Table 4. Fracture energy of interfacial roughness (MPa·mm)

	f0	f1	f2	f3	f4	f5
C30	0.565	0.817	1.142	0.541	0.166	0.024
C40	0.630	0.900	1.339	0.664	0.120	0.022
C50	0.709	1.413	1.630	1.115	0.120	0.024

It is acquired through the table 4 that before the f2 interface, the overall interface fracture energy is elevated as the concrete strength grade increases, and as the roughness increases. The maximum interface energy is reached at the f2 interface, and accordingly the influences of the concrete strength grade are respectively 14.7% and 42.7%. As the roughness exceeds the f2, and reaches between f3 to f5, the interface fracture energy shall be decreased with the increase of roughness. It shall be decreased by 98.5% to the largest extent. The concrete strength grade at the roughness section of this interface shall not exert the influence, bespeaking that while the roughness is relatively large, the huge failure of the colloid at the contact point between the fiber reinforced polymer and the concrete shall result in the loss of bonding property. The analysis result mentioned is consistent with the result that the average shearing strength tends to be in direct proportion of the failure time in the single shearing test.

Given the convergence of data, the G_f and τ_m are fitted. The G_f is fitted through adopting the method combining the Marquardt method and global optimization, and the τ_m is fitted through adopting the Guass2D. The fitting results are indicated as follow:

$$G_f = \frac{0.196 + 0.002 f_{co} - 0.325 f_i}{1 - 0.001 f_{co} - 3.684 f_i + 3.893 f_i^2} \quad (30 \le f_{co} \le 50; 0 \le f_i \le 1), \quad R^2 = 0.97 \tag{4}$$

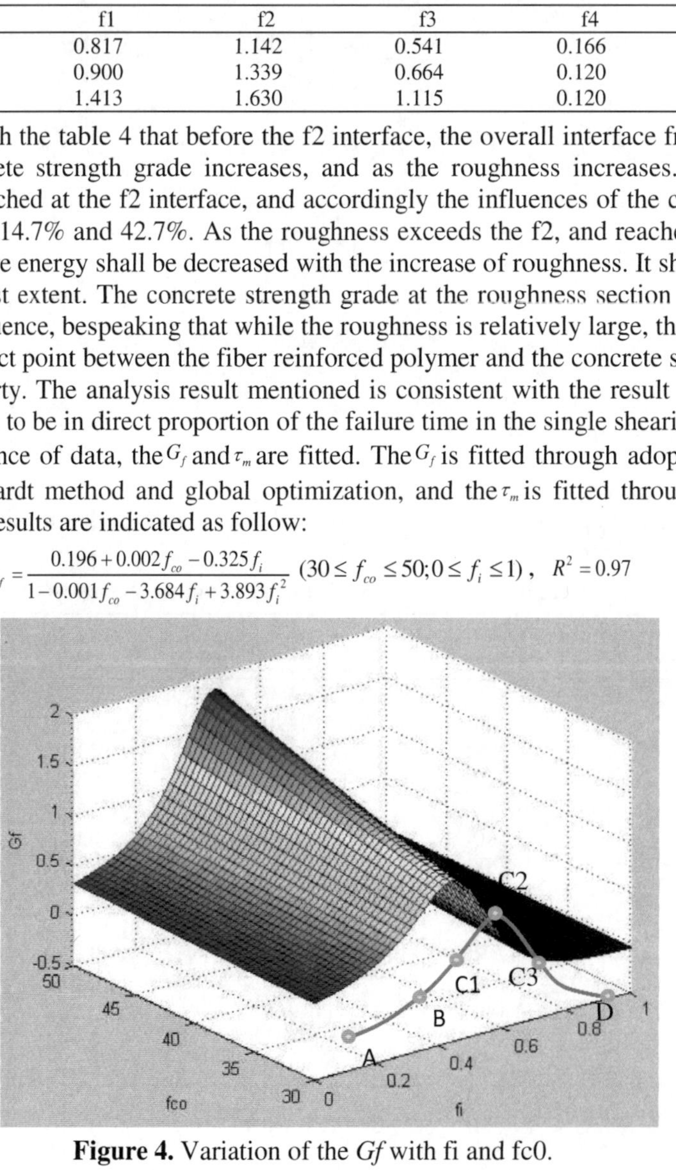

Figure 4. Variation of the *Gf* with fi and fc0.

It is acquired through viewing the Fig 4 that the *Gf* curve for different concrete strength grades can be approximately elevated in the preliminary stage and decreased in the follow-up stage. Three sections can be divided, viz. the A-B, B-C, C-D sections. In AB section, as the fi increases, the *Gf* increases. Moving into the BC section, from C1 to C2, *Gf* is significantly increased as the fi increases. In C2, the maximum *Gf* value is acquired, and the corresponding fi equals to 0.44. Moving into C2 to C3, the *Gf* turns out to be decreased. Reaching to D, the minimum value of *Gf* is acquired, and the corresponding fi equals to 0.88. The process mentioned above bespeaks that the bonding effect shall not be absolutely better with the increase of roughness on the interface, but there are an extreme value and a corresponding extreme value point. For this reason, in the engineering structure, the appropriate roughness shall reinforce the interface energy and be conducive to reinforcing the engineering structure.

Additionally, it is acquired from the Fig 4 that the concrete strength grade is higher than fi in the contribution to the *Gf*, which is consistent with the recognition made by other scholars in the research towards the contribution made by the concrete strength grade.

It is acquired from Fig 4 that as 0.34≤fi≤0.56, the fitting is favorably realized. As fi=0.88, the error is relatively bigger. The maximum error shall reach 212%, which also indicates the application range of the formula (4).

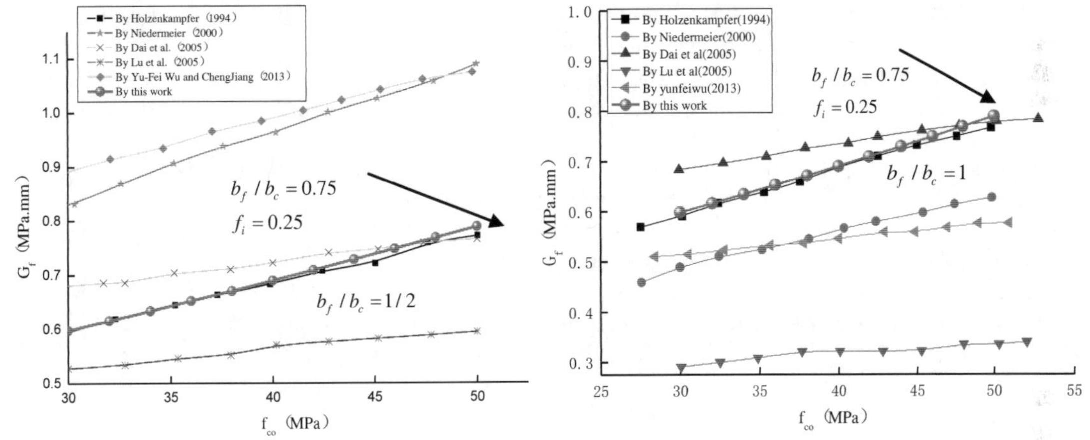

(a) (b)

Figure 5. Models of fracture energy

From Fig 5 we can draw the opinion that the existing model, except for the Holzenkampferand Dai et al models without considering the influence of b_f/b_c, the width ratio, on the interface energy, the other models all consider the influence of width ratio on the interface energy. As the width ratio $b_f/b_c=0.5$, the interface energies under Yunfei-Wu et al and Niedermeier shall be higher than that of other models to 28.5%, or even higher than Lu et al model to 70%. As $b_f/b_c=1$, the interface energies under Yun-FeiWu et al and Niedermeier shall be lower than that of other models to 20%, or even lower than Lu et al model to 40%. Hence it is bespoken that width ratio parameter exerts the larger influence to those two models with larger discrete distribution. If FRP width ratio is 0.5, as shown in Fig 5, the interface energy shall be overestimated, posing the potential safety hazard to the engineering structure. The model proposed in this paper can be well geared into the mean sections. Both the influence of concrete strength grade on the interface energy and the roughness of concrete surface are considered. The interface energy calculated under $b_f/b_c=0.75$, the width ratio adopted in this research, can better geared into three models, viz. Holzenkampfer, Lu et al and Dai et al. This is because the influence of width is not considered, and because the influence of concrete strength grade on the interface is approved.

4. Conclusion

(1) Surface roughness of concrete which impacts on fracture energy cannot be ignored, there exists one extremum points and extremum about fracture energy, when fi=0.44, Gf reaches its maximum which is 1.142 MPa. mm, 1.339 MPa. mm and 1.630 MPa. mm respectively.

(2) The new proposed models suggeststhat concrete compressive strength f_{c0} takes higher contribution to interface the interface roughness fi.

(3) For $0.34 \le f_i \le 0.56$, $30\text{MPa} \le f_{c0} \le 50\text{MPa}$is applicable for the closed-form solution of G_f based on the experimental data, whereas for more interfacial study about lower strength and higher strength of concrete,it remains further research.

Acknowledgments

This paper, its corresponding author is Yin Yu-shi, And this work was supported by National Natural Science Foundation of China (Grant Nos. 51578099) and Liaoning Provincial College of Communications's Technology Applied Research Projects(Grant Nos. lnccjyky201701)

References

[1] ZHANG Tingyi,GAODanying. Research on fracture energy of fiber high strength concrete[M].2010.
[2] Holzenkämpfer Holzenkampfer,p.1994.Ingenieurmodelle des verbundesgeklebterbewehrung fur betonbauteile.Ph.D.thesis,TUBraunschweig,Germany(in German).
[3] Neubauer and Rostasy, F.S.(1997).Design aspects of concrete structures strengthened with externally bonded CFRP-plates.Proc.,7thInt.Conf.on Structural Faults and Repair,M.C.Forde,ed.,Engineering Technics Press,Edinburgh,UK,109-118.
[4] Niedermeier, R.(2000).Envelope line of tensile forces while using externally bonded reinforcement.Ph.D.thesis,TU München ,Germany(in German).
[5] Dai,J.G.(2003).Interfacial models for fiber reinforced polymer(FRP)sheets externally bonded to concrete.Ph.D.thesis,HokkaidoUniv.,Hokkaido,Japan.
[6] Dai,J.G.,Ueda,T.,andSato,Y.(2005).Development of the nonlinear bond stress-slip model of fiber reinforced plastics sheet-concrete inerfaces with a simple method.J.Compos.Constr.,9(1),52-62.
[7] Lu Xinzheng. FRP and concrete interface behavior research [D]. Beijing: Tsinghua University, 2004.
[8] YuFei Wu and ChenJiang .Quantification of bond-sdlip relationship for externally bonded FRP-to-concrete joints. J. Compos. Constr. 2013.17:673-686.
[9] Chajes MJ, Finch WW, Januszka TF, Thomson TA. Bond and force transfer of composite material plates bonded to concrete. ACI Struct J 1996;93(2):208–17.
[10] Yao J, Teng JG, Chen JF.Experimental study on FRP-to-concrete bonded joints. Composites Part B 2005;36:99–113.
[11] Delaney J, Karbhari V. Defect criticality in FRP strengthening. In: Proc 8th IntSymp in fiber-reinforced (FRP) polymer reinforcement for concrete structures(FRPRCS-8) 2007 CD-ROM, Univ of Patras, Greece, Paper 3–20.
[12] Li Weiwen.Effect of concrete surface roughness on the bonding performance between the CFRP and concrete. Journal of Shenzhen university (version of technology).2007.01.
[13] The interior bureau of reclamation. Concrete handbook [M]. Wang Shengpei translation. Beijing: water conservancy electric power press, 1990.
[14] Highway Bridges reinforcement design specification, the ministry of transport of the People's Republic of China, JTG/T J22-2008.
[15] ZHAO Zhifang,ZHAOGuofan. Experimental research on treating interface of young-old concrete with high-pressure water jet method [J].Journal of Dalian University of Technology, 1999, 39(4):559-561. (in Chinese).
[16] SHANG Shouping,YUDejun,ZHANGRuiwen.Evaluation of surface roughness on strengthed RC structure[J].Journal of Building Structures. 2010,(10):120-124. (in Chinese).
[17] Wang Suyan,LiuGui,ZhouYingwu. GFR - high strength concrete interface bond slip relationship study [J]. Journal of civil engineering, 2009, and (6) : 8-13.

IOP Conf. Series: Materials Science and Engineering **389** (2018) 012017 doi:10.1088/1757-899X/389/1/012017

Oxidation Kinetics Researches under the Condition of Compressive Loading

C H Zhou[1,3] , R Y Pan[1], H T Ma[2] , H Zhang[1], X G Guan[1], N N Mao[1] and F J Sun[1]

[1] School of Materials Science and Engineering, Heilongjiang University of Science and Technology, Harbin 150022, P. R. China;
[2] School of Materials Science and Engineering, Dalian University of Technology, Dalian 116085, P .R. China

email: dlut2325@163.com (C.H. Zhou)

Abstract. High temperature oxidation behavior of alloys is mostly affected by an external loading. The growth of oxide scale, oxide products, interfacial reactions and other characteristics have be affirmed to be affected by the external loading. Here, it provided the oxidation kinetics of several metals under the compressive loading. And the growth of oxide scale was distinguished in the form of thin oxide layer and thick oxide layer. Different compressive effect on the growth of oxide scale was summarized. At the initial oxidation stage, the applied compressive stress induced more defects on the surface of alloy, supplying more oxide neclei, thereby increased the growth of the thin oxide layer. In the case of the growth of the thick oxide layer, it was attrbuted to the short circuit paths of the oxide grain boudary inherited from the initial oxidation stage and the cracks or other defects deduced by the applied compressive stress. Thus the oxide kinetics of alloys was promoted. However, considering the interfacial reaction at scale/metal, the coarsens of vacancies would occur after a long time of oxidation, decreasing the growth of oxide scale.

1. Introduction

The oxidation resistance of alloys is reflected by oxidation kinetics, and oxidation rate constant is close related to the application lifetime of alloys. Commonly, the growth of Cr, Al, Si-formed integrity oxide scale is slow, which can play a role of protection to alloys [1]. However, the alloys are always used in various serious environment in the practice application. Thus the oxide scale on the alloys will be destroyed, and cracking and even spallation are occurred. Once the further healing oxidation cannot supply the protective oxide scale, the alloys will be oxidized in a fast rate. Especially, the oxide scale is prone to spall under the external loading conditions, which brings a negative effect on the oxidation kinetics [2]. More and more investigations have been carried out to explore the effect mechanism of external loading on the growth of oxide scale. Nowadays, there are still controversy about the effect mechanism. Most studies suggested that the growth of oxide scale was increased by the external loading [3-5]. Several investigations reported that the oxidation rate was decreased by the loading [6,7]. And a few experiments affirmed that no effect of loading on the growth of oxide scale [8]. Undeniably, the external loading has a fatal effect on the brittle oxide scale. Once the oxide scale is broken, the slowly growth of integrity oxide scale is interrupted. Therefore, the explore on the oxidation kinetics of alloys under the external loading is very important. The authors reviewed the studies of oxidation kinetics of several metals under the compressive loading, and the effect mechanism was summarized.

2. Oxide-scale growth characteristics under compressive loading

Oxidation of a metal/alloy occurs owning to the reaction with an oxidizing atmosphere under the driving force of the free energy of the oxide formation. A continuous oxide scale as a barrier to decrease the fast depletion of metal can form on the surface of the metal. However, the growth of oxide scale will be affected when subjected to an external loading condition. The growth rate of oxide scale is depended on the oxidizing time, the integrity of oxide scale, and other factors. In order to clarify the effect of external loading on the growth of oxide scale, a schematic of oxide-scale evolution under the external loading is shown in Figure1. In the short-stage oxidation, a thin film will form on the surface of the metal shown in Figure1(a), and the growth of the thin film would be affected by the loading. In fact, it is indeed affected according to the previous studies [2]. With the time increasing, the oxide scale is gradually thickened, as indicated in Figure 1(b). As the PBR and the mechanical nature of oxide scale, the oxide scale would be bulking, cracking and spalling. The continuous growth of oxide scale is then damaged, as shown in Figure.1(c) and Figure.1(d). As for the destroyed oxide scale, its growth is depended on the healing behavior of further oxidation at the naked metal surface. It is concluded that the growth of a mental/alloy under the loading condition is consisted of two stages, i.e., the growth stage of thin oxide scale and the growth stage of thick oxide scale. And the growth behavior of thick oxide scale is depended on the mechanical nature of oxide scale, which can be comprised of two forms of buckling and spalling oxide scale. Therefore, the following summary of the effect of compressive stress on the growth of oxide scale of metal/alloy is based on this distinctive.

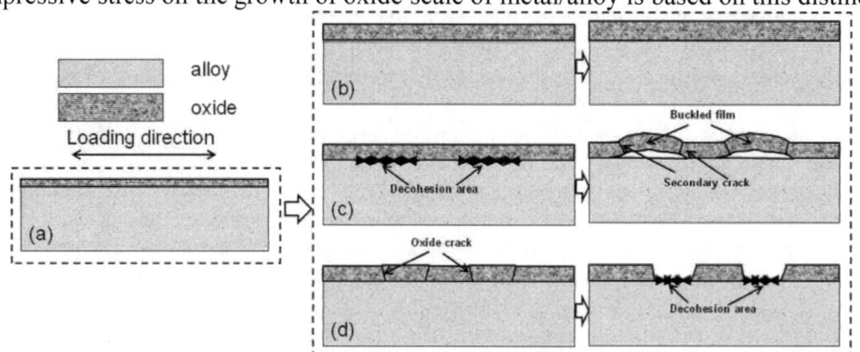

Figure 1. Oxide-scale growth characteristics under compressive loading: (a) thin oxide-scale formation; (b-d) thick oxide-scale continuous formation; (b) protective oxide-scale growth; (c)oxide-scale failure model I: decohesion and buckling precede through-scale cracking; (d)oxide-scale failure model II: through-scale shear cracking precedes interfacial decohesion [9]

3. Oxidation kinetics under compressive loading

3.1 Thin oxide-scale growth stage

Early stages of oxidation of a metal/alloy typical are consisted of nucleation and growth of nuclei and then a continuous thin layer of oxide. During this stage, the involved oxidation kinetics is close related to the surface condition of metals. Selective oxidation and short-circuit diffusion would occur at the local surface of metals where high subsurface dislocation density and lots of defects are existed [2]. And subsequently fined oxide particles would form, which enhances the effective diffusion according to the equation 1.

$$D_{eff} = \left(1 - f\right) D_b + fD_{gb} \tag{1}$$

Where D_{eff} is effective diffusion coefficient, D_b is diffusion in the bulk, D_{gb} is diffusion along grain boundaries, $D_{gb}>>D_b$ [10]. Thus, the induced dislocations and defects by some methods will supply more nuclei and increase the diffusion, and therefore increase the oxidation rate of a metal/alloy.

According to previous investigations, external loading, such as grinding, sandblasting and even directing loading, can introduce dislocations and defects [2]. Especially, the growth behavior of oxide scale is dependent on the mechanical loading.

Figure 2 show the evolution of instantaneous oxidation rate constant with increasing time of pure Ni and Fe-20Cr when subjected to compressive loading. It can be seen that compressive loading increase the initial oxidation rate for pure Ni oxidized at 800 ℃ and Fe-20Cr alloy oxidized at 900℃ respectively, as shown in Figure 2(a) and Figure 2(b). That is related to the short-circuit diffusion of grain boundaries, which can be supported by the evolution of oxide particles during such process.

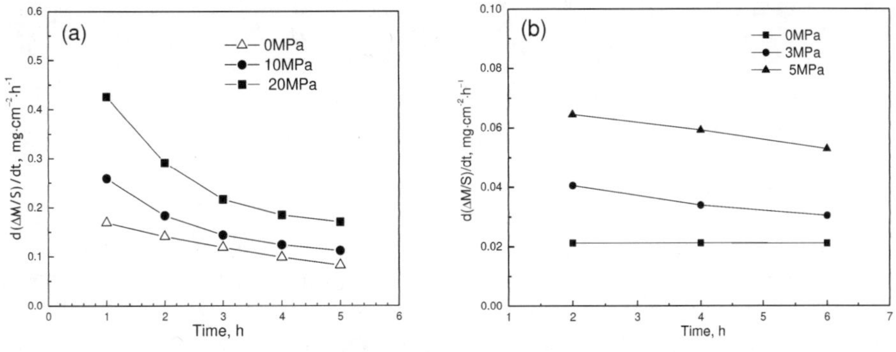

Figure 2. Instantaneous oxidation rate constant with increasing time of (a) pure Ni oxidized at 800 ℃ and (b) Fe-20Cr alloy oxidized at 900 ℃

Figure 3 displays the surface morphologies of pure Ni oxidized at 800 ℃ and Fe-20Cr oxidized at 900 ℃. As can be seen, compressive loading induced the change of surface morphology of pure Ni from classical ridge-like under unstressed condition displayed in Figure 3(a) to particle-like in the case of 20 MPa shown in Figure 3(b). As for Fe-20Cr alloy, the compressive loading made the oxide size more smaller (by comparing the oxide particles under 0 MPa stress in Figure 3(c) and 5 MPa stress in Figure 3(d)). Above evolution of surface morphologies of both metals implies that the instantaneous oxidation rate constants are related to the grain size of oxide. And the smaller grain size of oxide induced by the compressive loading can be as an ex situ evidence on the increasing nuclei at the introduced dislocations and defects. This phenomenon is also reflected in pure Ni oxidized at 700 ℃ and Fe-16Cr alloy oxidized at 900 ℃ under the condition of compressive loading [10-13].

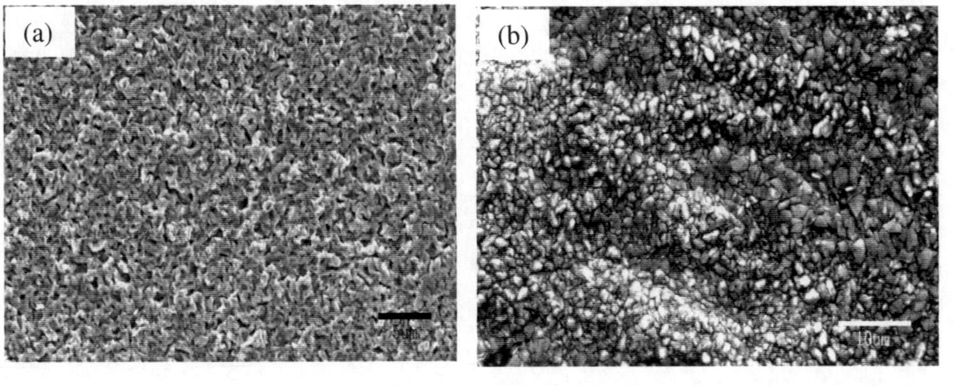

Figure 3. Surface morphologies of pure Ni oxidized at 800 ℃ under (a) 0 MPa , (b) 20 MPa for 20 h and Fe-20Cr alloy at 900 ℃ under (c) 0MPa, (d) 5 MPa for 10 h [10,13]

3.2 Thick oxide-scale continuous growth stage

After the formation of a thin layer of oxide, the continuous growth of oxide scale to form a thick layer is commonly controlled by the diffusion of cationic or anionic species through the oxide scale. The diffusion behavior through the oxide scale is strongly affected by the nature of oxide scale. The mechanical properties, microstructures, pores and defects of oxide scale is directly affect such diffusion process [1], which increased the difficulty for analysing the compressive loading on the growth of oxide scale. In order to simplify and scientize, the growth behavior of a metal/alloy under compressive loading is distinct according to Figure1.

3.2.1 growth of integrity oxide scale. For the growth of an integrity oxide scale, it is mainly closed to the diffusion of cationic or anionic species through the oxide scale and the interface reaction at the oxide/metal interface. In particular, the change of oxide products will occur owing to the external loading. That affects the diffusion behavior in the oxide scale. Though some different mechanisms has been proposed, it did not reach a consensus on the effect of external loading on the change of oxide products [14,15]. In the present work, pure Nickel, which have no change of oxide product during the oxidation and is widely used to investigate the oxidation mechanism under various conditions owing to classical and simply oxide product, were chosen to study the growth behavior under the compressive loading.

Figure 4 shows the parabolic oxidation curves of pure Ni oxidized at 700 ℃ under different compressive conditions. As can be seen, the applied compressive loading increase the growth of pure Ni. Besides, different laws was observed. It completely followed a parabolic law for conventional oxidation. However, it followed a two-stage parabolic law in the case of 10 MPa compressive loading. Interestingly, it did not follow parabolic law subjected to 20 MPa compressive loading. In order to clarify that, the cross-sectional morphologies were examined.

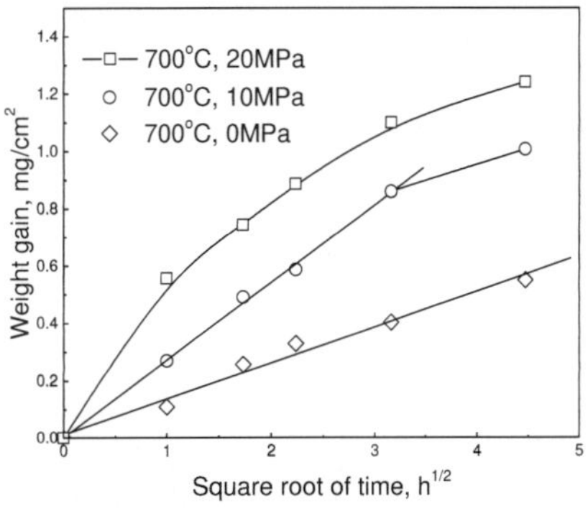

Figure 4. Weight gain as a function of the square root of time for Ni oxidized at 700℃ subjected to compressive loading [12]

Figure 5 shows the cross-sectional morphologies of pure Ni oxidized under different conditions. It can be seen, by comparing the cross-sectional morphologies of pure Ni oxidized at 700 ℃ and 10 MPa and 20 MPa for 20 h as shown in Figure 5(a) and Figure 5(b), that obvious coarsens of pores at the oxide/metal interface were generated. According to the relations between the growth of oxide scale and vacancy concentration proposed by Gibbs [16], which was proved by Stroosnijder [17].

$$dx/dt = B\Omega(p_{O_2})^{-1/n}[(V_L)_e/(V_L)^{2/n}](D_L/x) \tag{2}$$

Where Ω is volume of oxide per metal ion, $(V_L)_e$ is thermodynamic equilibrium concentration of lattice vacancies in a metal; V_L is concentration of lattice vacancies close to the scale/metal; D_L is cation diffusion coefficient in oxide; x is thickness of oxide scale; n and B are constants. It can be gained that coarsens of pores substantially increase the vacancy concentration at the oxide/metal interface, which decrease the oxidation rate. It can be explained that the parabolic oxidation rate at second stage is smaller than that at the initial one in the case of 10 MPa compressive loading. It is because that the growth of pores continuously decrease the oxidation rate. It displays more obvious in the case of 20 MPa, even no parabolic law was followed, and the oxidation rate gradually decrease with the time. So, it can be detected that the oxidation rate in the case of compressive loading will lower than that of the conventional test in the following oxidation time. Figure 6 shows the growth of oxide scale for pure Ni up to 140 h. It can be seen that the weight gain for the conventional test was higher than that under 10 MPa compressive loading after about 60 h. As for the increasing oxidation rate before 60 h, it was related to the nucleation, growth of nuclei and thin layer. The change of surface morphologies were approved that in the previous work [11-13]. Besides, the growth of integrity oxide scale on Fe-20Cr alloy was increased by compressive loading based on the short-circuit diffusion along oxide grain boundary [10].

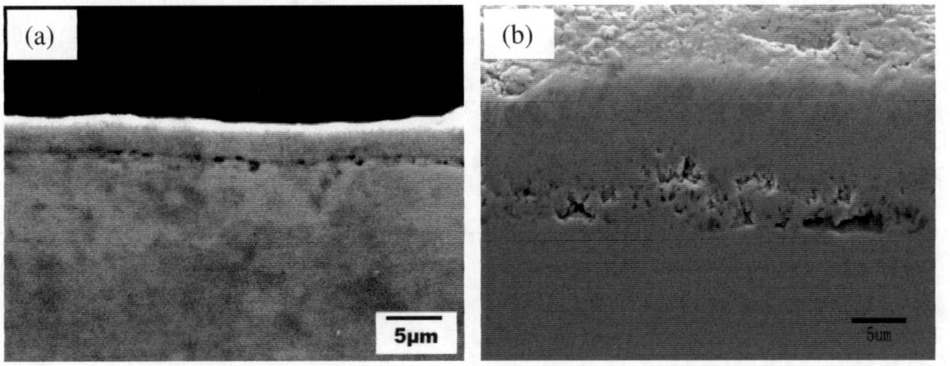

Figure 5. Cross-sectional morphologies of pure Ni oxidized at 700 ℃ under (a) 10 MPa and (b) 20 MPa for 20 h [11]

IOP Conf. Series: Materials Science and Engineering **389** (2018) 012017 doi:10.1088/1757-899X/389/1/012017

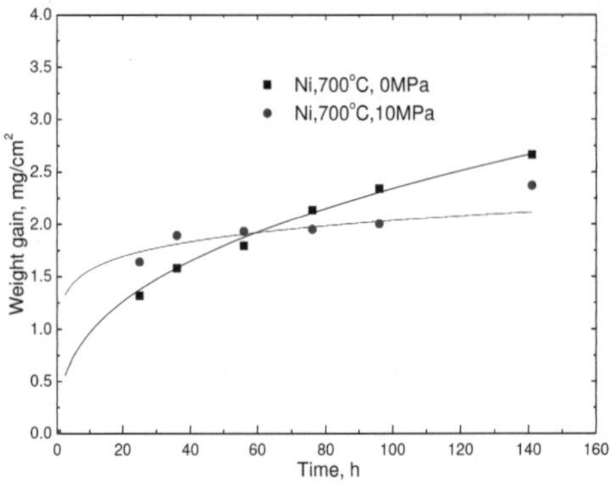

Figure 6. weight gain as a function of time for pure Ni oxidized at 700 ℃

3.2.2 growth of cracking oxide scale. The growth of cracking oxide scale under the compressive loading is dominated by short-circuit diffusion, such as grain boundary, cracks, pores and other defects. Such process shows that the growth of oxide scale was increased by compressive loading. As seen in Figure 7, the growth of oxide scale on pure iron was increased by compressive loading. Because the oxide products of pure iron at above 570 ℃ are consisted of haematite, magnetite and more amount of wustite phases, the oxide scale is prone to crack under the loading owing to the poor mechanical properties of wustite layer [1]. The short-circuit diffusion along the cracks and other defects becomes the main mechanism of oxidation of pure iron under the compressive loading. Figure 8 shows the surface morphologies of pure iron after 20 of oxidation under loading. And the surface morphology of pure iron oxidized at 650 ℃ for 20 h under 6MPa compressive stress is shown in Figure 8 (a). It can be seen that buckling, cracks and local spallation were displayed, which become the short-circuit diffusion paths. And high magnification surface morphologies show that a lot of pores in the oxide scale were generated in comparison with unstressed condition and 6 MPa compressive stress, as displayed in Figure 8(b) and Figure 8(c). Oxygen then can directly reach the reaction interface, rather than through the diffusion process. Those can greatly increase the growth of oxide scale, as shown in Figure 7, the weight gain substantially increases after 10 h of oxidation under 6 MPa loading.

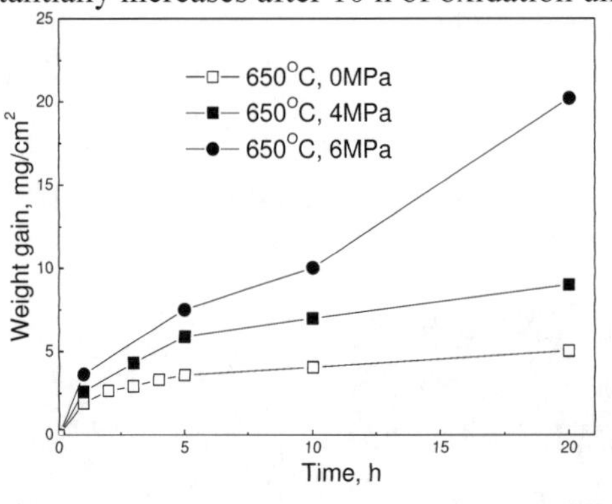

Figure 7. weight gain as a function of time for pure iron oxidized at 650 ℃ [18]

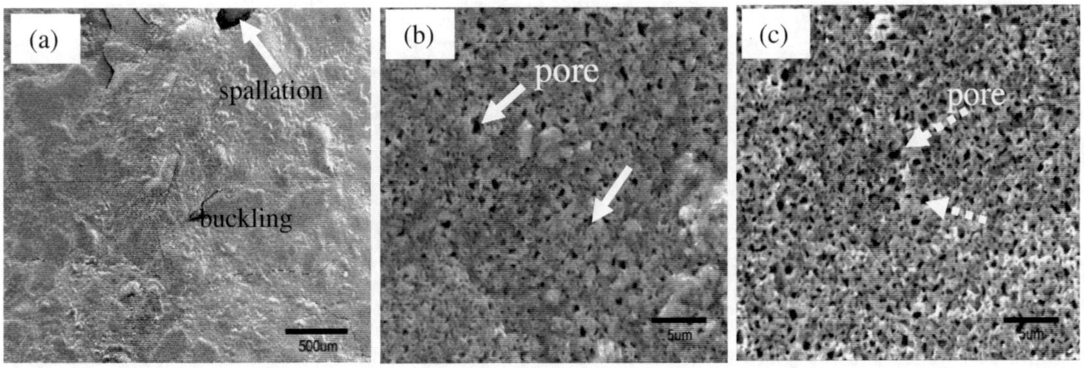

Figure 8. surface morphologies of pure iron oxidized at 650 ℃ for 20 h under compressive loading of (a) 6 MPa; (b) 0 MPa; (c) 6 MPa [18]

A Fe-16Cr alloy, which formed multi-layer oxide, was chosen to explain the effect of compressive loading on the growth of oxide scale. As can be seen, the growth of oxide scale on the Fe-16Cr alloy followed a parabolic law in the presence of compressive loading. And the compressive effect on the oxide scale induced its growth, as shown in Figure 9. According to the growth mechanism of Fe-Cr alloys with about 15% Cr content, commonly an outer Fe-oxide layer, a media $FeCr_2O_4$ layer and inner $Fe(Fe,Cr_2)O_4$ layer will form [1]. And the diffusion of cations or anions across the oxide scale is blocked by $FeCr_2O_4$ layer and inner $Fe(Fe,Cr_2)O_4$ layer. If the Fe-oxide layer is prone to crack, then more short-circuit diffusion paths will generate to increase the growth of oxide scale. Seen form Figure10, it can be gained that Fe-oxide multilayer and Cr-rich oxide layer was the critical characteristic of Fe-16Cr alloy. The previous investigation showed that the increased Fe-oxide layer by the loading was the important factor on the compressive effect on the whole growth of oxide scale [19]. Then, it comes a conclusion that the poor mechanical property of oxide scale is easily to induce the short-circuit diffusion paths by the compressive loading, which is one important mechanism for the growth of oxide scale.

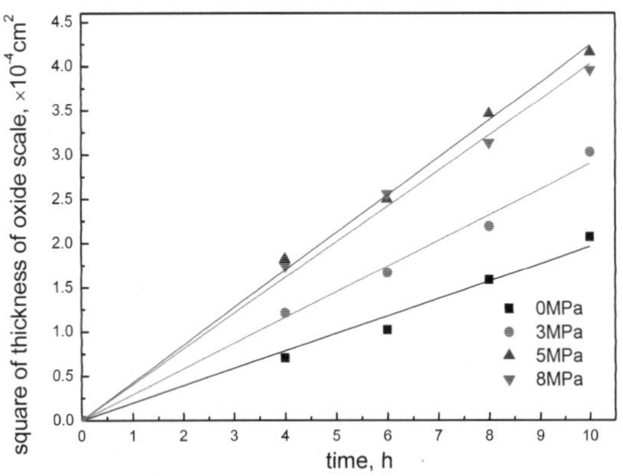

Figure 9. square of thickness of oxide scale as a function of time for Fe-16Cr alloy oxidized at 900 ℃ under compressive loading [19]

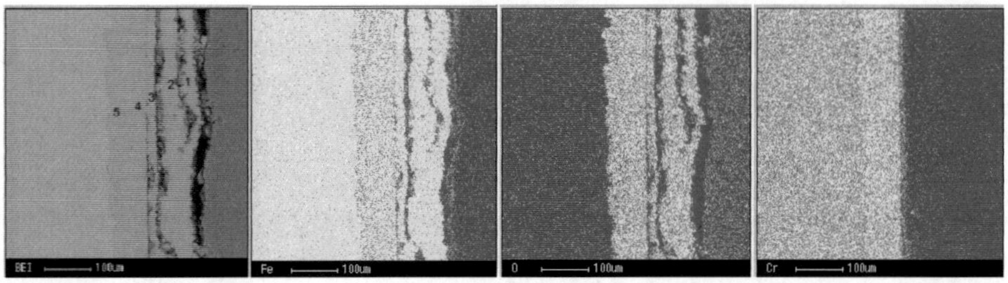

Figure 10. Elemental mapping of Fe-16Cr alloy oxidized at 900 ℃ under 5 MPa loading for 10 h [19]

Besides, the short-circuit diffusion paths in outer Fe-oxide layer can affect the growth of internal oxidation in a Fe-20Ni alloy. Figure 11 shows the growths of intragranular (as shown in Figure 11(a)) and intergranular (as displayed in Figure 11(b)) scale for the Fe-20Ni alloy oxidized at 800 ℃ and 900 ℃ for 20 h. As can be seen, the growths of intragranular and intergranular scale are both increased by the compressive loading for the Fe-20Ni alloy oxidized at 900 ℃. At 800 ℃, the compressive effect on the growths of both scales is not obvious. It is implied that the effect on the growths of intragranular and intergranular scale are depended on the applied loading and temperature.

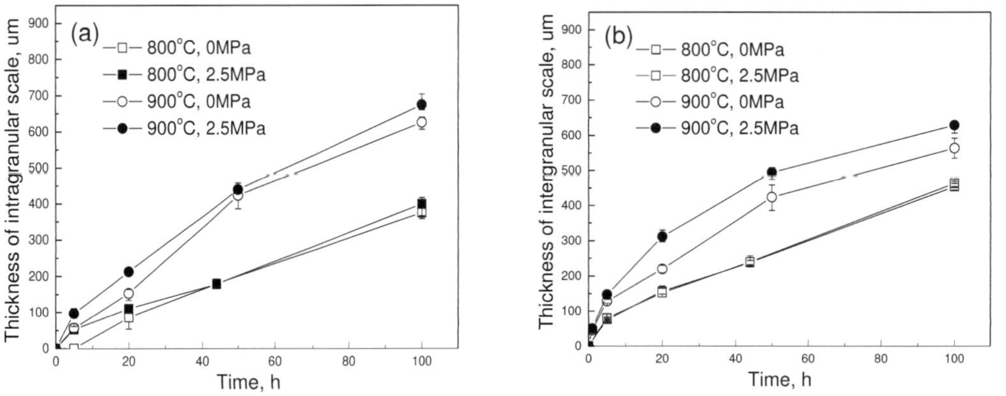

Figure 11. Effect of temperature on the thickness of internal scale on the Fe-20Ni alloy (a) intragranular scale; (b) intergranular scale

4. Summary

The growth of oxide scale of alloys under external loading is indeed effected, although different mechanism and effect are suggested. The present review only show the compressive loading on the growth of oxide scale on several metals. Both stages of growths of oxide scale by the effect of compressive loading are displayed. At the stage of growth of thin oxide layer, the loading introduces more defects on the subsurface of metals, which increases the oxide nucleation and thereby fined oxide grains form. Short-circuit diffusion along the oxide grain boundary is modified, and the growth of thin oxide layer is increased by the compressive loading. At the stage of growth of thick oxide layer, the mechanical properties of oxide scale become the critical factor on the compressive effect on the growth of oxide scale. For the oxide scale with good ductility, such pure Nickel, the compressive loading induces the pores generation at the oxide/metal interface, and finally decreases the growth of oxide scale of pure Nickel, although the growth of oxide scale is increased by compressive loading for a 20-hour oxidation. For the oxide scale with poor mechanical property, the oxide scale is prone to cracking, buckling and spallation by the compressive loading, short-circuit diffusion paths are more increased. And the growth of oxide scale is increased, even the internal oxidation is also increased.

Acknowledgments
The work is supported by National Natural Science Foundation of China (No. 51501058).

References

[1] Birks N, Meier G H, Pettit F S 2006 Introduction to the High-Temperature Oxidation of Metals(Cambridge: Cambridge University Press)
[2] Douglass D L, Kofstad P, et al 1996 *Oxid. Met.* **45** 529
[3] Rolls R and Shahhosseini M H 1982 *Oxid. Met.* **18** 115
[4] Calvarin-Amiri G, Molins R and Huntz A M 2000 *Oxid. Met.* **54** 399
[5] Zhou C H, Ma H T and Wang L 2008 *Oxid. Met.* **70** 287
[6] Roda G C, Santarelli F and Sarti G C 1984 *J. Electrochem. Soc.* **132** 1909
[7] Doremus R H 1984 *Thin Solid Film* **122** 191
[8] Qian Y H, Li M S and Zhang Y M 2001 *Corr. Sci. Protect. Technol.* **13** 342
[9] Nagl M M and Evans W T 1993 *J. Mater. Sci.* **28** 6247.
[10] Zhou C H, Ma H T and Wang L 2009 *Oxid. Met.* **71** 335
[11] Zhou C H, Ma H T and Wang L 2010 *Corros. Sci.* **52** 210
[12] Zhou C H, Ma H T and Wang L 2009 *Corros. Eng. Sci. Technol.* **44** 358
[13] Zhou C H, Ma H T and Wang L 2009 *The Chinese J. Nonferr. Met. 19* 2162
[14] Zhou C H, Ma H T and Wang L 2010 *J. Iron Steel Res. Inter* **17** 27
[15] Qian Y H, Li M S and Zhang Y M 2003 *Acta Metall. Sin.* **39** 989
[16] Gibbs G B 1968 *Philosophical Magazine* **18** 1175
[17] Stroosnijder M F, Guttmann V, et al 1989 *Mater. Sci. Eng. A* **121** 581
[18] Zhou C H, Ma H T and Wang L 2010 *J. Dalian Univer. Technol.* **50** 667
[19] Zhou C H, Ma H T and Wang L 2016 *Oxid. Me*t. **85** 537

Research on Preparation and Extrusion Process of Magnesium Calcium Alloy

Qiao Yang*, Yang Long, Wang Shouren, Guo Peiquan and Wang Yanan

School of Mechanical Engineering, University of Jinan, Jinan 250022, China

*Corresponding author. E-mail address: me_qiaoy@ujn.edu.cn

Abstract. Magnesium calcium alloy has attracted more and more attention in the medical metal field as it has the similar mechanical properties with human bone and biodegradable property. In this paper, the preparation and modification of the alloy were studied. It was found that, extrusion parameters had an important impact on the quality of product in extrusion process, like extrusion ratio, extrusion temperature and extrusion speed. Vickers hardness, and tensile mechanical properties test and bending mechanical properties were carried on as-cast and extruded alloy to obtain mechanical parameters, the mechanical property of MgCa0.8 alloy greatly improved after extrusion, and the extrusion process can be applied to other calcium content alloys.

1. Introduction

Magnesium (Mg) alloy attracts the attention of scholars because of its good mechanical and biodegradable properties [1]. Medical magnesium alloy applied to the field of bone fracture plate which replaces the original non degradable metal materials will effectively avoid the secondary damage caused by removing the plate after recovery [2].

Calcium (Ca) is one of a small number of alloying elements that can be included in implants. Binary magnesium calcium alloy has better corrosion resistance as the accession of calcium element, causing grain refinement. As the result of this similarity, it stimulates bone cells to attack the implant surface and make proper bonding [3, 4].

As a substitute of biomedical devices, magnesium calcium alloy is bound to be one of the main development directions of biomedical implants. According to the mechanical properties of magnesium calcium alloys and corrosion resistance, scholars carried out a series of studies [5-7]. These findings suggest that, we can improve the densification microstructure of alloy materials to control the content of Ca, thus improving the mechanical properties of magnesium alloy and corrosion resistance. It is found that MgCa0.8 will be the most suitable magnesium calcium alloy in the field of orthopedics implantation [8-10].

2. The preparation of magnesium calcium alloy

Magnesium ingot and MgCa15 master alloy were used for smelting in the experiment within the resistance furnace. The target alloys prepared of smelting were MgCa0.8. Follows were the steps:

(1) Calculated and prepared the raw materials for smelting.

(2) Raised the temperature of the furnace to 730℃ and stopped for half an hour as the temperature reached to 400℃and then protection gas was access into the furnace. The crucible with magnesium ingot was put into the furnace as the temperature reached to 500℃.

IOP Conf. Series: Materials Science and Engineering **389** (2018) 012018 doi:10.1088/1757-899X/389/1/012018

(3) MgCa15 master alloy was added after the magnesium ingot melted.

(4) Adjusted the temperature to 720℃ and stirred the crucible well after MgCa15 master alloy melt.

(5) Skimming was operated in the atmosphere of protection gas and the melted metal was poured into preheated graphite mold. The parts were taken out after the mold had colt in room temperature.

Since the chemical character of magnesium is very active and it was prone to oxidation and burning, which made the melting process must be under the environment of protection gas mixed with CO_2 and SF_6.

Inductively Coupled Plasma Emission Spectrometer (Optima 2100 DV) generated by Perkin Elmer was used to detect the composition of casting alloy. The protection gas mixed with CO_2 and SF_6 could stop the oxidation in the process of melting, but it could not stop evaporation of melted magnesium and calcium. So some difference appeared between tested value and calculated value. The right ratio of raw materials was continued to explore and then casted the alloys. Tested composition of Ca was 0.78 wt. %

3. Orthogonal Extrusion Experiment of magnesium calcium alloy

Extrusion parameters have an important impact on the quality of product in extrusion process, extrusion ratio, temperature and extrusion speed which had large influence on extrusion effect were chosen for orthogonal experiment.

The extrusion ratio was 8.1, 15.9, 29.5, extrusion temperature was 250℃, 300℃, 350℃, and extrusion speed was 10mm/s, 30mm/s. The level of 10mm/s was considered as a quasi-level in the orthogonal experiment which used L9 (3^3) orthogonal table. Table 1 shows the experiment plan, and the casting parts were treated at 370℃ for 6 hours for homogenization and turning to remove the oxide layer before experiment.

Table 1. Orthogonal experiment plan

No.	Ratio	Temperature /℃	Speed /mm • s-1	No. of product bar
1	8.1	250	10	A1
2	8.1	300	10	A2
3	8.1	350	30	A3
4	15.9	250	10	B1
5	15.9	300	30	B2
6	15.9	350	10	B3
7	29.5	250	30	C1
8	29.5	300	10	C2
9	29.5	350	10	C3

Phase analysis, mechanical property test and corrosion resistance test were carried on for a suitable extrusion process.

3.1. X-ray Diffraction Analysis

XRD analysis of the sample was used D8-ADVANCE type X-ray diffract meter generated by Brooks. Follows were its working parameters: copper target, operating voltage40kV, operating current 100mA, scanning speed of 4 °/min, scanning range 30°~ 90°.

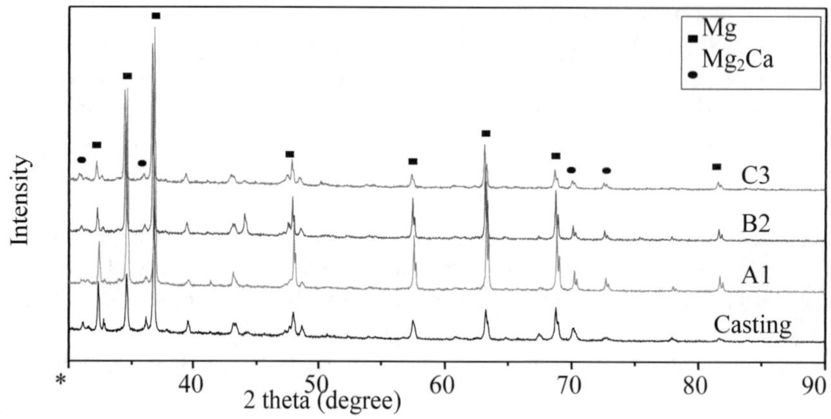

Figure1. X-ray diffraction pattern of as-cast and extruded Mg-Ca0.8 alloys

Figure 1 shows the X-ray diffraction pattern of MgCa0.8 alloy as-cast and extruded A1, B2 and C3. Compared with MDI Jade 6.0 software, MgCa0.8 alloy contains α-Mg and Mg_2Ca phases, and most of the alloy was α-Mg phase and Mg_2Ca phase content was relatively small according to the X-ray peak size. Compared with the X-ray of as-cast and extruded states, it was found that just the internal structure of MgCa0.8 alloy changed but no new phase appeared.

3.2. Mechanical Properties Tests

In order to study the effect of different extrusion process on the mechanical properties of the alloy, Vickers hardness, and tensile mechanical properties test and bending mechanical properties were carried on as-cast and extruded alloy to obtain mechanical parameters.

3.2.1. Vickers Hardness Test. Samples taken from different alloys respectively, were tested on digital micro hardness tester with applied pressure of 50N after mosaic, grinded and polished. The hardness would be the average of 5 points which tested on the same condition. Figure 2 shows the digital micro hardness tester.

Figure 3 shows the influence of different extrusion process on the Vickers hardness of MgCa0.8. It was found that the hardness decreased and then increased with the increase of extrusion temperature, and a sharp decrease of 7.2% with the temperature reached 300℃ from 250℃ according Figure 3(a). It was the same trend with the increase of extrusion ratio in the basis of Figure 3(b). The hardness reached the maximum of 67.4 HV as the extrusion ratio was 8.1, extrusion temperature was 250℃, and extrusion speed was 10mm /s, which was 21% higher than that of cast alloy.

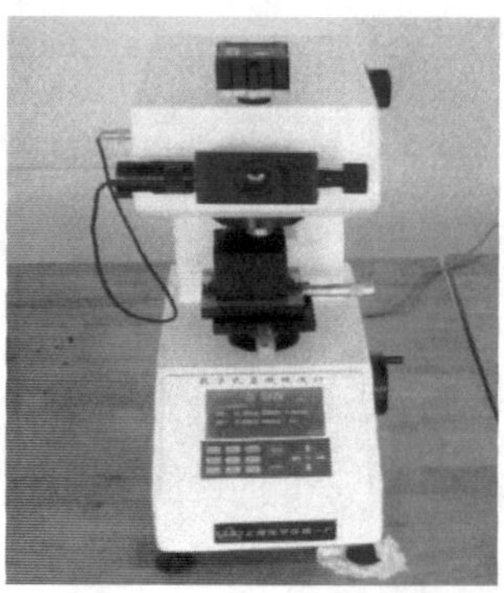

Figure 2. Digital micro hardness tester

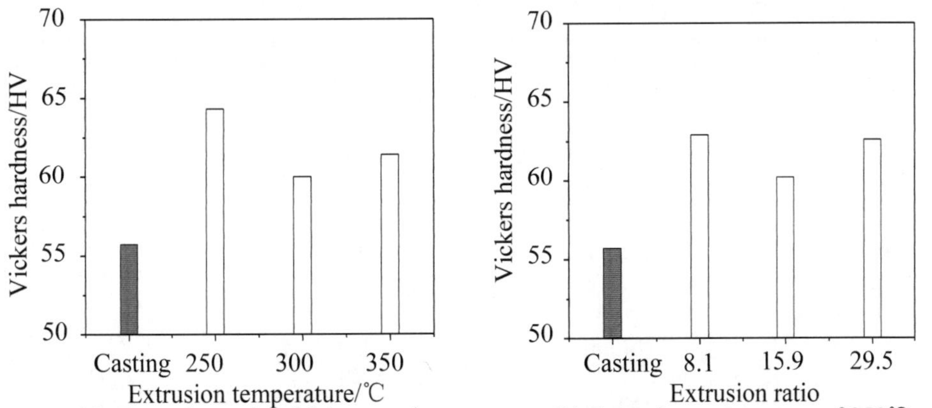

(a) Extrusion ratio of 8.1, extrusion speed of 10mm/s

(b) Extrusion temperature of 250℃, extrusion speed of 10mm/s

Figure 3. Influence of different extrusion process on Vickers hardness of MgCa0.8

3.2.2. Tensile and Bending Test. The tensile and bend test were carried out on WDW-100A type electronic universal testing machine. Figure 4 illustrated the dimension of tensile and bending samples, which were produced by wire cutting machine.

It was observed that the stress of casting alloy reached the yield point and fracture occurred abruptly, which showed a large brittleness. On the contrary the extruded alloy was broken after reaching yield point and then subjected to a certain length of elongation, which indicated the alloy obtained a certain degree of toughness during extrusion process. Table 2 listed the tensile parameters of as-cast and extruded alloys, from which it was found that the tensile strength of extruded alloy was significantly higher than that of the as-cast alloy. The maximum tensile strength was 245.42MPa and the maximum elongation was 12.32%, which was 477% and 254% higher than that of casting alloy respectively, resulted from the recrystallization and gain refinement and secondary phase refinement and distribution.

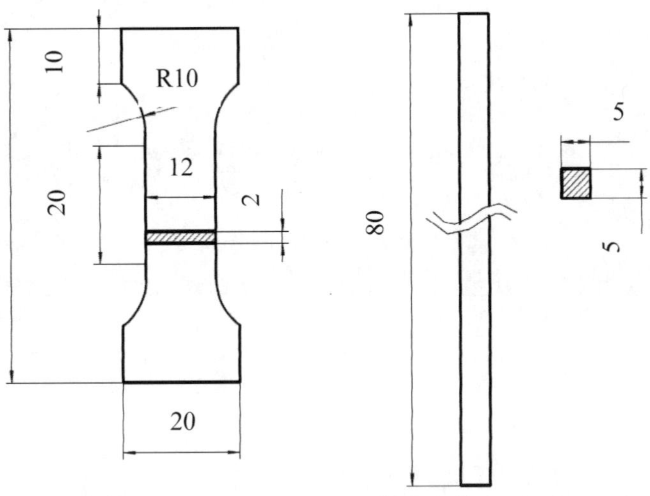

(a) Tensile sample (b) Bending sample

Figure 4. Samples of mechanical properties test

IOP Conf. Series: Materials Science and Engineering **389** (2018) 012018 doi:10.1088/1757-899X/389/1/012018

Table 2. Tensile mechanical parameters of MgCa0.8 alloys

Samples	Tensile strength σ_{bl}/MPa	Elongation δ/%	Elastic modulus E/GPa
casting	42.50	3.48	31.58
A1	212.71	8.54	45.22
A2	215.63	12.32	46.51
A3	182.78	10.38	42.76
B1	245.42	8.08	51.18
B2	198.13	11.60	41.35
B3	220.42	8.91	46.02
C1	185.00	10.44	42.54
C2	232.71	8.34	48.13
C3	234.17	7.44	35.77

Figure 5 shows the SEM pictures of casting MgCa0.8 alloy tensile fracture surface. As could be seen from figure 5(a), the surface was relatively rough as some short and curved tearing edge on. The gains were pulled up entirely in local area and crack expanded along gain boundary, with the characteristic of intergranular fracture. The comparison of matrix and precipitation phase of as-cast alloy could be clearly seen in figure 5(b). Precipitation phase Mg_2Ca was distributed along grain boundary and scattered within grain, as the white lines and sporadic white points in the photo were Mg_2Ca. The crack expanded along with the grain boundary revealing the brittleness of casting MgCa0.8 alloy.

Figure 6 shows the SEM pictures of extruded MgCa0.8, from which it was unable to intuitively distinguish the difference of matrix and precipitation phase, different from pictures of as-cast alloy, indicating that the precipitation phase experienced a greater degree of refinement along with the recrystallization. The pictures also showed the deformation of the grain after extrusion. Precipitation phases scattered were pulled off the matrix entirely in the process and displayed as micro-spherical morphology.

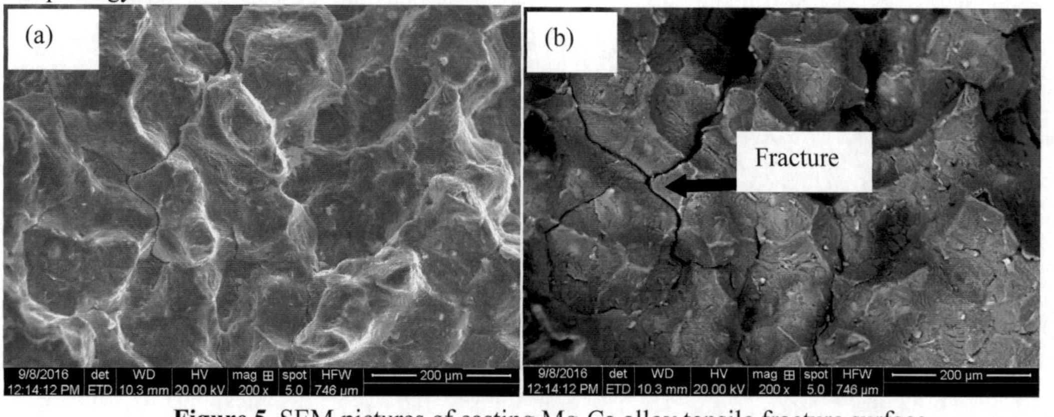

Figure 5. SEM pictures of casting Mg-Ca alloy tensile fracture surface ((a) Electronic scanning pictures (b) Back scattered pictures)

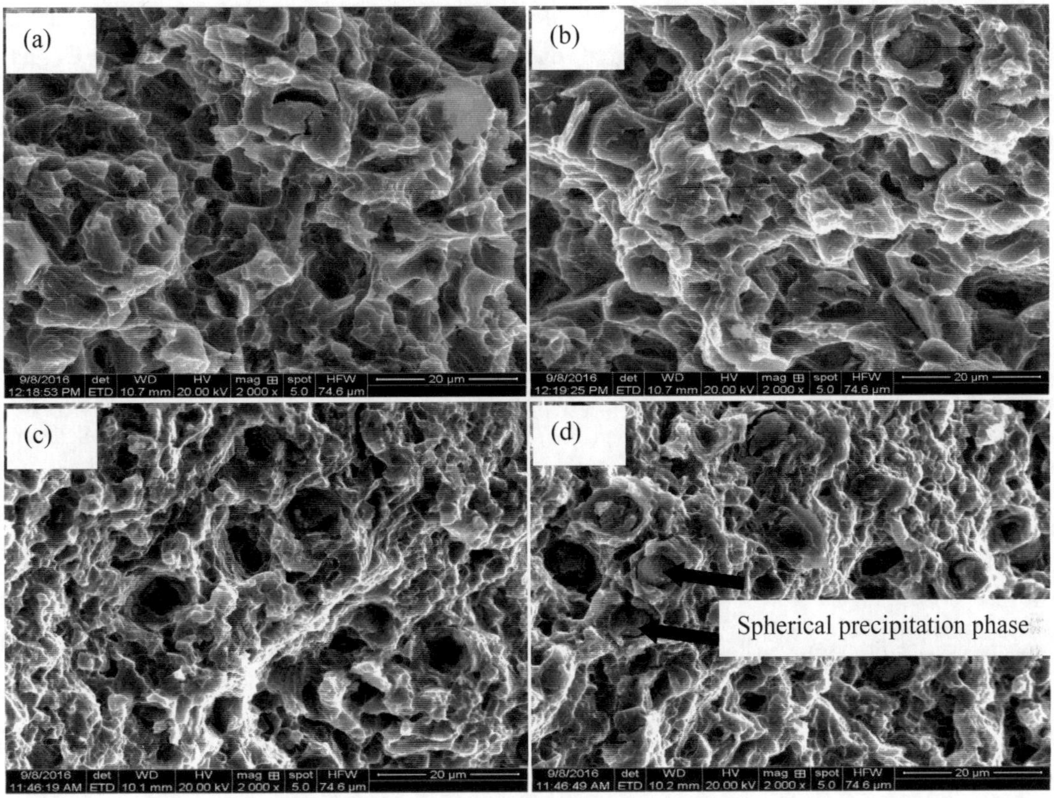

Figure 6. SEM pictures of extruded MgCa0.8 alloy tensile fracture surface

((a) (b) Alloy A3, (c) (d) Alloy B1)

Table 3 listed the bending parameters of as-cast and extruded alloys, from which it was found that the bending strength of extruded alloy was significantly higher than that of the as-cast alloy. The maximum bending strength was 379.2MPa, which was 405.6% higher than that of casting alloy, resulted from the recrystallization and gain refinement and secondary phase refinement and distribution. There was some difference between bending parameters of different extruded alloys because of the different degree of matrix and phase refinement due to dissimilar extrusion process.

Table 3. Bending mechanical parameters of MgCa0.8 alloys

Samples	Bending strength σ_{b2}/MPa	Maximum bending force /KN
Casting	67.50	0.13
A1	379.20	0.79
A2	360.02	0.75
A3	292.80	0.61
B1	384.04	0.80
B2	314.40	0.66
B3	362.44	0.76
C1	285.62	0.60
C2	348.04	0.73
C3	364.80	0.76

4. Conclusions

In this paper, alloys of MgCa0.8 were prepared, extruded and tested for mechanical properties and corrosion resistance. Conclusions can be summered as follows:

(1) It was found that mechanical properties of extruded alloy increased greatly in the research of extrusion process of MgCa0.8 alloy. The maximum tensile strength reached to 245.42MPa, maximum elongation to 12.32%, maximum bending strength to 379.20MPa, and different extrusion parameters had significant influence on mechanical properties of alloys.

(2) Compared with traditional medical metal materials, the mechanical properties of extruded alloy MgCa0.8 were similar to human bone. From the viewpoint of mechanical properties, MgCa0.8 was suitable for implant material within the quantity range of element Ca mentioned this paper.

Acknowledgments

This work was supported by Taishan Scholar Engineering Special Funding, the Development Program of Science and Technology of Shandong (2017GGX30143), and Shandong Natural Science Foundation of China (ZR2017LEE015).

References

[1] Zhang E, Yang L. Microstructure,mechanical properties and bio-corrosion properties of Mg-Zn-Mn-Ca alloy for biomedical application. Materials Science and Engineering A, 2008, 497(1-2):111-118

[2] N. Von Der Hoh, D. Bormann, A. Lucas, B. Denkena, C. Hackenbroich, A. Meyer-Lindenberg. Influence of different surface machining treatments of magnesium-based resorbable implants on thedegradation behavior in rabbits. Adv. Eng. Mater, 2009, 11(5):B47-54

[3] B. Aksakal, C. Hanyaloglu. Bioceramic dip-coating on Ti-6Al-4V and 316L SS implantmaterials. J. Mat. Sci. Materials in Medicine, 2008, 19(5):2097-2104

[4] S.E. Harandi, M.H. Idris, H. Jafari. Effect of forging process on microstructure, mechanical and corrosion properties of biodegradable Mg–1Ca alloy. Materials & Design, 2011, 32(5):2596-2603

[5] M.I. Jamesh, G. Wu, Y. Zhao, D.R. McKenzie, M.M.M. Bilek, P.K. Chu. Electro-chemicalcorrosion behavior of biodegradable Mg-Y-RE and Mg-Zn-Zr alloys in Ringer's solution andsimulated body fluid. Corros. Sci, 2015, 91:160-184

[6] N. Ishida, Z. Abidin, A.D. Atrens, D. Martin, A. Atrens. Corrosion of high purity Mg, Mg2Zn0.2Mn, ZE41 and AZ91 in Hank's solution at 37℃. Corros. Sci, 2011, 53(11):3542-3556

[7] M. Ascencio, M. Pekguleryuz, S. Omanovic. An investigation of the corrosion mechanisms of WE43 Mg alloy in a modified simulated body fluid solution: the influence of immersion time. Corros. Sci, 2014, 87(5):489-503

[8] Denkena B, Lucas A. Biocompatible magnesium alloys as absorbable implant materialsadjusted surface and subsurface properties by machining processes. Ann. CIRP, 2007, 56(1):113-116

[9] G.Y. Yuan, J. Zhang, W.J. Ding. Research progress of biodegradable medical magnesium-based biomaterials. Materials China, 2011, 30(2):44-50

[10] M Salahshoor, Y B Guo. Surface integrity of biodegradable Magnesium-Calcium orthopedic implant by burnishing. Journal of the mechanical behavior of biomedical materials, 2011, 4(8):1888-1904

IOP Conf. Series: Materials Science and Engineering **389** (2018) 012019 doi:10.1088/1757-899X/389/1/012019

Luminescence Properties Of Dy^{3+} Doped $Gd_2(WO_4)_3$ Phosphor Prepared by Hydrothermal Method

Zhengyang Zhang†,*, Jinkai Li †, Wenzhi Wang, Guangbin Duan, Weilin Zhao, Zongming Liu*

School of Materials Science and Engineering, University of Jinan, Jinan, Shandong 250022, China

* Corresponding author:

† The two authors contributed equally to this paper.

E-mail: mse_zhangzy@163.com; ost_liuzm@ujn.edu.cn

Abstract. The Dy^{3+} doped $Gd_2(WO_4)_3$ phosphors have been successfully obtained through calcining the precursor prepared via hydrothermal method at 900℃. The structure, morphology and luminescent properties of the resultant products were characterized by X-ray diffraction (XRD), field emission scanning electron microscopy (FE-SEM), photoluminescence excitation/photo-luminescence spectroscopy (PLE/PL), and fluorescence decay analysis. The $(Gd_{1-x}Dy_x)_2(WO_4)_3$ (x=0.01-0.10) solid solutions, exhibit strong yellow luminescence at 575 nm (the $^4F_{9/2}\rightarrow{}^6H_{13/2}$ electric dipole transition of Dy^{3+}) optimal UV excitation into the $^6H_{15/2}\rightarrow{}^4H_{11/2}$ of Dy^{3+} at ~452 nm. The quenching concentration was determined to be ~7at.%, and the quenching was due to the exchange reaction between Dy^{3+}. The emission intensity decreased monotonically with the used temperature increasing while the fluorescence lifetime (0.5±0.05 ms) does not alter appreciably. All the $(Gd_{1-x}Dy_x)_2(WO_4)_3$ phosphors have similar CIE chromaticity coordinates and color temperatures of (~0.43±0.02, ~0.47±0.02) and ~3550 K respectively, but the lifetime for 575 nm emission decreased with the Dy^{3+} increasing ascribed to the energy transfer between Dy^{3+}. The $Gd_2(WO_4)_3$:Dy^{3+} yellow phosphors are expected to be widely used in white light LED and display areas.

1. Introduction

Tungstate $Ln_2(WO_4)_3$ (Ln=rare earth) is one of the most classical inorganic luminescent materials which could be widely used in lighting and display areas due to its good physical and chemical properties [1]. $Gd_2(WO_4)_3$ is a bottom centered monoclinic structure [2]. Ln^{3+} ion in $Ln_2(WO_4)_3$ could be replaced by various activators and exhibit colorful emission. Zhang et al. prepared Eu^{3+} doped Gd_2WO_6 and $Gd_2(WO_4)_3$ nanophosphors via co-precipitation method [3]; Liu et al. prepared Eu^{3+} doped $Gd_2(WO_4)_3$ and studied energy transfer [4]; Zeng et al. prepared Eu^{3+}/Tb^{3+} doped $Gd_2(WO_4)_3$ nanophosphors by hydrothermal method [5].

Dy^{3+} doped $Gd_2(WO_4)_3$ system was chosen in the present work based on the following three reasons: (1) WO_4^{2-} ion itself can effectively absorb the spectrum of blue violet LED emission, and transfer to the rare earth ions doped in tungstate matrix [6]. (2) The characteristic emission peak of Gd^{3+} at 275 nm overlapped with the excitation band of many activator ions, and thus the Gd^{3+} could be usually served as a kind of effective sensitizer [7-8]. The Dy^{3+} emission in $Gd_2(WO_4)_3$ system could be further improved through the $Gd^{3+}\rightarrow Dy^{3+}$ energy transfer. (3) The Dy^{3+} mainly consisted two emission bands at 480 nm (blue emission) and 575 nm (yellow emission) corresponding to the $^4F_{9/2}\rightarrow{}^6H^{15/2}$ and $^4F_{9/2}\rightarrow{}^6H_{13/2}$ transitions of Dy^{3+}, respectively. The Dy^{3+} emission was effected significantly by the matrix material, and the emission color of Dy^{3+} could be adjusted by modulation the coordination environment [9-11]. This makes Dy^{3+} doped phosphors possess multi-functional performance, and thus have been extensively studied in the recent years.

Rare earth ions doped $Gd_2(WO_4)_3$ phosphors usually were synthesized, however, there are few reports about hydrothermal synthesis which can control the morphology and size of resultant products easily. In this work, we synthesized $Gd_2(WO_4)_3$:Dy^{3+} nanoscale phosphor by hydrothermal method. The resultant products were systematically studied using the combined techniques of XRD, FE-SEM, PLE/PL spectroscopy and fluorescence decay analysis. In the following, we report the synthesis, phase evolution, and the optical properties of $Gd_2(WO_4)_3$:Dy^{3+} samples in detail.

2. Experiment Section

2.1. Materials

The started chemicals required in this work mainly include gadolinium oxide (Gd_2O_3, 99.99%, Huizhou Ruier Rare Chemical Hi-Tech Co. Ltd., Huizhou, China), dysprosium oxide (Dy_2O_3, 99.99%, Huizhou Ruier Rare Chemical Hi-Tech Co. Ltd., Huizhou, China), sodium tungstate dihydrate ($Na_2WO_4\cdot 2H_2O$, 99.5%, Sinopharm Chemical Reagent Co. Ltd., Shanghai, China), sodium hydroxide (NaOH, 96%, Sinopharm Chemical Reagent Co. Ltd., Shanghai, China) and nitric acid (HNO_3, Sinopharm Chemical Reagent Co. Ltd., Shanghai, China). All chemical reagents are used as received without further purification.

2.2. Preparation procedure

The $RE(NO_3)_3$ (RE=Gd and Dy) solutions were prepared via dissolving Gd_2O_3 and Dy_2O_3 in hot nitric acid. The mother salt solution was prepared by $RE(NO_3)_3$ and $Na_2WO_4\cdot 2H_2O$ according to the chemical formula of $(Gd_{1-x}Dy_x)_2(WO_4)_3$

After the above process is completed, a certain proportion of $Na_2WO_4\cdot 2H_2O$ white particles dissolved in 50 ml water, then take the appropriate amount of nitrate solution dissolved in 20 ml water, respectively, placed in a collector type thermostatic magnetic stirrer, stirring about 5 min; When the two raw materials are fully dissolved and stirred in deionized water, the resulting nitrate aqueous solution is uniformly dropped into the sodium tungstate solution and treated at the end of the titration for 30 min, add NaOH solution to the mixture and adjust the pH until pH=7, The resulting reaction solution is then moved to a stainless steel autoclave in the capacity of 100 ml; The reaction kettle is placed in the oven and the oven is set at a temperature of 120℃, and the reaction time is 24 h. When the reaction is finished, the reaction kettle is cooled to room temperature and the hydrothermal product is removed; The obtained hydrothermal product is washed repeatedly by deionized water and once cleaned by anhydrous ethanol, and the product is dispersed into alcohol and dried in a drying box (80℃, dried 6 h) to obtain precursor; Finally, the precursor was calcined in an air atmosphere at high temperature (900℃ 2 h), and finally the target phosphor was obtained

2.3. Characterization

The XRD patterns were collected at room temperature by diffractometer (Model D8 ADVANCE, BRUKER Co., Germany) using nickel-filtered CuK_α radiation in the 2θ range 10-70° at a scan rate of 4.0° 2θ/min. The morphologies of products were observed by the FE-SEM with an acceleration voltage of 10 kV (Model QUANTA FEG 250, FEI Co., America). The PL/PLE spectra of the $(Gd_{1-x}Dy_x)_2(WO_4)_3$ phosphors were studied by a fluorescence spectrophotometer (FP-6500, JASCO, Tokyo, Japan) equipped with a Φ60-mm intergating sphere (Model ISF-513, JASCO, Tokyo, Japan) and a 150-W Xe-lamp as the excitation source.

IOP Conf. Series: Materials Science and Engineering **389** (2018) 012019 doi:10.1088/1757-899X/389/1/012019

3. Results and discussion

The XRD patterns of the $(Gd_{1-x}Dy_x)_2(WO_4)_3$ phosphors calcined at the same temperature 900℃ with different Dy^{3+} contents (x=0.01, 0.05 and 0.10) are shown in Fig. 1. It is well known that the pure phase is beneficial to the luminescent properties of phosphors. As shown in Fig. 1, all the Dy^{3+}-free samples can be well indexed to $Gd_2(WO_4)_3$ with the bottom centered monoclinic structure (space group: C_2/c, JCPDS: no.23-1076) [12]. No additional diffraction peaks were observed on the XRD spectra indicating that the pure phase $(Gd_{1-x}Dy_x)_2(WO_4)_3$ has been formed and Dy^{3+} doping does not change the crystal structure.

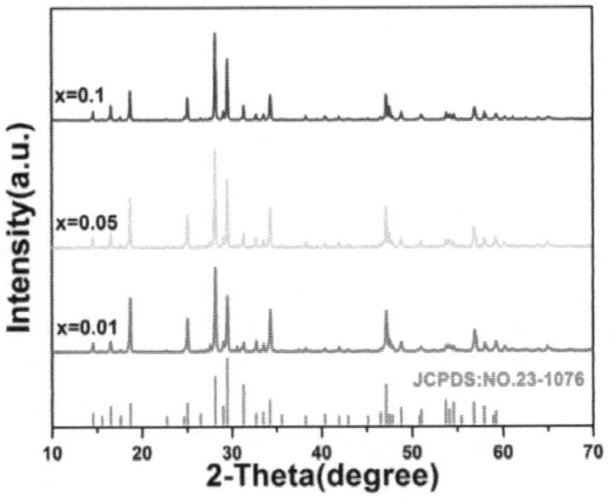

Figure 1. XRD spectra of $(Gd_{1-x}Dy_x)_2(WO_4)_3$ calcined 900 °C at as function of Dy^{3+} content (x value)

Figure 2. FE-SEM of the $(Gd_{1-x}Dy_x)_2(WO_4)_3$ precursors, x=0.01 (a), x=0.10 (b), and the precursors calcined at 900 °C, x=0.01 (c), x=0.1 (d)

Fig. 2 shows FE-SEM micrographs of $Gd_2(WO_4)_3$ with different Dy^{3+} content. From which it can be seen that the particle shape of samples does not change significantly with the Dy^{3+} incorporation, and all the precursors show needle-like appearance. The products with good dispersion calcined at 900℃ maintain the morphology of precursor.

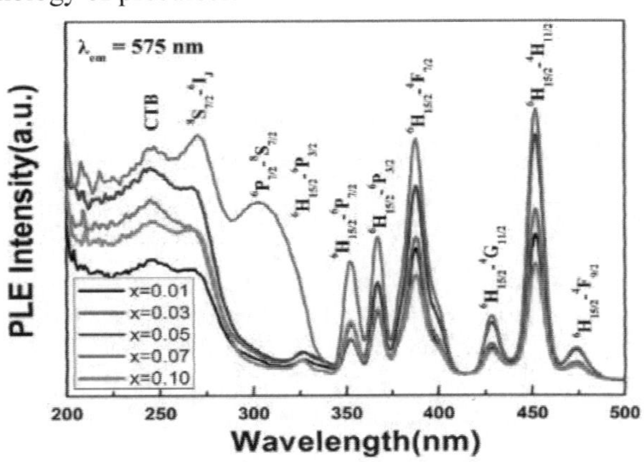

Figure 3. PLE spectra of the $(Gd_{1-x}Dy_x)_2(WO_4)_3$ phosphors calcined at 900 °C. The PLE spectra were taken by monitoring the 575 nm emission

Fig. 3 shows the excitation spectra of Dy^{3+} doped $Gd_2(WO_4)_3$ phosphor as a function of Dy^{3+} content. The PLE spectra all mainly are composed of five excitation bands and the excitation intensity of the 452 nm (Dy^{3+}: ${}^6H_{15/2}\rightarrow{}^4H_{11/2}$) wavelength is the strongest. The excitation peak at 245 nm originates from the charge transfer band transition of W-O proving direct evidence of $WO_4^{2-} \rightarrow Dy^{3+}$ energy transfer. The $^8S_{7/2} \rightarrow ^6I_J$ transition of Gd^{3+} appeared at ~270 nm [13-14] indicating the efficient energy transfer from Gd^{3+} to Dy^{3+}. As shown in Fig. 3, the other bands observed in the longer wavelength region at ~352, ~366, ~386, ~427 and ~470 nm are corresponded to the $^6H_{15/2} \rightarrow ^6P_{7/2}$, $^6H_{15/2} \rightarrow ^6P_{3/2}$, $^6H_{15/2} \rightarrow ^4F_{7/2}$, $^6H_{15/2} \rightarrow ^4G_{11/2}$ and $^6H_{15/2} \rightarrow ^4F_{9/2}$ transitions of Dy^{3+}, respectively.

Fig. 4 shows the PL properties obtained under UV excitation at 452 nm of the $(Gd_{1-x}Dy_x)_2(WO_4)_3$ phosphors calcined at 900℃. The PL spectra for all the samples with different Dy^{3+} concentrations include three emission banks at ~480 nm (blue), ~575 nm (yellow, the strongest) and ~663 nm (red), corresponding to the $^4F_{9/2} \rightarrow ^6H_{15/2}$, $^4F_{9/2} \rightarrow ^6H_{13/2}$ and $^4F_{9/2} \rightarrow ^6F_{11/2}$ transitions of Dy^{3+}, respectively [15]. Further observation is that the yellow emission is much stronger than the blue one in each case, and the integrated intensities of the yellow to blue emission bands have ratios of ~3.91, 4.13, 4.18, 4.30 and 4.33 for the different Dy^{3+} contents studied in this work. The reason is that the Judd-Ofelt parity law [16, 17] predicts that the magnetic dipole transition of $^4F_{9/2} \rightarrow ^6H_{15/2}$ is allowed while the electric dipole transition of $^4F_{9/2} \rightarrow ^6H_{13/2}$ is forbidden, the electric dipole transition is permitted only when the Dy^{3+} ions occupy a low symmetry site without an inversion center. $Gd_2(WO_4)_3$ is a bottom centered monoclinic structure,, the Gd^{3+} mainly occupies the low symmetry site without an inversion center [18]. The Dy^{3+} ions are expected to partially substitute for Gd^{3+}, and thus the emission spectra are thus dominated by the $^4F_{9/2} \rightarrow ^6H_{13/2}$ forced electric dipole transition at 575 nm rather than $^4F_{9/2} \rightarrow ^6H_{15/2}$ magnetic dipole transition at 480 nm. The Dy^{3+} addition does not alter the emission peak position and shape significantly, but the intensity varied. The inset in Fig. 4 shows intensity of the 575 nm emission (λ_{ex}=452 nm) plotted against Dy^{3+}concentration. It is clearly seen that the intensity steadily improves with increasing Dy^{3+} doping up to 7.0 at% (x=0.07) and then rapidly deteriorates owing to concentration quenching. The optimal Dy^{3+} content is thus determined to be 7.0 at% (x=0.07).

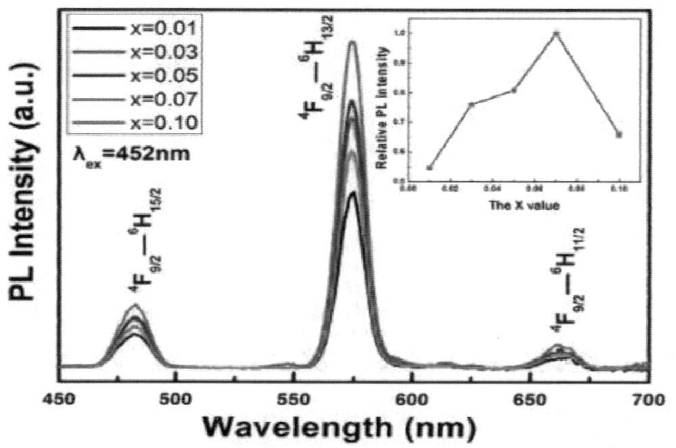

Figure 4. PL spectra of the $(Gd_{1-x}Dy_x)_2(WO_4)_3$ phosphors calcined at 900 °C. The PL spectra were measured under UV excitation at 452 nm. The inset is the relative intensity of the 575 nm emission, normalized to that of the $(Gd_{0.93}Dy_{0.07})_2(WO_4)_3$ phosphors, as a function of the Dy^{3+} content

The interaction type of luminescence quenching in solid phosphors can be concluded by analyzing the constants according to the following equation [19]:

$$\log(I/c) = (-s/d)\log c + \log f \tag{1}$$

where I is the emission intensity, c the activator concentration, d the sample dimension (d=3 for regular sample), f is a constant, and s is the index of electric multipole. The s values of 6, 8, and 10 represent the dipole-dipole, dipole-quadrupole, and quadrupole-quadrupole electric interactions, respectively, whereas s=3 corresponds to exchange interaction. The log(I/c)-log(c) plot of the 575 nm emission is given in Fig. 5. The slope (-$s/3$) was determined to be -0.933 yielding an s value of ~ 2.8

for the Dy^{3+} doped $Gd_2(WO_4)_3$ systems. This indicates that the concentration quenching is mainly due to the energy transfer between Dy^{3+}-Dy^{3+} [20].

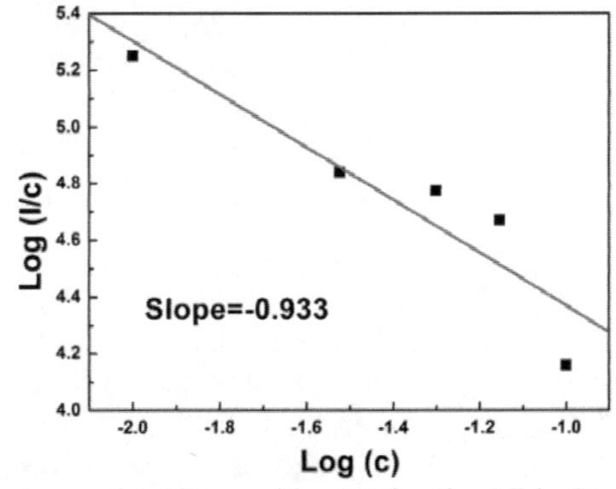

Figure 5. The relationship between log(I/c) and log(c) for the $(Gd_{1-x}Dy_x)_2(WO_4)_3$ samples calcined at 900 °C

Fig. 6 shows the decay kinetics for the 575 nm emission of $(Gd_{0.93}Dy_{0.07})_2(WO_4)_3$ calcined at 900 ℃ for 2 h. The decay curve can be fitted via the equation:

$$I = A\exp\left(-t/\tau_R\right) + B \tag{2}$$

where I is the relative fluorescence intensity, t is the decay time, τ_R is the fluorescence lifetime. A and B are constants [21]. The fitting results are A=7.19±0.07, B=-0.03±0.07, τ_R=0.55±0.004 ms. The inst (a) in Fig. 6 displays the lifetime value against the Dy^{3+} concentration. The lifetime shortens from 0.56 ms to 0.48 ms when the content of Dy^{3+} increases from x=0.01 to x=0.10. The shortened lifetime can be explained as follows: when the Dy^{3+} content is low, the distance among the Dy^{3+} ions is relatively long and the interaction among luminescent centers can be neglected. A higher Dy^{3+} concentration would lead to the formation of a resonant energy transfer net, which can act as an additional channel to the nonradiative centers of surface and therefore shorten the lifetime. The inset (b) in Fig.6 shows the lifetime of the $(Gd_{0.93}Dy_{0.07})_2(WO_4)_3$ at the temperature of 423 K. The fitting results are A=1.73±8.45, B=-0.03±0.27, τ_R=0.45±0.001 ms. The used temperature does not affect the fluorescence lifetime appreciably, and the lifetime in the temperature range 298-573 K was similar to be ~0.5±0.05 ms.

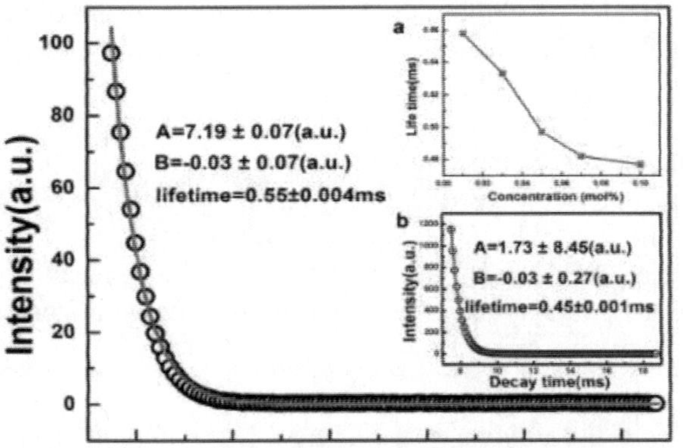

Figure 6. Fluorescence decay curve for the 575 nm emission of $(Gd_{0.93}Dy_{0.07})_2(WO_4)_3$ phosphors calcined at 900 °C for 2 h. The inset (a) in the figure is the lifetime values as function of Dy^{3+} content, (b) is the fluorescence decay curve for the 575 nm emission of $(Gd_{0.93}Dy_{0.07})_2(WO_4)_3$ at used temperature of 423 K

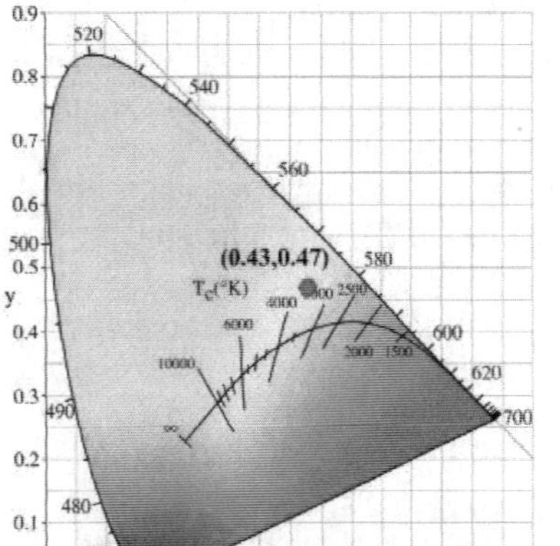

Figure 7. CIE chromaticity diagram for the $(Gd_{0.93}Dy_{0.07})_2(WO_4)_3$ system under 452 nm excitation

Moreover, the color chromaticity coordinates of phosphor has been calculated by using the CIE 1931 color matching functions, and the result was shown in Fig. 7. Fig. 7 shows the CIE chromaticity diagram of $(Gd_{0.93}Dy_{0.07})_2(WO_4)_3$ phosphor under 452 nm light excitation. The CIE chromaticity coordinate of sample $(Gd_{0.93}Dy_{0.07})_2(WO_4)_3$ is determined to be (0.43, 0.47), indicating that the samples emit a vivid yellow color. The corresponding color temperature was calculated to be ~3550 K according to the following formulas [22]:

$$T = -437n^3 + 3601n^2 - 6861n + 5514.31 \tag{3}$$

$$n = (x - 0.332)/(y - 0.1858) \tag{4}$$

It should be noted that all the samples studied in this work have almost identical CIE chromaticity coordinate and color temperature of (0.43±0.02, 0.47±0.02), and ~3550 K, respectively.

The CIE coordinate (x, y) of $(Gd_{0.93}Dy_{0.07})_2(WO_4)_3$ is (0.4300, 0.4700). The dominant wavelength of $(Gd_{0.93}Dy_{0.07})_2(WO_4)_3$ is determined to be 575 nm, and we can obtain the (x_d, y_d) chromaticity coordinate of $(Gd_{0.93}Dy_{0.07})_2(WO_4)_3$ is (0.5096, 0.4904). By substituting the coordinates of (x, y), (x_i, y_i), and (x_d, y_d) in equation (6), the color purities of $(Gd_{0.93}Dy_{0.07})_2(WO_4)_3$ is calculated to be 71%. This result indicates that $(Gd_{1-x}Dy_x)_2(WO_4)_3$ phosphor is a deep yellow phosphor and features higher color purity.

4. Conclusions

In the experiment, the $Gd_2(WO_4)_3$ doped Dy^{3+} phosphors were successfully obtained via hydrothermal method. The phase-pure $(Gd_{1-x}Dy_x)_2(WO_4)_3$ with good dispersion and uniform particle size formed at 900 ℃. The $(Gd_{1-x}Dy_x)_2(WO_4)_3$ phosphors with needle-like feature exhibit strong yellow emission at 575 nm ($^4F_{9/2}\rightarrow{}^6H_{13/2}$ transition of Dy^{3+}) under UV excitation into the $^6H_{15/2}\rightarrow{}^4H_{11/2}$ of Dy^{3+} at ~452 nm. The currence of charge transfer band of W-O and $^8S_{7/2}\rightarrow{}^6I_J$ transition of Gd^{3+} provide the direct evidence of the $WO_4^{2-}\rightarrow Dy^{3+}$ and $Gd^{3+}\rightarrow Dy^{3+}$ energy transfer. The quenching concentration of $(Gd_{1-x}Dy_x)_2(WO_4)_3$ phosphors was determined to be 7 at% (x=0.07) and the luminescence quenching is dominated by exchange interactions resulting from the energy transfer among Dy^{3+} ions. All the $(Gd_{1-x}Dy_x)_2(WO_4)_3$ phosphors have the similar CIE chromaticity coordinates, and color temperatures of (~0.43±0.02, ~0.47±0.02), and ~3550 K, respectively. By analyzing the fluorescence decay kinetics of $Gd_2(WO_4)_3:Dy^{3+}$ phosphor, it is found that the fluorescence lifetime of best $(Gd_{0.93}Dy_{0.07})_2(WO_4)_3$ sample was determined to be 0.55±0.004 ms, and the lifetime shortens with the increase of Dy^{3+} concentration while changes little with the temperature increasing.

5. Acknowledgments

This work was supported in part by the National Natural Science Foundation of China (No. 51402125, and 51602042), China Postdoctoral Science Foundation (No. 2017M612175), the Special Fund for the Postdoctoral Innovation Project in Shandong Province (No. 201603061), the Research Fund for the Doctoral Program of University of Jinan (No. XBS1447), the Natural Science Foundation of University of Jinan (No. XKY1515), the Science Foundation for Post Doctorate Research from the University of Jinan (172335).

References

[1] Brixner L H, Sleight A W. Crystal growth and precision lattice constants of some $Ln_2(WO_4)^{3-}$ type rare earth tungstates. Mater. Res. Bull., 1973, 8(10):1269.
[2] Keve E T, Abrahams S C, Bernstein J L. Ferroelectric paramagnetic beta-$Gd_2(MoO_4)_3$ crystal structure of the transition-metal molybdates and tungstates. J. Chem. Phys., 1971, 54(7): 3185.
[3] Nassau K, Levinstein H J, Loiacono G M. A comprehensive study of trivalent tungstates and molybdates of the type $L_2(MO_4)_3$. J. Phys. Chem. Solids, 1965, 26: 1805.
[4] Liu Z, Meng Q, Liu H, Yao C, Meng Q, Liu W, Wang W. Energy transfer and electron-phonon coupling properties in $Gd_2(WO_4)_3$:Eu phosphor. Opt. Mater., 2013, 36(2): 384.

[5] Zeng Y, Li Z, Wang L, Xiong Y. Controlled synthesis of $Gd_2(WO_4)_3$ microstructures and their tunable photoluminescent properties after Eu^{3+}/Tb^{3+} doping. CrystEngComm, 2012, 14(20): 7043.

[6] Kodaira C A, Brito H F, Felinto M C F C. Luminescence investigation of Eu^{3+} ion in the $RE_2(WO_4)_3$ matrix (RE=La and Gd) produced using the Pechini method. J. Solid State Chem., 2003, 171(1): 401.

[7] Li J K, Teng X, Wang W Z, Zhao W L, Liu Z M. Investigation on the preparation and luminescence property of $(Gd_{1-x}Dy_x)_2O_3$ (x = 0.01-0.10) spherical phosphors. Ceram. Int., 2017, 43: 10166.

[8] Li J G, Sakka Y. Recent progress in advanced optical materials based on gadolinium aluminate garnet ($Gd_3Al_5O_{12}$). Sci. Technol. Adv. Mat., 2015, 16(1): 14902.

[9] Wang W, Li J K, Duan G, Zhao W, Cao B, Liu Z. Morphology/size effect on the luminescence properties of the $[(Y_xGd_{1-x}0.98Dy_{0.02}]_2O_3$ phosphor with enhanced yellow emission. J. Lumin., 2017, 192: 1056.

[10] Wang J, Zhang Z J, Zhao J T, Chen H H, Yang X X, Tao Y, Huang Y. Luminescent metastable $Y_2WO_6:Ln^{3+}$ (Ln=Eu, Er, Sm, and Dy) microspheres with controllable morphology via self-assembly. J. Mater. Chem., 2010, 20(48): 10894.

[11] Ding B, Han C, Zheng L, Zhang J, Wang R, Tang Z. Tuning oxygen vacancy photoluminescence in monoclinic Y_2WO_6 by selectively ℃cupying yttrium sites using lanthanum. Sci. Rep-UK, 2015, 5: 9443.

[12] Lu H, Hao H, Gao Y, Li D, Guang G, Song Y, Wang Y, Zhang X. Optical sensing of temperature based on non-thermally coupled levels and upconverted white light emission of a $Gd_2(WO_4)_3$ phosphor co-doped with in Ho (III), Tm (III), and Yb (III). Micr℃him. Acta, 2017, 184(2): 641.

[13] Mahlik S, Lazarowska A, Grobelna B, Grinberg M. Luminescence of Gd2(WO4)3:Ln^{3+} at ambient and high hydrostatic pressure. J. Phys-Condens. Mat., 2012, 24(48): 485501.

[14] Kim J. The role of auxiliary alkali metal ions on scheelite structure double molybdate and tungstate phosphors. Inorg. Chem., 2017, 56(14).

[15] Li J K, Li J G, Liu S H, Sun X D, Sakka Y. Greatly enhanced Dy^{3+} emission via efficient energy transfer in gadolinium aluminate garnet ($Gd_3Al_5O_{12}$) stabilized with Lu^{3+}. J. Mater. Chem. C, 2013, 1(45): 7614.

[16] Kaczmarek A M, Hecke K V, Deun R V. Enhanced luminescence in Ln^{3+}-doped Y_2WO_6 (Sm, Eu, Dy) 3D microstructures through Gd^{3+} codoping. Inorg. Chem., 2014, 53(18): 9498.

[17] Sun J, Yu T, Li X, Zhang J, Cheng L, Zhong H, Tian Y, Hua R, Chen B. Co-precipitation synthesis, structural and morphological characterization, and luminescence properties of Dy^{3+} doped nanocrystal $Gd_2(WO_4)_3$ and Gd_2WO_6 phosphors. J. Phys. Chem. Solids, 2012, 73(3): 465.

[18] Li J K, Li J G, Zhang Z, Wu X, Liu S, Li X, Sun X, Sakka Y. Gadolinium aluminate garnet ($Gd_3Al_5O_{12}$): crystal structure stabilization via lutetium doping and properties of the $(Gd_{1-x}Lu_x)_3Al_5O_{12}$ solid solutions (x=0–0.5). J. Am. Ceram. Soc. 2012, 95(3): 931.

[19] Li J K, Li J G, Li X, Sun X. Tb^{3+}/Eu^{3+} codoping of Lu^{3+}-stabilized $Gd_3Al_5O_{12}$ for tunable photoluminescence via efficient energy transfer. J. Alloy. Compound., 2016, 670: 161.

[20] Shukla R, Ningthoujam R S, Tyagi A K, Vatsa R K. Luminescence properties of Dy^{3+} doped Gd_2O_3 nanoparticles prepared by glycine route: annealing effect. Int. J. Nanotechnol., 2010, 7: 843.

[21] Li J, Wu Z, Sun X, Zhang X, Dai R, Zuo J, Zhao Z. Controlled hydrothermal synthesis and luminescent properties of $Y_2WO_6:Eu^{3+}$ nanophosphors for light-emitting diodes. J. Mater. Sci., 2017, 52(6): 3110.

[22] Mccamy C S. Correlated color temperature as an explicit function of chromaticity coordinates. Color Res. Appl., 1992, 17(2): 142.

IOP Conf. Series: Materials Science and Engineering **389** (2018) 012020 doi:10.1088/1757-899X/389/1/012020

Morphology of Strontium Carbonate Particle Adjusted by Phthalic Acid and Isophthalic Acid

Yuxi Liu, Xianghai Jing, Xianhong Hu and Gang Wu

College of Material and Chemical Engineering, Chu Zhou University, Chu Zhou 239012, P. R. China

liuyuxiemail@126.com

Abstract. $SrCO_3$ crystals with various morphologies were prepared using precipitation method by the reaction of Na_2CO_3 and $SrCl_2$ in aqueous solution in the presence of isophthalic acid and phthalic acid, respectively, which were used as crystal growth adjusters. When isophthalic acid was used as the adjuster, the major morphologies of $SrCO_3$ were center-bundled and two ends emission growing clusters formed by the self-assembling of short rod whiskers. While using phthalic acid as the adjuster, flower-like $SrCO_3$ clusters were assembled by needle-shaped whiskers when the concentration was low. But the morphologies changed from flower shape to sphere shape with the increasing amount of phthalic acid. The results indicate that the morphology of $SrCO_3$ may be adjusted by the structure of carboxylic acid.

1. Introduction

The development of specific microstructure morphologies and controllable morphosynthesis of inorganic materials with specific size, shape and phase etc. by using diverse synthetic methods has attracted research interest, because shape control and structural speciality have significant effects for the properties in various fields [1,2]. Aragonite-type $SrCO_3$, strontianite, is widely used as an additive in the glass of color television tubes to block X-rays and improve glass quality. $SrCO_3$ is also used in ferrite magnets, pigments, iridescent materials, pyrotechnics, a precursor for various strontium compounds, and as a catalyst [3-6]. Therefore, $SrCO_3$ is a very important inorganic material and the development of specific microstructure morphologies has attracted more and more attention because of their novel properties and potential applications [7,8]. Many methods for the fabrication of $SrCO_3$ (micro/nanostructures) have been reported, including refluxing process [9], catalytic decomposition [10], self-assembled monolayers [11], hydrothermal method [12,13], microemulsion mediated solvothermal method [14], homogeneous precipitation by enzyme-catalyzed reaction [10], biological synthesis using the fungus fusarium oxysporum [15] and so on. Strontianite with different morphologies such as spheres and ellipsoids [14,16], olive-like and flower-like nanostructures [17,18], and 1D nanostructures have been synthesized [15,19]. In addition, a variety of $SrCO_3$ rods, wires, ribbons, hexagonal prisms et al. micro- and nano-structures have been synthesized [20-22]. The effects of the concentration of PSMA(poly-(styrene-alt-maleic acid)) on the morphologies and phase structures of strontium carbonate particles were investigated by Yu et al. The results showed that $SrCO_3$ particles with various shapes, for example, bundles, dumbbells, irregular aggregates and spheres could be obtained by varying the concentration of PSMA [23]. Huang et al. synthesized and reported $SrCO_3$ in the presence of nonionic/anionic surfactants F127/SDS in the water or alcohol. $SrCO_3$ particles change from pancakes to flowers, and then to irregular flakes [24].

It is well known that phthalic acid (PA) and isophthalic acid (IPA) can coordinate with Sr(II) ion and therefore, may control the initial growth and further assemblies of $SrCO_3$ particles, at micro- and nanoscale level, into target structured materials bearing the controlled size and tailored morphology [25]. Therefore, in this work, we reported a facile and direct method for hierarchical $SrCO_3$ in presence of appropriate amount of PA and IPA as morphology adjust control agent, using strontium chloride hexahydrate ($SrCl_2 \cdot 6H_2O$) and anhydrous sodium carbonate (Na_2CO_3) as the starting materials. The influence of structure of carboxylic acid on morphology was investigated.

2. Experimental

2.1. Materials

All the chemicals used to prepare reactant solutions, $SrCl_2 \cdot 6H_2O$, Na_2CO_3, phthalic acid (PA), isophthalic acid (IPA) were analytical grade, commercially obtained and used without further purification. Ultrapure water was used throughout the experiment.

2.2. Preparation of $SrCO_3$

The 0 mg (0 mmol), 34.2 mg (0.205 mmol), 62.0 mg (0.369 mmol), 88.5 mg (0.533mmol) IPA were added into five 50 mL beakers with 10 mL ultrapure water, respectively. Then, 0 mg (0 mmol), 22.0 mg (0.205 mmol), 39.0 mg (0.369 mmol), 56.8 mg (0.533 mmol) Na_2CO_3 were added into each beaker under stirring. After the mixture solutions were stirred for 10 min. 111.0 g (1.0 mmol) $SrCl_2 \cdot 6H_2O$ was added into each beaker under stirring. Then, 0.1060 g (1.0 mmol) Na_2CO_3 was added into each beaker. The precipitates produced were separated by centrifugation, washed three times with ultrapure water and anhydrous ethanol, and then vacuum dried at 40℃ for about 48 h. The size and morphologies of the particles were examined by SEM. The crystal phases were determined by FT-IR spectroscopy and powder X-ray diffractometer. Using PA as crystal growth adjuster to preparing $SrCO_3$, procedure and the amount of reactants are the same as using IPA as adjuster expect that IPA was replaced by PA.

2.3. Characterization

The as-prepared samples were characterized by X-ray diffraction (XRD) analysis using an X-ray diffractometer (Model Bruker D8 ADVANCE) with Cu-Kα1 (λ=1.54178 Å) monochromatized radiation at 40 kV and 40 mA. The samples were scanned over a 2θ range of 20-70° at a step size of 0.02°. The size and morphologies of the samples were examined by a scanning electron microscope (SEM) (JEOL JSM 5600LV) with gold coating. The Fourier transform infrared (FT-IR) spectroscopy on a Nicolet 6700 FT-IR spectrophotometer by using KBr pellet in the range of 4000~400 cm^{-1} and 32 scans were collected at a resolution of 4 cm^{-1}.

3. Results and discussion

The products prepared using $SrCl_2 \cdot 6H_2O$ and Na_2CO_3 as precursor in the 10.0 mL water solution were first examined by XRD, and the XRD patterns are shown in figure 1 and figure 2. All diffraction peaks of the product can be indexed as the (111), (002), (112), (220), (221), (132), (113) reflections of a pure strontianite-type crystal [9], indicating that $SrCO_3$ crystal was formed. The samples consisted of a single phase of well-crystalline $SrCO_3$ with an orthorhombic structure (PDF 05-0418) and with lattice parameters of a) 5.107 Å, b) 8.414 Å, c) 6.029 Å. No peaks from impurities were observed, showing pure strontianite-type crystal according to the XRD standard spectrum of $SrCO_3$ crystal. But in figure 1 and figure 2, compared with sample 1, the diffraction peaks of samples 2, 3, 4, 5, 6, 7 broaden obviously. The broadening features of the samples implied that the crystals could be composed of primary nanoparticles [26].

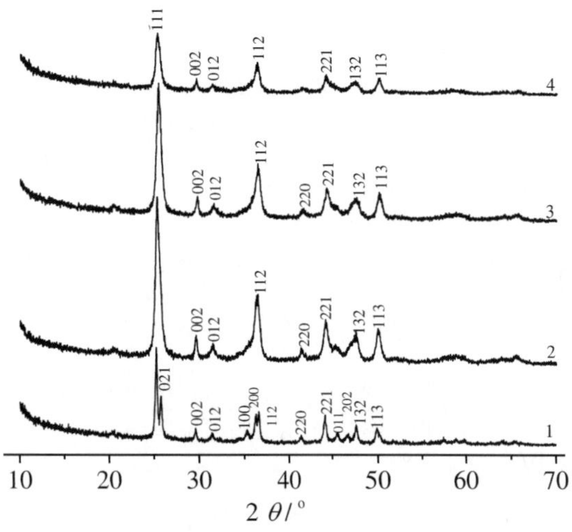

Figure 1. XRD patterns of the as-synthesized samples in the presence of IPA (mg): 1: 0.0; 2: 34.2; 3: 62.0; 4: 88.5.

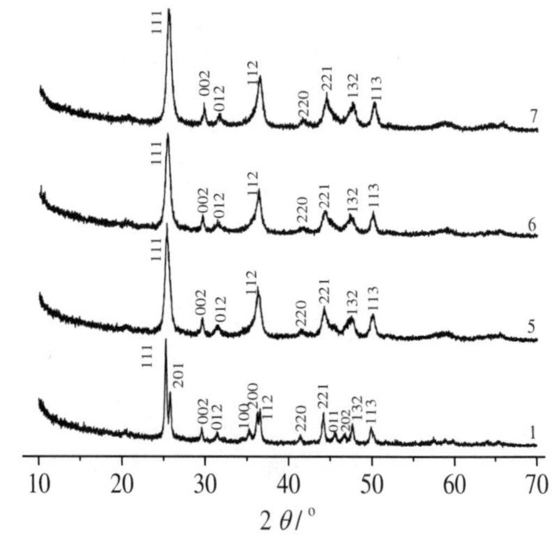

Figure 2. XRD patterns of the as-synthesized samples in the presence of PA (mg): 1: 0.0; 5: 34.2; 6: 62.0; 7: 88.5.

The FT-IR spectrum is a strong tool for characterizing inorganic minerals. The IR spectra of the samples are shown in figure 3 and figure 4. Isolated planar CO_3^{2-} anion has a D_{3h} symmetry [27]. The absorption bands attributed to the vibrations in CO_3^{2-} anion are located range from 1800 to 400 cm^{-1}. The IR spectra of the samples represent the characteristic absorbance of $SrCO_3$ particles. The strong broad absorption centered at about 1465 cm^{-1} is assigned with the asymmetric stretching vibrations, and the strong sharp absorption band at 856 cm^{-1} and 706 cm^{-1} can be assigned to the bending out of plane vibrations and in plane vibrations [28], respectively, which provide further evidence for the pure $SrCO_3$ composition of the product. This conclusion is in agreement with the result from the XRD pattern shown in figure 1 and figure 2.

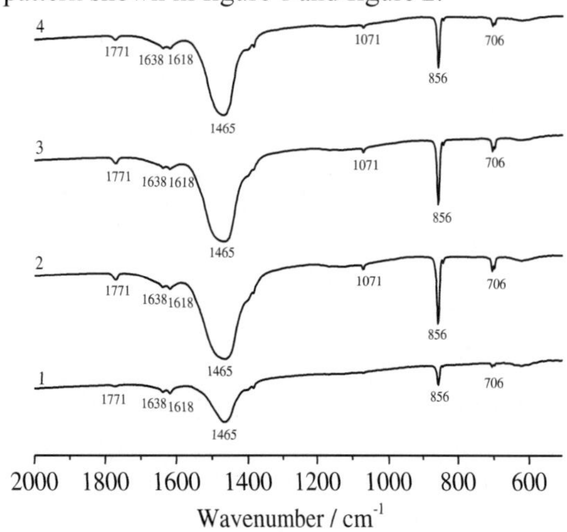

Figure 3. The FT-IR spectra of the as-synthesized samples in the presence of IPA (mg): 1: 0.0; 2: 34.2; 3: 62.0; 4: 88.5.

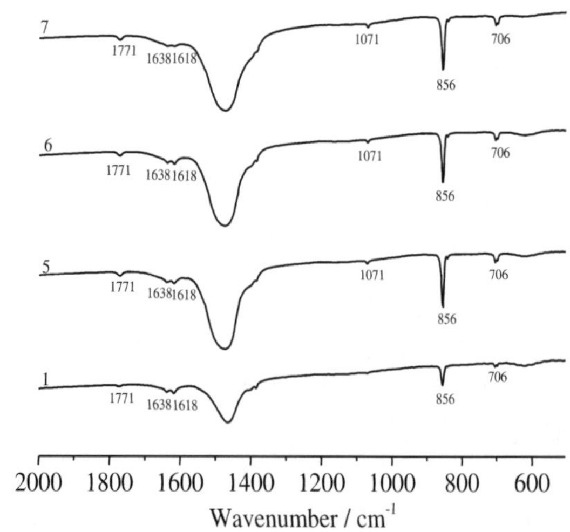

Figure 4. The FT-IR spectra of the as-synthesized samples in the presence of PA (mg): 1: 0.0; 5: 34.2; 6: 62.0; 7: 88.5.

The morphologies of all samples were further investigated by SEM and compiled in figure 5 (without and with the exist of IPA) and figure 6 (using PA as adjuster). The as-obtained $SrCO_3$ particles without IPA or PA (shown in figure 5-1) appear the uniformed morphology of spindle shape

IOP Conf. Series: Materials Science and Engineering **389** (2018) 012020 doi:10.1088/1757-899X/389/1/012020

with big length/diameter ratio of 3/1 and have good dispersion. When different amounts of IPA were added into the liquid phase preparation system, the morphologies of $SrCO_3$ particles are flower-like clusters formed by the assembling of short rod whisker (figure 5-2). The majority of clusters are center-bundled and two ends emission growing shape which are like two beams of flower bundled reversely together. Some clusters are like a beam of flower. In addition, the sizes of clusters and short rods are the smallest when the addition amount of IPA is 62.0 mg (figure 5-3).

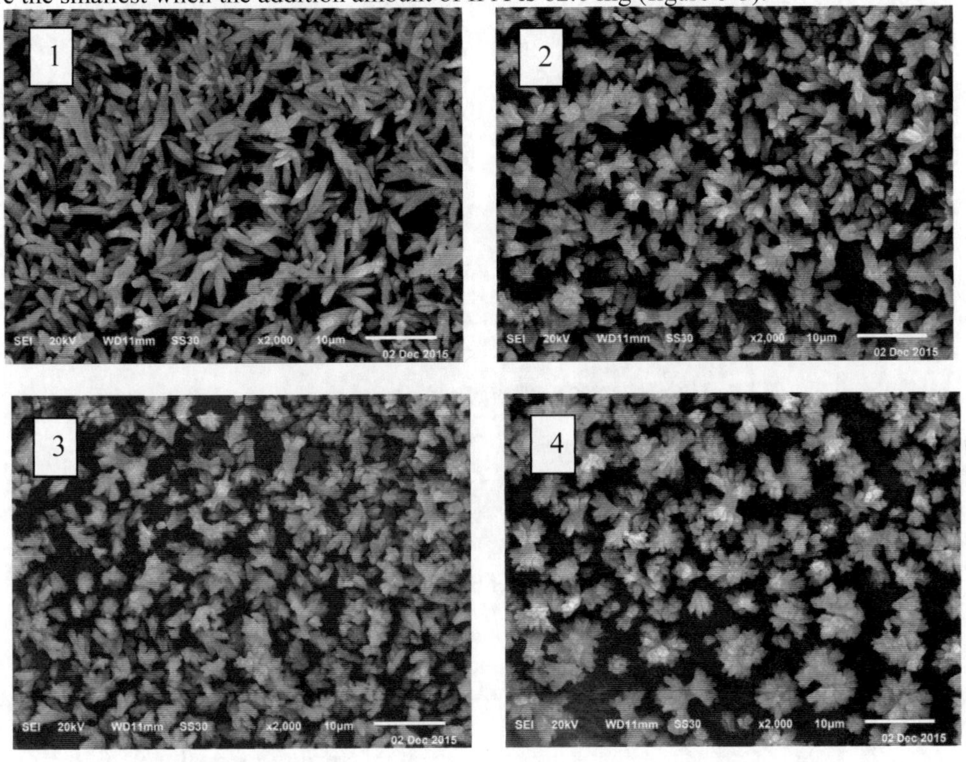

Figure 5. SEM images of the as-synthesized samples in the presence of IPA (mg): 1: 0.0; 2: 34.2; 3: 62.0; 4: 88.5.

Figure 6 shows the SEM images of $SrCO_3$ particles prepared in the presence of PA. When the adding amount of PA is 34.2 mg, figure 6-5 shows that $SrCO_3$ particles prepared are frost flower-like clusters. The magnified SEM image in Fig.6-5 (insert) indicates that each cluster is assembled with numerous well-aligned needle-like whisker, which is different from the self-assembling of short rod whiskers (figure 5-4). When the amount of PA is elevated to 62.0 mg, it is obvious that $SrCO_3$ particles have smaller size and better dispersion by contrasting to figure 6-5 and figure 6-6. The amplified SEM image (insert in figure 6-6) reveals that most $SrCO_3$ clusters are carnation flower-shaped clusters assembled with needle-like whisker and the overall appearance of clusters has two types. One is like two beams of carnation flower bundled reversely together, the other is like a beam of carnation flower. figure 6-7 shows the SEM image of $SrCO_3$ prepared with 88.5 mg of PA, which can be seen that the particles size become big again and the morphology change into irregular sphere without exist of needle-like whisker on the outside surface. To sum up, the adding amount of PA has noteworthy effects on the size, morphology and dispersion of $SrCO_3$ particles.

IOP Conf. Series: Materials Science and Engineering **389** (2018) 012020 doi:10.1088/1757-899X/389/1/012020

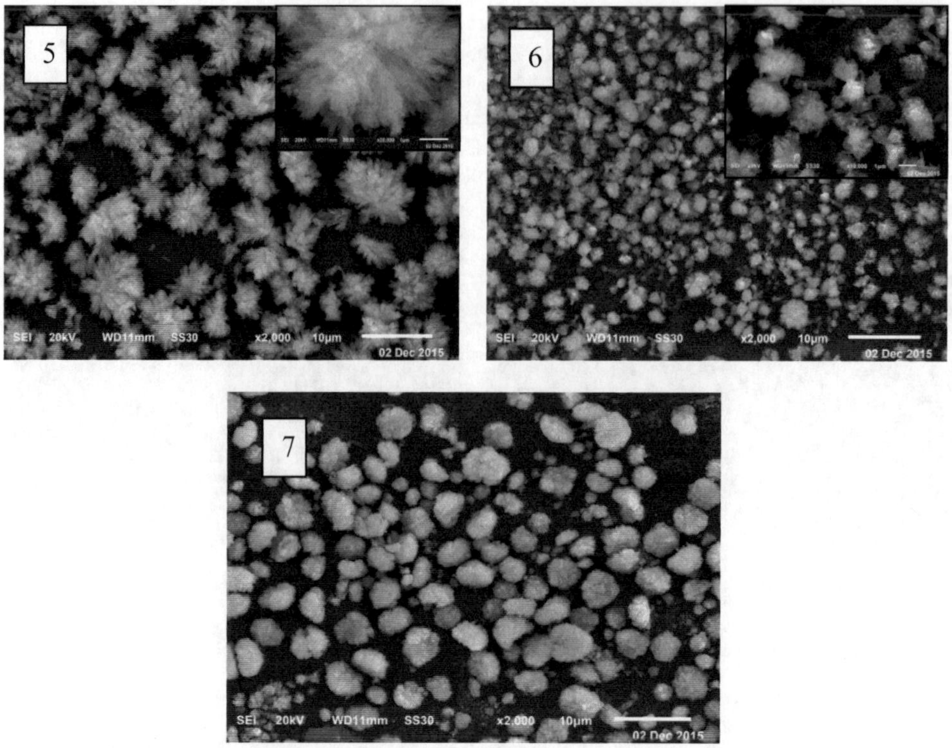

Figure 6. SEM images of the as-synthesized samples in the presence of IPA (mg): 1: 0.0; 5: 34.2; 6: 62.0; 7: 88.5.

It is well known that IPA and PA have strong coordination ability with Sr^{2+} ion. In the presence of IPA and PA, the complexes of $Sr(IPA)_m^{n-}$ or $Sr(PA)_m^{n-}$ formed due to the strong coordination ability of IPA and PA. But when Na_2CO_3 is added, dissociated Sr^{2+} can react with CO_3^{2-} species spontaneously to form the SrCO3 nuclei at room temperature because of small solubility product of $SrCO_3$ (at 25 °C, $K^{\Theta}_{sp} = 1.1\times10^{-10}$). Then the $SrCO_3$ nuclei grew to form nanocrystals.

The negatively charged groups of $-COO^-$ of PA or IPA molecules can bind Sr^{2+} strongly. Therefore, PA and IPA molecules can act as capping agents which selectively bind to facet(s) on a nanocrystal or microcrystal to alter their specific surface free energies. At the same time, the capping agent chemisorbed on a facet will hinder or prevent the deposition of ions onto this facet. The facet covered by the capping agent will take a slower growth rate and thus a greater proportion on the surface. This in turn changes both the edge-step free energies and the shape of the steps. Changes in step shape directly modify the macroscopic crystal shape by inducing the formation of new crystal faces resulting in macroscopic morphological modifications [29]. The distances between two carboxylate groups of PA and those of IPA are different, which mean that carboxylate groups of PA and IPA may bind different crystal face(s) of $SrCO_3$ particles, which result in the formation of crystals with different morphologies.

4. Conclusions

$SrCO_3$ crystals with various morphologies were prepared using PA and IPA as crystal growth adjusters, respectively. When IPA was used as adjuster, the major morphologies of $SrCO_3$ are center-bundled and two ends emission growing clusters formed by the self-assembling of short rod whiskers. While using PA as adjuster, flower-like $SrCO_3$ clusters are assembled by needle-shaped whiskers and their morphologies change from flower shape to sphere shape with the increasing of PA amount. The result indicates that the morphology of $SrCO_3$ may be adjusted by the structure of carboxylic acid.

Acknowledgements

This work was financially supported by, The Natural Science General Subject of An Hui Education Department (KJ2015B14), Scientific Research Starting Foundation (2016qd04) and National Training Program of Innovation and Entrepreneurship for Undergraduates (201710377023) for financial support of this work.

References

[1] Bao S P, Chen X Y, Li Z, Yang B J and Wu Y C 2011 *Cryst. Eng. Comm.* **13** 2511
[2] Li Y, Cao M H and Feng L Y 2009 *Langmuir* **25** 1705
[3] Zhang Y Y, Liu J L, Zhu Y X, Shang Y, Yu M and Huang X J 2009 *Mater. Sci.* **44** 3364
[4] Suryanarayanan V, Nair A S and Tom R T 2004 *J. Mater. Chem.* **14** 2661
[5] Homeijer S J, Barrett R A and Gower L B 2010 *Cryst. Growth Des.* **10** 1040
[6] Wu W K, Ma Y R, Xing Y, Zhang Y Z, Yang H, Luo Q, Wang J, Li B and Qi L M 2015 *Cryst. Growth Des.* **15** 2156
[7] Oldenberg S J, Averitt R D, Westcott S L and Halas N J 1998 *Chem. Phy. Lett.* **288** 243
[8] Liz-Marzan L M, Giersig M and Mulvaney P 1996 *Langmuir* **12** 4329
[9] Du J M, Liu Z M, Li Z H, Han B X, Huang Y and Zhang J L 2005 *Microporous and Mesoporous Mater.* **83** 145
[10] Sondi I and Matijević E 2003 *Chem. Mater.* **15** 1322
[11] Küther J, Bartz M, Seshadri R, Vaughane G B M and Tremela W 2001 *J. Mater. Chem.* **11** 50
[12] Li S Z, Zhang H, Xu J and Yang D R 2005 *Mater. Lett.* **59** 420
[13] Huang Q, Gao L, Cai Y and Aldinger F 2004 *Chem. Lett.* **33** 290
[14] Cao M H, Wu X L, He X Y and Hu C W 2005 *Langmuir* **21** 6093
[15] Rautaray D, Sanyal A, Adyanthaya S D, Ahmad A and Sastry M 2004 *Langmuir* **20** 6827
[16] Balz M, Therese H A, Kappl M, Nasdala L, Hofmeister W, Butt H J and Tremel W 2005 *Langmuir* **21** 3981
[17] Ma M G and Zhu Y J 2007 *J. Nanosci. Nanotechno.* **7** 4552
[18] Xu S J and Wu P Y 2014 *Cryst. Eng. Comm.* **16** 1311
[19] Rautaray D, Sainkar S R and Sastry M 2003 *Langmuir* **19** 888
[20] Lu Y, Yin Y D, Li Z Y and Xia Y N 2002 *Nano Lett.* **2** 785
[21] Caruso R A and Antonietti M 2001 *Chem. Mater.* **13** 3272
[22] Schuetzand P and Caruso F 2002 *Chem. Mater.* **14** 4509
[23] Yu J G, Guo H and Cheng B 2006 *J. Solid State Chem.* **179** 800
[24] Guo G L, Yan G W, Wang L and Huang J H 2008 *Mater. Lett.* **62** 4018
[25] Zhu W C, Zhang G L, Li J, Zhang Q, Piao X L and Zhu S L 2010 *Cryst. Eng. Comm.* **12** 1795
[26] You C, Zhang Q, Jiao Q Z and Fu Z D 2009 *Cryst. Growth Des.* **9** 4720
[27] Miller F A, Carlson G L, Bentley F F and Jones W H 1960 *Spectrochimica Acta* **16** 135
[28] Alavi M A and Morsali A 2010 *Ultrason. Sonochem.* **17** 132
[29] Orme C A, Noy A, Wierzbicki A, McBride M T, Grantham M, Teng H H, Dove P M and DeYoreo 2001 *J .J. Nature* **411** 775

IOP Conf. Series: Materials Science and Engineering **389** (2018) 012021 doi:10.1088/1757-899X/389/1/012021

Effect of Applied Pressure on Densification and Mechanical Properties of Sialon-ZrN Composites Fabricated Through a Reaction Bonding /Gas Pressure Sintering Process

L Yin[1] and M I Jones[1,*]

[1] Department of Chemical and Materials Engineering, Faculty of Engineering, University of Auckland, Auckland 1023, New Zealand

[*]mark.jones@auckland.ac.nz

Abstract. Sialon-ZrN composites have been produced by reaction bonding and gas pressure sintering, and the effects of applied pressure on the sintering and properties has been investigated. The results showed that composites were mainly composed of β-Sialon and ZrN, with intermediate phases (including Y_2SiAlO_5N, Al_2SiO_5 and $Y_2Si_2O_7$) for 0.4 MPa and $Y_5Si_4Al_2O_{17}N$ for 1.0 MPa. The mass loss decreased with pressure increase due to suppression of thermal decomposition of the Sialon. Composites sintered under 1.0 MPa had better densification than those sintered under 0.4 MPa, and presented greater mechanical properties. Under 1.0 MPa, compared to the monolithic Sialon, the addition of ZrN resulted in an increased hardness up to 30% ZrN, with the highest hardness of 16.16 GPa being observed with 20% ZrN. The hardness decreased with higher ZrN content due to the weak bonding between the ZrN particles and the Sialon matrix. However this resulted in a continuous increase in fracture toughness, reaching a maximum of 5.41MPa·m1/2 with 50% ZrN.

1. Introduction

Sialon is a group of compounds in the Si-Al-O-N system with a structure based on Si_3N_4, where Si-N bonds are partially replaced by Al-O bonds[1]. Sialon has been a promising ceramic material because of its excellent properties, such as high thermal shock resistance, high hardness, chemical inertness, good erosion resistance and low creep property[2,3]; and is widely used for bearings, heat exchangers, cutting tools, and refractories. Sialons are normally produced from expensively commercial Si_3N_4, Al_2O_3 and AlN powders. In order to reduce the raw material cost, a reaction bonding method using less inexpensive Si powders to replace the costly commercial Si_3N_4 powders has been employed to fabricate Sialon materials. Conventional liquid phase reaction sintering, spark plasma sintering (SPS) and hot pressing sintering (HP) are the most common methods to fabricate high dense Sialon materials[4-6]. However, these methods require high temperature, high applied pressure and complicated equipment. SPS and HP are suitable for fabricating high density Sialon materials at a relatively low temperature in a short holding time, but they are not suitable for samples with complex shapes and large sizes. In recent research works gas pressure sintering (GPS) was shown to be suitable for mass production of Sialon/Sialon-based ceramics and other ceramics as a laboratory and industrial technique[7,8]. During the densification of Sialon and Sialon-based ceramics, the decomposition of Sialon can result in negative effects on the sintering behavior and properties of the materials. For the GPS technique, a high nitrogen gas pressure can prevent the decomposition of Sialon and also provide

IOP Conf. Series: Materials Science and Engineering **389** (2018) 012021 doi:10.1088/1757-899X/389/1/012021

an extra driving force for the sintering which helps to lower the sintering temperature and the amount of sintering additives required. The sintering of Sialon and Sialon-based ceramics produced by GPS, is not ascertained by a single factor, but is influenced by several factors[9,10], such as sintering temperatures, gas pressures, compositions of the gas and sintering additive contents.

In previous work[11], we have discussed the effect of sintering temperature on the sintering behavior and properties of Sialon-ZrN composites. In the current work, we have investigated the effect of the applied nitrogen gas pressure on the densification and mechanical properties of the β-Sialon-ZrN composites. The scope of this work was to determine an available pressure range to obtain high density Sialon-ZrN composites with satisfactory hardness and fracture toughness.

2. Experiment Work

The starting composition of the composites consisted of pre-synthesized β-Sialon (z=1) powders (details of the preparation process have been described in previous work[11]), and commercial ZrN powders (Sigma-Aldrich Ltd., D50≈3 μm). The ZrN content was varied from 0% to 50 wt% in increments of 10 wt%. Sintering additives were composed of Y_2O_3 and Al_2O_3 with a molar ratio of 3:5 (i.e. YAG composition), and a total amount of 8 wt%. The powder mixtures were ball-milled in isopropanol for 4h with Si_3N_4 balls, then dried in a rotary evaporator and passed through a 250 μm sieve. The sieved powders were uniaxially pressed into pellets, and cold isostatically pressed at 200 MPa for 5 min. The green compacts were placed into a BN crucible and sintered in a high temperature graphite furnace at 1700 ℃ for 6h under an nitrogen pressure of 0.4 and 1.0 MPa, respectively.

Following sintering, bulk densities and open porosities of the ground and polished samples were determined by the Archimedes' method. Relative density was determined by comparison to the theoretical density of the composites determined from a rule of mixtures for each composition.

The phase assemblages of the sintered samples were analyzed by XRD diffraction. The morphologies were observed with a field-emission gun scanning electron microscope (ESEM, FEI QUANTA 200, USA). Vickers hardness (according to ASTM C1327-15 [12]) and indentation fracture toughness were measured using a Vickers hardness tester equipped with a diamond indenter on the polished surface with a 98 N load for 15 s. Fracture toughness was calculated by the following equation as described by G. R. Anstis et. al[13] :

$$K_{IC}=(0.016\pm0.004)\times\left(\frac{E}{H_V}\right)^{0.5}\times\left(\frac{P}{c^{3/2}}\right) \quad (1)$$

where Hv is the Vickers hardness, c is the crack length, and E is the Young's modulus for the composites calculated by rule of mixtures.

3. Results and discussions

3.1. Influence of applied pressure on phase assemblage and microstructure of the sintered composites

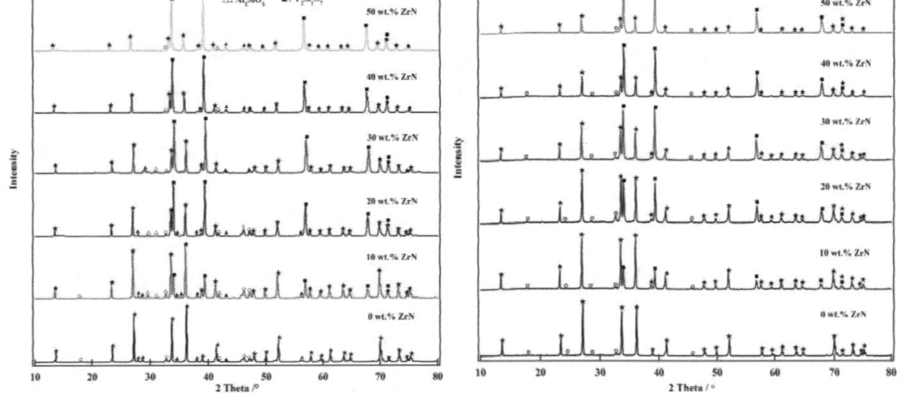

Figure 1. The XRD patterns of the composites sintered at 1700 °C under nitrogen pressure of (a) 0.4 and (b) 1.0 MPa for 6 h, respectively.

Figure 1 shows the phase assemblage of the composites sintered at 1700 °C under applied nitrogen gas pressure of 0.4 and 1.0 MPa, respectively. According to the XRD results, under these two pressures, the composites were mainly composed of β-Sialon and ZrN, with secondary phases of Y_2SiAlO_5N, Al_2SiO_5 and $Y_2Si_2O_7$ for 0.4 MPa (shown in figure 1(a)), while only $Y_5Si_4Al_2O_{17}N$ was observed in samples produced under 1.0 MPa shown in figure 1(b). These results show that variation in applied pressure had no influence on the major phases, while it affected the formation of minor phases. For this sintering process, the nitrogen pressure mainly served as an external pressure to assist the densification of composites, and it had limited influence on their phase assemblage.

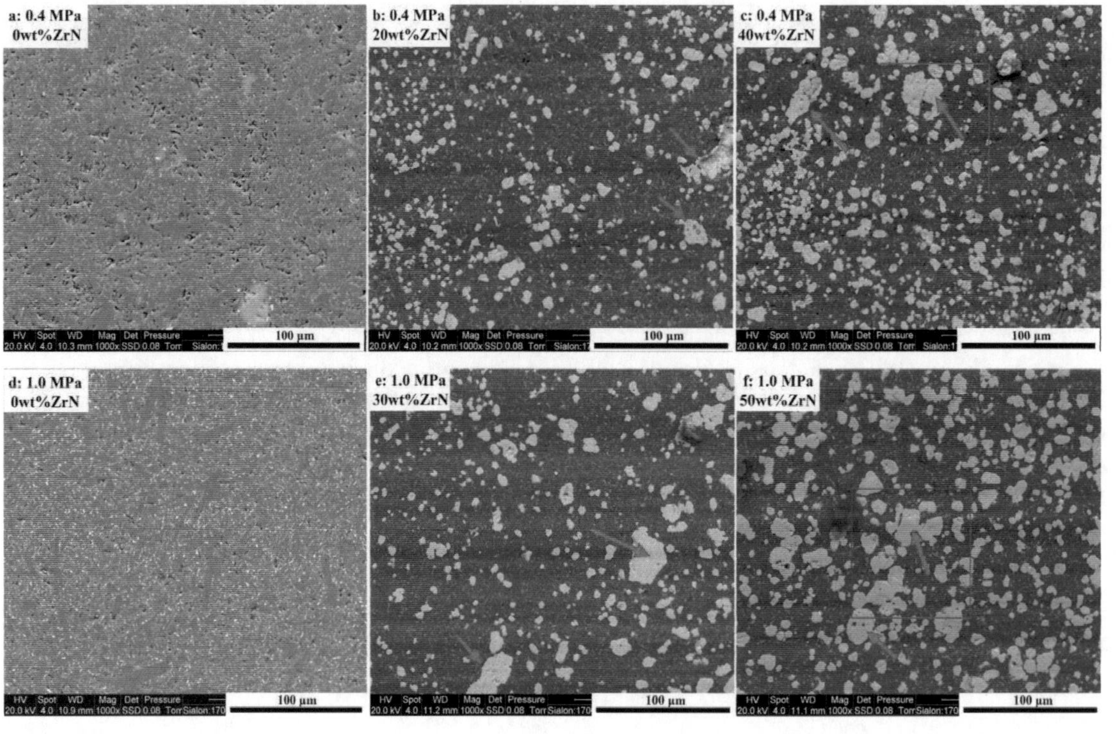

Figure 2. The morphologies of the composites sintered at 1700 °C under nitrogen pressure of 0.4MPa and 1.0 MPa, respectively.

During gas pressure sintering, pressure also plays an important role on grain growth, pore elimination and grain boundary mobility, leading to differences in the microstructure of the composites, as shown in Figure 2. As can be seen in Figures 2a and 2d, the monolithic Sialon sintered under 0.4 MPa had more residual pores with larger sizes than those under 1.0 MPa. As shown in Figure 2d, a larger number of high aspect ratio β-SiAlON (Z=1) grains were observed under a nitrogen gas pressure of 1.0 MPa. For all of the samples, the amount of pores decreased with increase in ZrN content due to the ZrN particles distributing into the grain boundaries to affect the boundary mobility. For the composites sintered under these two applied pressures, abnormal grain growth of the ZrN particles (red arrows) was found and some pores became trapped within the surface of ZrN particles.

3.2. Influence of applied pressure on densification behaviour of the sintered composites

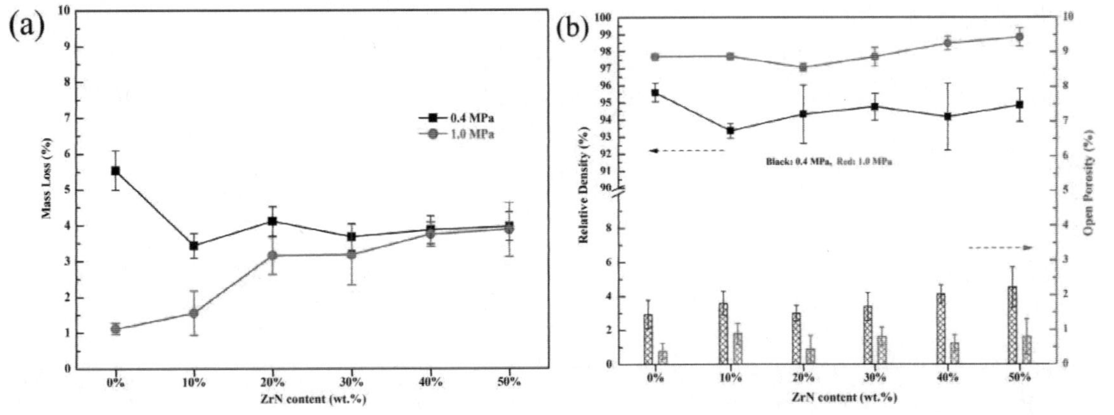

Figure 3. Effect of nitrogen pressures on (a) mass loss, (b) relative density and open porosity of composites with different ZrN contents sintered at 1700 °C for 6h.

Figure 3a shows the mass loss of composites sintered at 1700 °C under a nitrogen pressure of 0.4, and 1.0 MPa, respectively. The results indicated that composites sintered under low pressure (0.4 MPa) had the highest mass loss, and this decreased with increase in nitrogen pressure. For Sialon-ZrN composites, the thermal decomposition of Sialon matrix was the predominant reason for mass loss. The results in Figure 3a show that this thermal decomposition of β-SiAlON is affected by the magnitude of the gas pressure. It has been shown that the partial pressure of SiO is important for the stability of β-SiAlON[14]. If it is assumed that $P_{N2}+P_{SiO}=1$ atm, at 1800 °C, the single phase β-SiAlON would be stable when the SiO partial pressure was between 0.39 and 0.60 atm. During the decomposition process of β-SiAlON, SiO gas could escape from the bulk sample under a low sintering atmosphere pressure. This could decrease the partial pressure of SiO in the bulk sample, leading to further thermal decomposition of Sialon. Increasing pressure makes it more difficult for SiO gas to diffuse out the sample, causing the increased SiO partial pressure to halt thermal decomposition of Sialon, finally resulting in less mass loss. High pressure is believed to result in suppression of the thermal decomposition of the Sialon matrix, resulting in the decreased mass loss with pressure. Under a nitrogen pressure of 0.4 MPa, the mass loss of composites was lower than that of monolithic Sialon and varied slightly with increase in ZrN content. However, under high nitrogen pressure of 1.0 MPa, the mass loss increased with increasing ZrN content, reaching around 4% for a ZrN content of 50%. For the commercial ZrN powders, ZrO_2 is an unavoidable oxide existing on the surface of the ZrN particles. A reaction between ZrO_2 and ZrN could take place by the following equation, leading to the slight increase in mass loss with ZrN content.

$$2ZrO_2+N_2 \rightarrow 2ZrN+O_2 \quad (2)$$

The relative density and open porosity of the composites sintered at 1700 °C under applied pressure of 0.4 and 1.0 MPa are presented in Figure 3b. The results show that the composites sintered under low pressure had poorer densification behavior with relative density around 93% and open porosity between 1.5 to 3%. The composites sintered under high pressure (1.0 MPa) had greater densification behavior, where the relative densities reached 98% and open porosities were less than 1.5%, indicating that increasing pressure was beneficial for the densification. This was because the densification occurred by diffusional processes under the influence of surface energy as a driving force. The driving force increased with applied pressure and resulted in higher relative density and low open porosity. However, during the sintering, it was impossible to avoid thermal decomposition of the Sialon matrix, the formation of intermediate phases and the existence of residual closed pores located at grain-boundary intersections, indicating the difficulty in obtaining 100% full dense composites.

3.3. Influence of applied pressure on the hardness and fracture toughness of the sintered composites

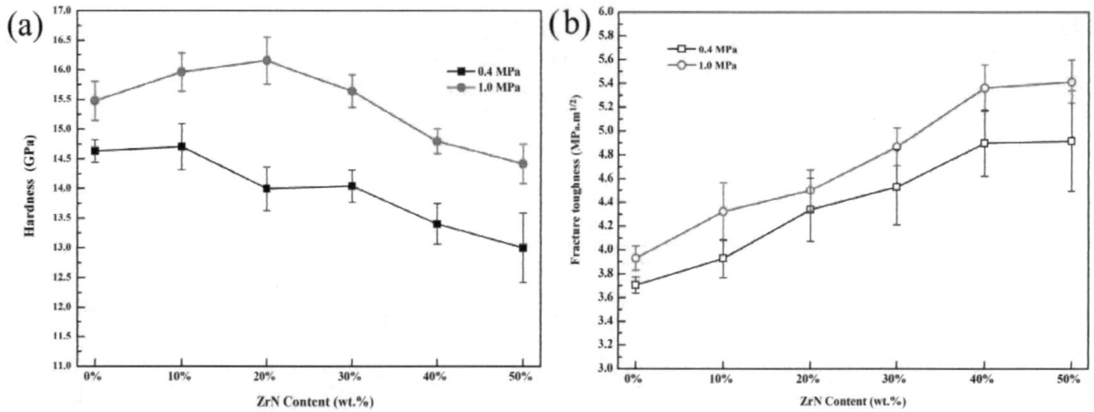

Figure 4. The (a) hardness and (b) fracture toughness of the β-Sialon -ZrN composite with different ZrN contents sintered at 1700 °C under a nitrogen gas pressure of 0.4 and 1.0 for 6h.

Figure 4 presents the hardness (HV10) and indentation fracture toughness of the β-Sialon-ZrN composite with different ZrN contents sintered at 1700 °C under a nitrogen gas pressure of 0.4 and 1.0 for 6 h. The results in this figure demonstrate that it is possible to improve the mechanical properties of the Sialon matrix composites through a combination of increase in applied pressure and amount of reinforcing phase. As shown in figure 4(a), the highest hardness for samples sintered under a high gas pressure (1.0 MPa) regardless of ZrN content, thought to be due to the higher density of these samples (shown in figure.3b). Under 1.0 MPa, compared to the sample with no reinforcement, the addition of ZrN resulted in an increased hardness up to a ZrN content of 30 wt%, with the highest hardness around 16.10 GPa being observed with 20 wt% ZrN; The hardness decreased for higher ZrN content, but this was not related to density since they were similar for all samples. The similar effect of secondary phase on hardness of Sialon-matrix composites has been observed in Y-α-Sialon-cBN composites, and was attributed to weak bonding between the matrix grains and the reinforcing particles [15]. In the present work, such weakening may also have been observed, due to the differences in thermal expansion coefficients of the two materials and/or the residual stresses due to the differences in sinterability of the two materials, normally densification of ZrN being obtained at higher than 2000 °C (much higher than fully densification of Sialon matrix)[16]. This weakening is increased with increasing amounts of ZrN, since the distribution of ZrN grains tends towards a continuous network, softening the composite and resulting in the decrease in hardness. Again as can be seen in figure 4(b), the samples sintered under high pressure (1.0 MPa) had the greater fracture toughness due to the higher density and lower amount of porosity. Unlike hardness, which went through a maximum with increasing ZrN content, the fracture toughness increased continuously up to a value of 5.40 MPa· m1/2 for 50 wt% ZrN. The crack paths and toughening mechanisms of the composites are shown in Figure 5, where all of the toughening mechanisms known in ceramic materials can be observed (i.e. crack branching, crack deflection and crack bridging) [17,18].

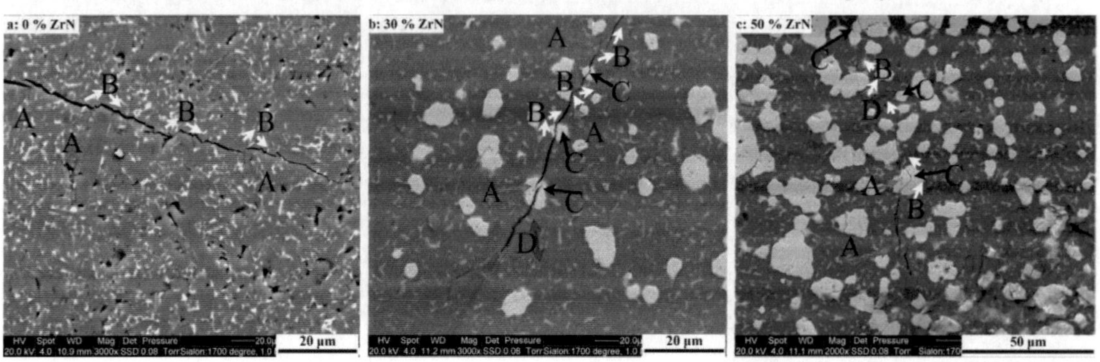

Figure 5. The crack paths and toughen mechanisms being observed in the selected composites, including the crack deflection (B-white arrows), crack bridging (A-red arrows), transgranular fracture (C-black arrows) and crack branching (D-blue arrows)

Generally, for Si_3N_4 and Si_3N_4-based materials, the composition and microstructure (i.e grain size and shape, grain boundary phases and reinforcing phases) have great effect on the fracture toughness, as described by numerous works[19-21]. As can be seen in Figure 5, for the monolithic Sialon material, the fracture toughness was predominantly influenced by crack bridging and crack deflection. With increasing in ZrN content, more crack deflection and crack branching were observed due to ZrN particles distributed into the grain boundaries and triple junctions and the more weaker interfaces between the matrix and the reinforcing phase. Also it needed more energy to break the higher ZrN particles than the Sialon matrix. These phenomena all consume energy and/or make the propagation of cracks more difficult, and lead to the increased toughness. Although the weaker interface with increasing ZrN results in decreasing hardness when the amount of reinforcement is increased, it allows for more crack bridging and crack deflection, and thus results in a higher toughness.

4. Conclusions

The influence of applied nitrogen gas pressure on the densification and properties of Sialon-ZrN composites have been investigated through a reaction bonding /post sintering process. Composites sintered under nitrogen pressures of 0.4 and 1.0 MPa were composed of an elongated β-Sialon (Z=1) matrix, and equiaxed ZrN particles, with intermediate phases (including Y_2SiAlO_5N, Al_2SiO_5 and $Y_2Si_2O_7$) for 0.4 MPa and $Y_5Si_4Al_2O_{17}N$ for 1.0 MPa. The composites sintered under higher nitrogen gas pressure (1.0 MPa) had less mass loss due to the suppression of the thermal decomposition of β-Sialon, and also showed greater densification with a relative density around 98% and small open porosity less than 1.5%. These samples achieved the higher hardness and fracture toughness, which were both affected by the ZrN content. The hardness of the composites reached a maximum (16.16 ±0.40 GPa) at 20 wt% ZrN and decreased for higher contents, due to weakly bonded interfaces between the Sialon matrix and the ZrN particles. The reinforcing ZrN phase helped to toughen the composite by different toughening mechanisms, such as crack deflection, crack bridging, transgranular fracture and crack branching. The addition of ZrN promoted a continuous increase in fracture toughness of the composites, where the maximum fracture toughness of 5.41±0.18 MPa·m1/2 was observed with 50 wt% ZrN. This work indicates that increase in the applied gas pressure in GPS technique was positive for the densification and mechanical properties of Sialon and Sialon-based ceramics and an appropriate content of a reinforcing phases was beneficial for the densification and properties of Sialon materials.

Acknowledgements

Li Yin would like to acknowledge the China Scholarship Council (CSC) for providing a doctoral scholarship. The authors are grateful to technical staff in the Chemical & Materials Engineering Department at the University of Auckland for their assistance with preparation, property testing and data analysis.

References

[1] Liu H Tao, Meng F Rong, Li Q, Huang Z Hui, Luo S Qin, Yin L, Fang M Hao, Liu Y Gai and Wu X Wen 2015 *Cryst. Eng. Comm* **17** 1591
[2] Rueanngoen A, Kanazawa K, Akiyoshi M, Imai M, Yoshida K and Yano T 2013 *J. Nucl. Mater* **442** S394
[3] Li Y jun, Yu H Liang, Jin H Yun, Liu D Hua, Qiao G Jun, Jin Z Hao and Shi Z Qi 2014 *J. Alloy. Compd* **616** 639.
[4] Khan M A Raja, Malki M A Moath, Hakeem S Abbas, Ehsan A Muhammad and Laoui T 2016 *Mater. Sci. Eng. A* **673** 243
[5] Nekouee A Kh and Khosroshahi R A 2016 *Int. J. Refract. Met. H* **61** 6

IOP Conf. Series: Materials Science and Engineering **389** (2018) 012021 doi:10.1088/1757-899X/389/1/012021

[6] Abo-Naf S M, Dulias U, Schneider J, Gahr K H Z, Holzer S and Hoffmann M J 2007 *J. Mater. Process. Tech* **183** 264
[7] Liu S Chao, Ye F, Hu S Qing, Yang H Xia, Liu Q and Zhang B 2015 *J. Alloy. Compd* **647** 686
[8] Rocha V Cláudio and Costa A Célio 2006 *Materials Research.* **9** 143
[9] Hyuga H, Yoshida K, Kondo N, Kita H, Sugai J, Okano H and Tsuchida J 2009 *Ceram. Int* **35** 1927
[10] Li Y, Liu D, Zeng C, Shi Z and Jin Z 2016 *J. Mater. Eng. Perform* **25** 1143
[11] Yin L and Jones I Mark 2018 *Ceram. Int.* In press
[12] ASTM C 1327-15, ibid, vol. **15**.01
[13] Anstis G R, Chantikul P, Lawn B R and Marshall D B 1981 *J. Am. Ceram. Soc* **64** 533
[14] Mitomo M, Kuramoto N and Yajima Y 1980 Journal of the Ceramic Association, Japan. **88** 49
[15] Garrett J C, Sigalas I, Herrmann M, Olivier E J and O'Connell J H 2013 J. Eur. Ceram. Soc. **33** 2191
[16] Li D, Yang Z, Jia D, Hu C, Liang and Zhou Y 2015 *J. Eur. Ceram. Soc* **35** 4399
[17] Steinbrech R W 1992 *J. Eur. Ceram. Soc* **10** 131
[18] Ritchie O Robert 1988 *Mater. Sci. Eng. A* **103** 15
[19] Björklund H, Falk L K L, Rundgren K and Wasén J 1997 *J. Eur. Ceram. Soc* **17** 1285
[20] Becher F Paul, et al 1998 *J. Am. Ceram. Soc* **81** 2821
[21] Blugan G, Hadad M, Janczak-Rusch J, Kuebler J and Graule T 2005 *J. Am. Ceram. Soc* **88** 926

IOP Conf. Series: Materials Science and Engineering **389** (2018) 012022 doi:10.1088/1757-899X/389/1/012022

Microstructure and Properties of Ni/Al_2O_3 Composites Prepared by Powder Injection Molding

Feng Shang[1,2,3*], Zhenwei Cao[4], Peng Zhang[1], Bin Qiao[1] and Huaqiang Li[1]

[1]School of Mechanical Engineering, Huaihai Institute of Technology, No.59 Cangwu Road Haizhou District , Lianyungang, Jiangsu, 222005, China
[2]Marine Resources Development Institute of Jiangsu, No.108 Tongguan South Road Haizhou District, Lianyungang, Jiangsu, 222001, China
[3]Jiangsu Key Laboratory of Large Engineering Equipment Detection and Control, Xuzhou Institute of Technology, No.1 Fuchun Road Xincheng District, Xuzhou, Jiangsu, 221111, China
[4]School of Mechanical and Electrical Engineering, China University of Mining and Technology, No.1 Daxue Road, Xuzhou, Jiangsu, 221116, China

*E-mail: shangfeng@hhit.edu.cn (Feng Shang); Tel.: +86-518-85895323 (Feng Shang)

Abstract. Ni/Al_2O_3 composite was prepared by powder injection molding, the effects of Ni content and different sintering process on the microstructure and mechanical properties of Ni/ Al_2O_3 composites were studied .The results show that when the Ni content is 10%, the comprehensive performance of Ni/ Al_2O_3 composite is better. The optimum vacuum sintering process of 10%Ni/ Al_2O_3 composite is as follows: sintering temperature of 1500 ℃ for 1 hour, with the relative density of 96.7%, bending strength of 312 MPa, Vickers hardness of 1528HV and fracture toughness of 5.3 MPa•$m^{1/2}$.

1. Introduction

With characteristics of high hardness, high tensile, high temperature resistance, corrosion resistance, etc., Al_2O_3 ceramic is one of the most commonly applied ceramic materials at present. However, with the deficiencies of great fragility, mechanical processing difficulty, etc., its application is restrained [1]. Metal has good heat stability and ductility, but it may be easily oxidized at high temperatures. Therefore, the composites combined with metal and ceramic have not only the high tensile and high hardness of ceramic, but also the ductility and good toughness of metal [2].

The Al_2O_3 based cermets are mainly manufactured by vacuum sintering, hot isostatic pressing, self-propagating high-temperature synthesis, sol-gel method, etc [3-5]. In recent years, laser near net shaping, pseudo-semi-soid thixoforming, gel casting, powder injection molding and other powder near net shaping techniques have all been applied in the manufacturing of alumina ceramic matrix composites [6-7]. Powder injection molding (PIM) is a new near net shaping technology of metal/ceramic components combining injecting molding with traditional powder metallurgy processing [8]. PIM overcomes the deficiencies of traditional powder metallurgy processing, including large density gradient [9]. With this technology, near net shape components can be manufactured with high material utilization rate and excellent mechanical property. Therefore, it is widely applied in automobile, medicine, military, aerospace, etc. due to automation and bulk manufacturing. This work

IOP Conf. Series: Materials Science and Engineering **389** (2018) 012022 doi:10.1088/1757-899X/389/1/012022

manufactured Ni/Al_2O_3 composites by powder injection molding and studied the effects of vacuum sintering on the microstructure and properties of Ni/Al_2O_3 composites with different Ni contents.

2. Experimental

Table 1 shows the basic performance parameters of Al_2O_3 powder and Ni powder used in the experiment and figure 1 shows the powder morphology. By mechanical alloying, powder of Ni/Al_2O_3 composites with Ni content of 0%, 5%, 10% and 15% (mass fraction) can be manufactured. For convenience of sintering, a trace of Y_2O_3 and MgO with mass fraction of 0.5% is added into powder of Ni/Al_2O_3 composites as sintering additive. Binding agent is wax based with components of paraffin wax (PW), low-density polyethylene (LDPE), EVA ethylene (EVA) and stearic acid (SA). The mass fractions are 65%, 20%, 10% and 5%, respectively and the powder loading is 58% (volume fraction). The injection processing parameters obtained by orthogonal optimizing experiment are: injection pressure 90MPa, injection temperature 140 ℃ and injection time 8s. The degreasing of injection green parts is conducted by solvent degreasing + thermal debinding where the solvent is n-heptane. Thermal debinding process: heat for 30min to 100 ℃ at room temperature (insulation for 30min) and then heat for 60min to 200 ℃ (insulation for 30min), heat for 90min to 350 ℃ (insulation for 40min), heat for 120min for 550min (insulation for 60min) and heat for 60min to 650 ℃ (insulation for 30min). Then frozen to room temperature in the furnace.

Table 1. Basic characteristics of powder

Powder	True Density ($g \cdot cm^{-3}$)	Average particle size (μm)	Apparent density ($g \cdot cm^{-3}$)	Tap density ($g \cdot cm^{-3}$)
Al_2O_3	3.92	0.52	0.7	2.12
Ni	8.8	4.5	2.1	3.13

Figure 1. SEM morphology of powders (a) Al_2O_3; (b) Ni

HAAKE PolyLab torque rheometer is used to prepare feedstock, HAAKE MiniJet trace injection molding machine used to injection molding, WZDS-20 vacuum sintering furnace used for vacuum thermal debinding and sintering, DH-300X electronic hydrometer used for measuring the density of sintering sample, HV-30 Vickers hardness tester used for measuring the hardness of sintering sample, CMT-5205 electronic universal testing machine used for measuring the three-point bending strength, Axio vert.A1 microscope used for observing the microstructure of samples, and HELIOS NanoLab 600i scanning electron microscope of U.S. FEI Company and Japan JSM-6390LV scanning electron microscope used for observing the powder morphology and fracture morphology of sintering pieces.

3. Results and discussion

3.1. Effect of vacuum sintering process on microstructure and properties

Ni/Al_2O_3 composite with Ni content of 10% is used for vacuum sintering. Table 2 shows the properties of pieces sintered at different temperatures with the insulation time of 1h. According to Table 2, the density, bending strength and hardness will reduce after increase with the increase of sintering temperature in the same insulation time. When the sintering temperature is slightly lower than Ni melting point (1453 ℃), the sintering is solid phase sintering. The binding force of Ni/Al_2O_3 composite in solid phase sintering without pressure action is not as the binding force between powder in liquid phase sintering. The sintering temperature of Al_2O_3 is about 1600 ℃, so the relative density is not high, the pores are many, and bending strength and hardness are low. When sintering temperature is slightly higher than Ni melting point, Ni is in liquid state which is liquid phase sintering. In sintering, Ni may evaporate, causing uneven distribution of Ni in Al_2O_3 matrix with increasing pores and reducing relative density, bending strengh and hardness. Therefore, the sintering temperature should be controlled and sintering pressure be increased in vacuum sintering to improve the microstructure and properties of Ni/Al_2O_3 composites [10].

Table 2. Properties of sintered samples with different sintering temperatures

Sintering temperature (℃)	Density ($g{\cdot}cm^3$)	Bending strength (MPa)	Hardness (HV)
1400	4.013	293	1443
1500	4.053	312	1528
1600	4.028	300	1426

Figure 2 shows the microstructure at different sintering temperatures. The distribution of Ni is relatively even in Al_2O_3 matrix. When sintering at 1400 ℃ and 1500 ℃, respectively, the combination of Ni and Al_2O_3 matrix is good with few Ni reunion. When sintering at 1600 ℃, there will be hole in Ni and Al_2O_3 matrix with Ni reunion. The main reason is the thermal expansion coefficients of Ni and Al_2O_3 mismatch. The higher the sintering temperature is, the worse the binding capability will be and the easier Ni reunion may occur [11]. Figure 3 shows the fracture morphology of bending sample. When sintering at 1400 ℃ and 1600 ℃, there are many pores between Ni and Al_2O_3 matrix, affecting the mechanical property of sintering pieces. When sintering at 1500 ℃, the fracture has fewer pores and better mechanical properties.

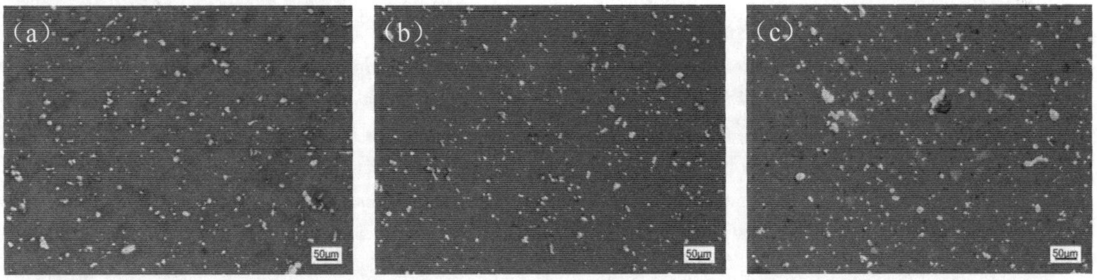

Figure 2. Microstructure of sintered samples at different sintering temperatures (a) 1400 ℃; (b) 1500 ℃; (c) 1600 ℃

3.2. Effect of Ni content on microstructure and properties

Table 3 shows the mechanical properties of Ni/Al_2O_3 composites sintering sample with different Ni contents (0%, 5%, 10% and 15%) in vacuum sintering at 1500 ℃ with thermal insulation for 60min.

IOP Conf. Series: Materials Science and Engineering **389** (2018) 012022 doi:10.1088/1757-899X/389/1/012022

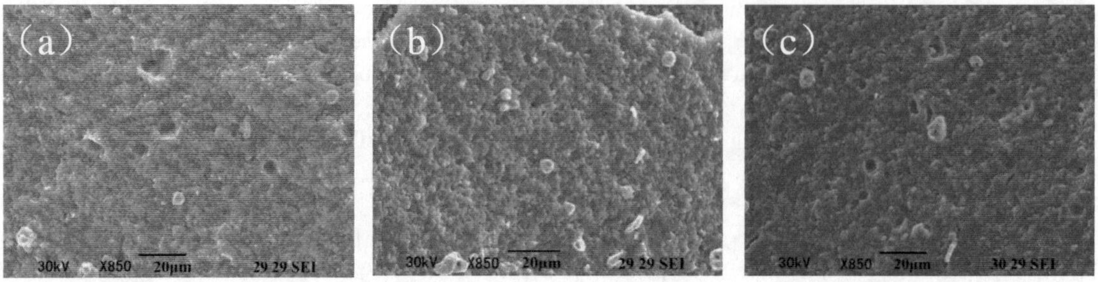

Figure 3. Fracture morphology of bending specimens at different sintering temperatures (a) 1400 ℃; (b) 1500 ℃; (c) 1600 ℃

With the increase of Ni content, the relative density of sintering sample gradually increases. When Ni content reaches 15%, the relative density of sintering sample will reduce. When Ni content is small, active sintering cannot be realized. When Ni content is large, Ni reunion may occur at crystal boundary with the intragranular Ni content reducing and Ni content at crystal boundary increasing due to the viscous flow and plastic flow of powder. Therefore, too large Ni content may be disadvantageous against the improvement of relative density [12]. Figure 4 shows the microstructure of Ni/Al_2O_3 composites with different Ni contents.

Table 3. Mechanical properties of sintered samples with different Ni content

Sintering process parameters	Ni content (%)	Relative density (%)	Bending strength (MPa)	Vickers hardness (HV)	Fracture toughness ($MPa{\cdot}m^{1/2}$)
1500 ℃ for 60 min	0	96.1	291	1648	3.9
	5	96.4	298	1573	4.6
	10	96.7	312	1528	5.3
	15	95.9	303	1427	4.6

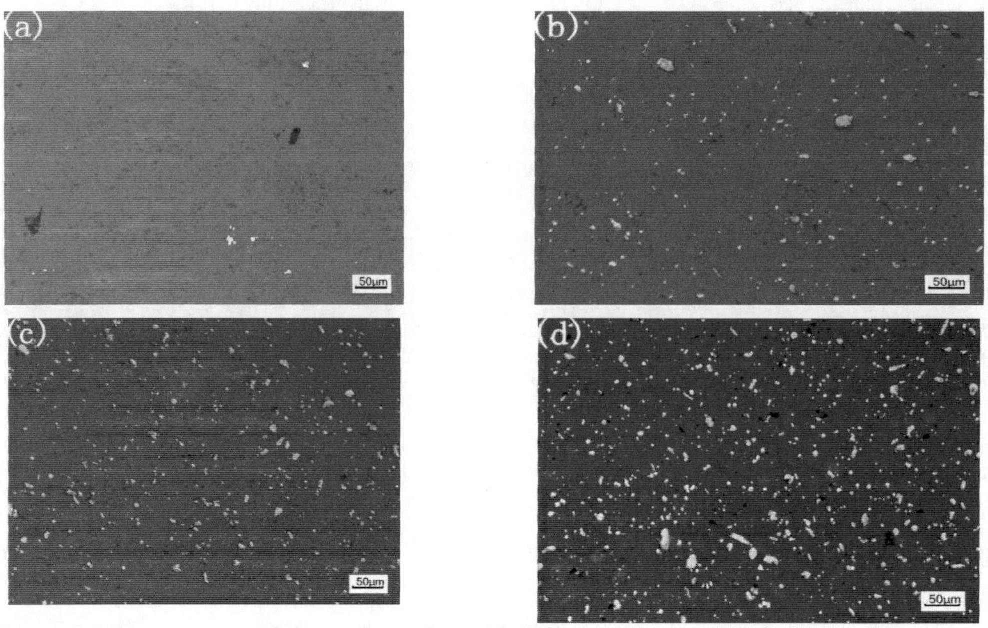

Figure 4. Microstructure of sintered samples with different Ni content (a) 0%; (b) 5%; (c) 10%; (d) 15%

Figure 4 shows that the distribution of Ni is relatively even in Al_2O_3 matrix. With the same sintering process, the evenness of microstructure in figure 5 (c) is good. When Ni content is low, the active sintering is not good enough. When Ni content is large, Ni reunion may occur with particle size increasing, causing adverse effects on sintering densification. Moreover, due to the mismatch of thermal expansion coefficients, pores may gradually increase, while the relative density decreases[13].

With the increase of Ni content, the bending strength of sintering sample will reduce after increasing. Ni can inhibit the growth of Al_2O_3 grain, so the higher the Ni content is, the better the inhibition effect and bending strength will be. When Ni content is larger than 10%, the thermal expansion coefficient of Ni (15×10^{-6}/℃) will be the twice of Al_2O_3 (8.4×10^{-6}/℃). When sintering cools down, the thermal expansion coefficients may mismatch to cause pores at crystal boundary. Meanwhile, with the growth of Ni, Ni may preferentially absorb the neighboring Ni to cause reunion. When the content of Ni is larger than 10%, the bending strength may reduce. Li et al.[14] used homogeneous deposition and coating technology to study the mechanical properties of Ni/Al_2O_3 composites with different Ni contents. With the increase of Ni content, the bending strength tends to reduce after increasing. This is consistent with the research results of this work. Figure 5 shows the fracture morphology of sintering sample.

According to figure 5, the fracture morphology shows the change from neat and leveling to irregular with the increase of Ni content. It is mainly because Ni is liquid phase at the sintering temperature of 1500 ℃. It is easy to find pores at fracture which are mainly caused by the mismatch between the thermal expansion coefficients of Ni and Al_2O_3. Pores at the crystal boundary in fracture morphology (d) are significantly more than those in fracture (c), showing the reason of bending strength reduction.

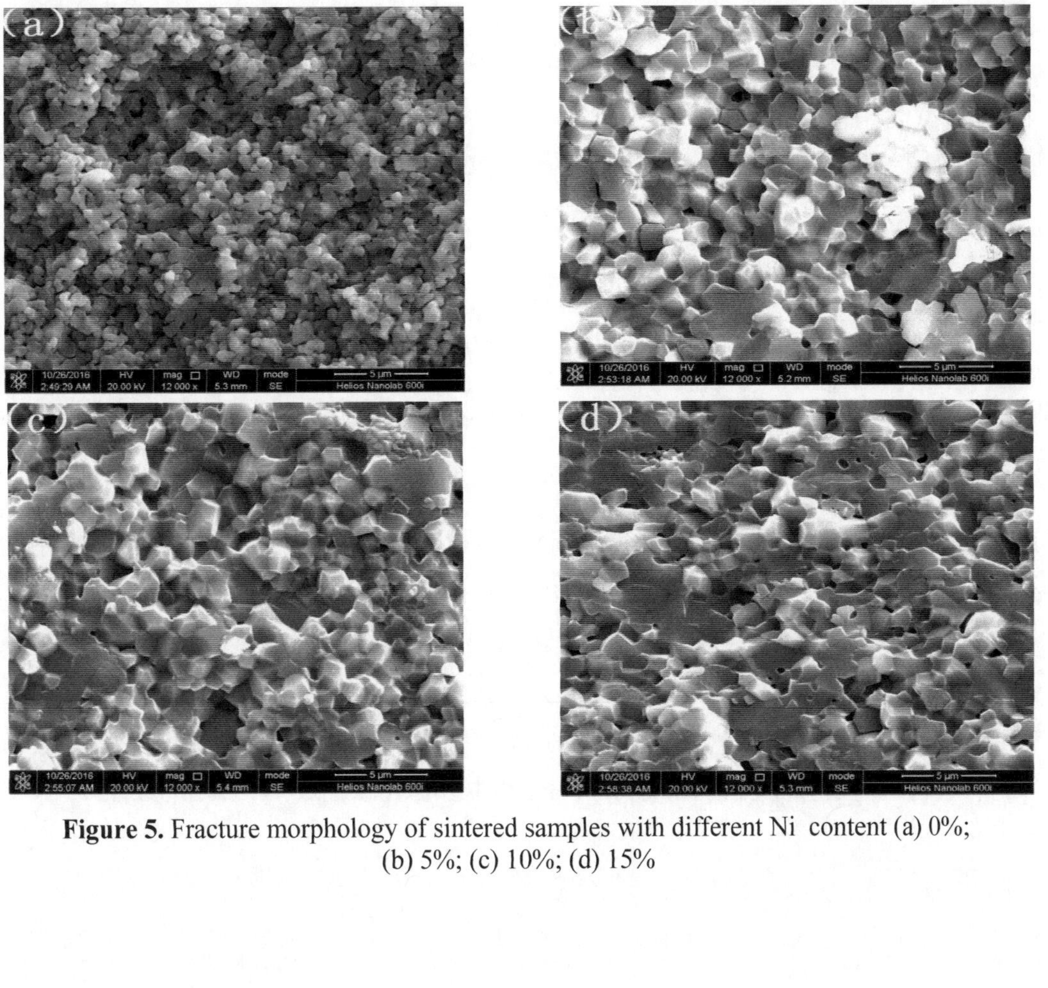

Figure 5. Fracture morphology of sintered samples with different Ni content (a) 0%; (b) 5%; (c) 10%; (d) 15%

The type and performance of added phase have great impacts on the hardness of Al_2O_3-based composites. Table 3 shows that with the same sintering process, the hardness of sintering sample may gradually decrease with the increase of Ni content. This rule meets with the rule of metal as the second phase toughening and the hardness of Ni/Al_2O_3 composite is related to that of Ni and Al_2O_3. With good ductility and toughness, Ni can toughen the matrix. Therefore, with the increase of Ni content, the overall hardness of composite will gradually decrease.

With the increase of Ni content, the fracture toughness of Ni/Al_2O_3 decreases after increasing. Ni is introduced to Al_2O_3 matrix. Compared with Al_2O_3, the composite has its fracture toughness improved. The toughness will reach the maximum when Ni content is 10%. The research proves that the introduction of hard particle into Al_2O_3 matrix as the second phase can improve the fracture toughness of composites [15].

4. Conclusions

When Ni content is 10%, Ni/Al_2O_3 composite has better comprehensive properties. The optimal vacuum sintering process for 10% Ni/Al_2O_3 composite is as follows: sintering temperature 1500 ℃, insulation for 1 hour, relative density 96.7%, bending strength 312 MPa, Vickers hardness 1528 HV and fracture toughness 5.3 MPa•$m^{1/2}$.

Acknowledgments

The research described in this publication was made possible by financial support of the open Foundation of Jiangsu Key Laboratory of Large Engineering Equipment Detection and Control(No.JSKLEDC201506) , Natural Science Foundation of Lianyungang (No. CG1514).

References

[1] Chou T C, Nieh T G, Mcadams S D, Phar G M and Oliver W C 1992 *Journal of Materials Research* **7(10)**, 2774-2784
[2] Evans A G, Lu M C, Schmauder S and Rühle M 1986 *Acta Metallurgica* **34(8)**, 1643-1655
[3] Wang J, Zhao N, Nash P, Liu E, He C, Shi C and Li J 2013 *Journal of Alloys & Compounds* **578**, 481-487
[4] Bhaduri S B, Somani M C and Rao R M 2014 *Transactions of the Indian Ceramic Society* **47(1)**, 1-4
[5] Zhu G, Wang X, Feng P, Li D, Yang T and Akhtar F 2016 *Journal of Alloys & Compounds* **688**, 870-877
[6] Cole and Harold E 2016 *Analytical Chemistry*, **21(4)**, 669-679
[7] Verma V, Kumar B V M 2017 *Materials Today* **4(2)**, 3062-3071
[8] Ko T J, Kim J H and Yoon I J 2009 *International Journal of Precision Engineering & Manufacturing* **10(4)**, 13-17
[9] Sommer F, Talpeanu D, Kern F, Gadow R and Heise U 2015 *International Journal of Applied Ceramic Technology* **12(1)**, 147-156
[10] Isobe T, Daimon K, Sato T, Matsubara T, Hikichi Y and Ota T 2008 *Ceramics International* **34(1)**, 213-217
[11] Li G J, Huang X X, Guo J K and Chen D M 2003 *Acta Materiae Compositae Sinica* **20(2)**, 58-63
[12] Li G J, Huang X X and Guo J K, 2004 *Journal of Inorganic Materials* **19(3)**, 546-552
[13] Li G J, Chen D M, Huang X X and Guo J K 2002 *Journal of Aeronautical Materials* **22(1)**, 1-5
[14] Gao L, Hong J S, Miyamoto H and Torre S D D L 2002 *Journal of the European Ceramic Society* **20(12)**, 2149-2152
[15] Jia X J, Zhang X B, Yu J Z, Fan D, Xu F and Qi F 2015 *Materials Review* **29(s2)**, 524-528

IOP Conf. Series: Materials Science and Engineering **389** (2018) 012023 doi:10.1088/1757-899X/389/1/012023

Rapid Detection of Cracks in Turbine Blades Using Ultrasonic Infrared Thermography

Yong Jia[1,*], Rui-min Zhang[2], Wei Zhang[1], Guo-feng Jin[3], Ji-fang Liu[2], Gan Tian[1] and Zheng-wei Yang[1]

[1]Xi'an Research Institute of High-Tech, Xi'an, 710025, P.R.China
[2]High-Tech Institute, Qing Zhou, 262500, P.R.China
[3]96879 Unit, Baoji, 721012, P.R.China

*jiayong0718@163.com

Abstract. Ultrasonic infrared thermography was used to detect the crack in the turbine blades in this work. In order to explore the effect of couplant on the detection results, three groups of comparative test were carried out respectively by using no couplant, sheet metal and medical adhesive plaster as couplant. On this basis, infrared image sequences and temperature data were analyzed. The results obtained show that couplant is the key factor to the detection efficiency of ultrasonic infrared thermography, and the medical adhesive plaster is a type of better couplant compared to the sheet metal. Additionally, ultrasonic infrared thermography can effectively detect the crack in the blade with the detection time of 1s, which demonstrates that the ultrasonic infrared thermography is very suitable for the situation of rapid detection.

1. Introduction

Blades are the key components in the turbine to realize the energy conversion. During the working process, the blades are subjected to a variety of forces such as static centrifugal stress, bending stress, torsion force and exciting force, and are prone to cracking [1]. Therefore, it is of great significance for improving the safety and reliability of the turbine to detect the crack in the blade accurately and effectively. Many techniques have been applied to detect the crack of blade, among which the most common detection methods are based on penetration inspection [2] and ultrasonic inspection [3]. However, the penetration inspection can only detect the surface opening defects, and the penetrant is toxic and volatile. The ultrasonic inspection has low detection efficiency for complex structures, and the detection results are not intuitive. So it is necessary to explore the effective detection method.

Ultrasonic infrared thermography possesses superior performances such as high-speed, great structural adaptability, large area inspection and intuitional results in detecting the internal defect of object and attracts extensive attention in the world [4-9]. In the 1970s, E. G. Henneke et al [10] found that the defect of the sample may generate the thermal mode when the high frequency vibration excitation was applied to the sample. On this basis, Adams R. D et al [11] proposed the ultrasonic infrared thermography detection method. However, the simulation calculation and experimental verification were difficult to carry out due to the limitations of computer performance and infrared camera capability. Afterwards, with the improvement of computer performance and the measurement accuracy of infrared camera, the research of ultrasonic infrared thermography detection entered a completely new stage. In 2000, L. D. Favro et al [12] detected micro-cracks of aluminum alloy sample using ultrasonic infrared thermography and pointed out its ability to inspect contact interface type of

IOP Conf. Series: Materials Science and Engineering **389** (2018) 012023 doi:10.1088/1757-899X/389/1/012023

defects with high sensitivity and great speed. In 2007, X. Han et al [13] studied the relationship between the crack surface closed state and the crack heat generation, and explored the detectability of ultrasonic infrared thermography. In 2013, Guo-feng Jin et al [14] used numerical method to study the feasibility of detecting cracks in sample with curvature using ultrasonic infrared thermography. Therefore, ultrasonic infrared thermography can effectively detect the micro-cracks in metal materials and has great complex structure adaptability. At present ultrasonic infrared thermography technology is mostly used for the sample-level inspection. Involving the advantages of the ultrasonic infrared thermography technology, this work tries to use this method to detect of blade with crack.

Firstly, this work analyzes the non-destructive testing principle of ultrasonic infrared thermography and the heat generation law in the internal damage region of component. On this basis, a set of experiments were carried out to detect the crack in the blade by using no couplant, sheet metal and medical adhesive plaster as couplant. Then, the effect of couplant on the test results was analyzed by comparing the experimental results.

2. The detection principle of ultrasonic infrared thermography

Ultrasonic infrared thermography detection system consists of ultrasonic excitation source, infrared camera and computer processing system. Its detection principle and composition can be showed in Figure 1. Ultrasonic infrared thermography is an active detection technology with ultrasonic pulse excitation. The ultrasonic wave enters the detected object through the tip of the ultrasonic probe, and it will produce scattered reflection when encounters discontinuity or interface, while it makes the damage contact interface to generate heat that the particles rub each other due to the viscosity of the propagation medium. The mechanical energy is converted into heat, which is transmitted to the surface of the object in the form of thermal waves, the changeable temperature of the surface of the object will be generated and recorded by the infrared camera. Detection of the defect inside the object can be realized after the infrared image sequences collection and analysis.

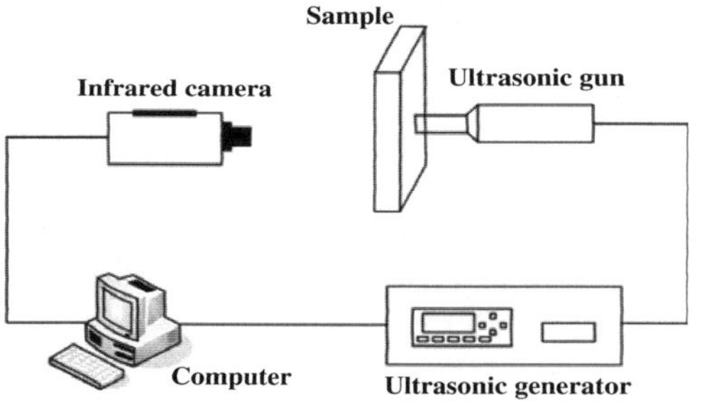

Figure 1. The detection system of the ultrasonic infrared thermography.

It can be seen from the above analysis that the damage contact interfaces are collided during the ultrasonic excitation, where can be regarded as the multiple-degree freedom vibration system. And the ultrasonic mechanical energy is converted into heat. The control equation of forced vibration under ultrasonic stimulation can be described as follow [15]:

$$M\frac{d^2U}{dt^2}+C\frac{dU}{dt}+KU=F+R \tag{1}$$

According to Newton third law and the first law of thermodynamics, it is known that the heat flux density at the damage is:

$$q(t)=\gamma\cdot\left[\mu_{\mathrm{d}}+(\mu_{\mathrm{s}}-\mu_{\mathrm{d}})e^{-\beta|v|}\right]R_{\mathrm{N}}(t)\cdot v_{\mathrm{T}}(t) \tag{2}$$

The meaning of each physical quantity in the above formula is shown in Table 1.

Table 1. The meaning of the physical quantities.

Symbol	Name	Unit
U	Nodal displacement matrix	m

IOP Conf. Series: Materials Science and Engineering **389** (2018) 012023 doi:10.1088/1757-899X/389/1/012023

M	Mass matrix	kg
C	Damping matrix	kg/s
K	Stiffness matrix	kg/s^2
F	External load matrix	N
R	Contact force matrix	N
t	Time	s
$R_N(t)$	Normal friction	N
$V_T(t)$	Tangential relative speed difference	m/s
γ	Friction heat transfer ratio coefficient	
β	Static and dynamic friction conversion factor	
μ_s	Static friction coefficient	
μ_d	Dynamic friction coefficient	

3. Experimental investigation

3.1. Test Overview

The sample studied in this work is a turbine blade made of alloy steel, as shown in Figure 2.

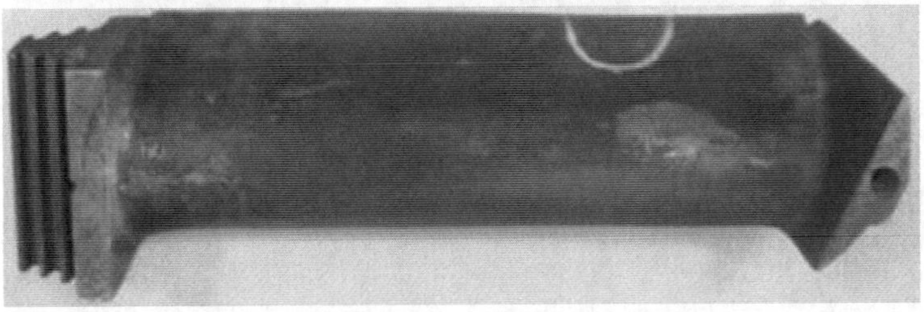

Figure 2. The turbine blade used in the test.

Experimental ultrasonic infrared thermography detection system, as shown in Figure 3, consists of ultrasonic excitation system, sample fixture, infrared camera and computer processing system. The ultrasonic excitation system is mainly composed of an ultrasonic gun and a test parameter control panel. The infrared image collection equipment is the Vhr680 uncooled infrared camera manufactured by InfraTec. Its temperature measurement range is from -40 to 1200℃ and the image resolution is 320×240.

In the experiment, the ultrasonic excitation power is set to 800w, the amplitude of vibration output is 70%, the pre-tightening force between the ultrasonic gun and the blade is 440N, and the infrared camera acquisition frequency is 50Hz. In addition, in order to explore the effect of couplant on the detection results, three groups of comparative test were carried out respectively by using no couplant, sheet metal and medical adhesive plaster as couplant.

Figure 3. The detection system of the ultrasonic infrared thermography.

3.2. Results and analysis

Figure 4 shows the detection results of blade using ultrasonic infrared thermography. The bright spots on the upper and right of blade are all surface reflection, the hot spots on the lower right corner are the excitation position of the ultrasonic gun, and the hot spots on the upper left are the damage region. The results show that there is no hot spot on the surface of the blade without the couplant. When the couplant is used, there are hot spots in the damage region and no hot spot in the non-damage region. Therefore, the couplant is the key factor to the detection efficiency of ultrasonic infrared thermography. Ultrasonic infrared thermography can effectively detect the crack in the blade with the characteristics of selective heating.

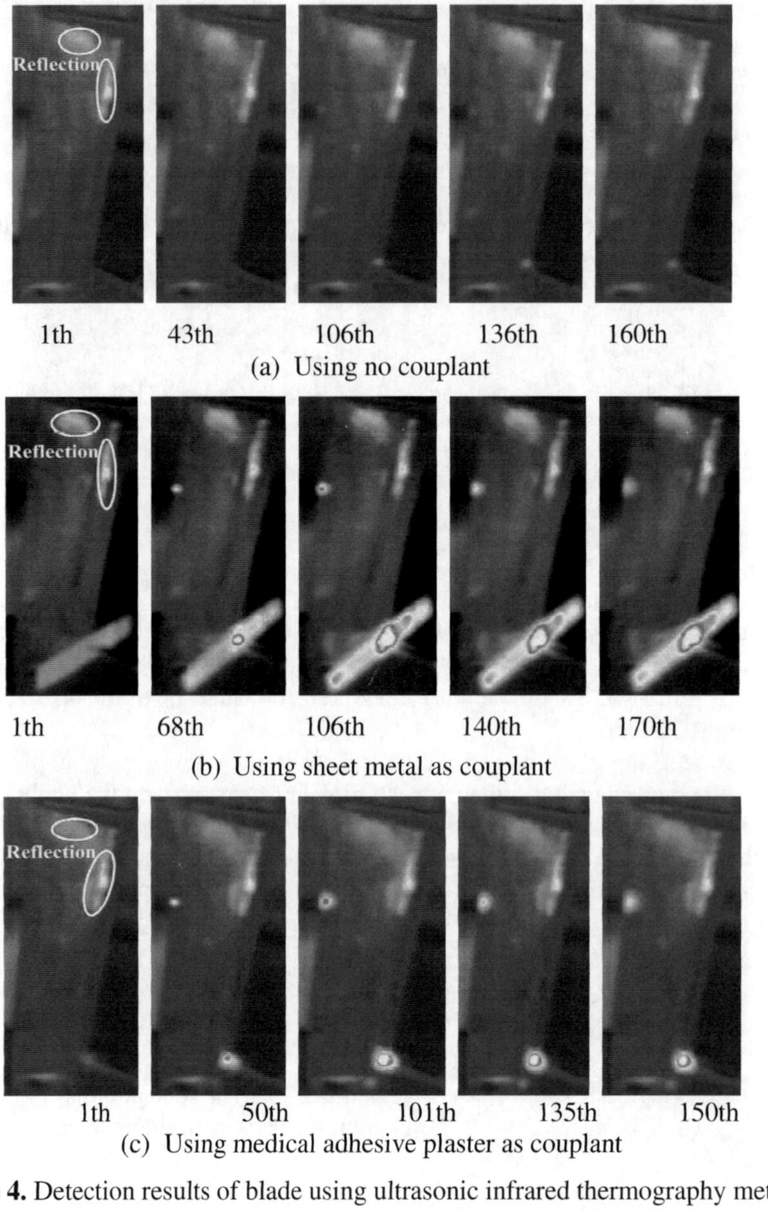

Figure 4. Detection results of blade using ultrasonic infrared thermography method.

In Figure 4 (a), the ultrasonic wave cannot be effectively injected into the blade, as a result, the damage region cannot be selectively heated. So there is no hot spot in the infrared image sequences collected. Figure 4 (b) and (c) show the detection results by using sheet metal and medical adhesive plaster as couplant respectively. After ultrasonic excitation, hot spots appear in the upper left region of the blade, then extend around. It is related to the process of the vibration friction heat generation and heat conduction that hot spots appear from dark to bright and then to dark due to the effect of thermal diffusion. The reason why the hot spots turn from dark to bright is that friction occurs in the damage region due to the ultrasonic vibration and a great deal of heat is produced, leading to temperature rise. The reason why the hot spots turn from bright to dark is that heat dissipation leads to a gradual decrease of temperature during heat conduction process. The bright spots have no obvious change in the excitation process, which are marked in the infrared images, so they are the surface reflection. Comparing Figure 4 (b) and (c), the hot spots are larger and brighter in Figure 4 (c), so the medical adhesive plaster is more effective than the sheet metal.

Additionally, the hot spots in Figure 4 (c) appear in the 50th frame among the infrared image sequences, i.e., the crack in the blade was detected. According to the acquisition frequency of 50Hz in the experiment, it is concluded that ultrasonic infrared thermography can effectively detect the crack of the blade with the detection time of 1s.

3.3 Data processing

To directly analyze the detection results, we use maximum surface temperature difference to weigh the detection effect of ultrasonic infrared thermography in this section.

The surface temperature difference parameters, defined in Eq. (3), were used to investigate the effect of couplant on the detection results. The point P was selected in the damage region of blade and the point Q was selected in the non-damage region, as shown in Figure 5. The temperature data of P and Q were extracted, which represented by T1 and T2 respectively. The temperature difference between damage region and non-damage region is expressed byΔ T:

$$\Delta T = T_1 - T_2 \tag{3}$$

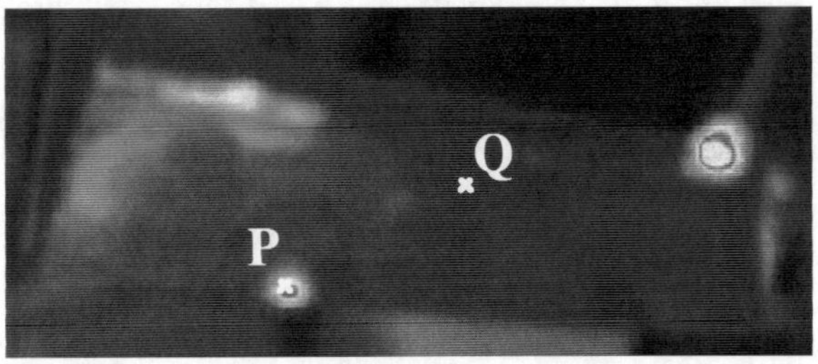

Figure 5. Selection of non-damage region and damage region of blade.

In the three groups of comparative test, the temperature data T1 and T2 were respectively extracted, and the temperature difference Δ T was obtained according to formula (3). The time curvest of T1, T2 and Δ T were drawn, as shown in Figure 6 and 7.

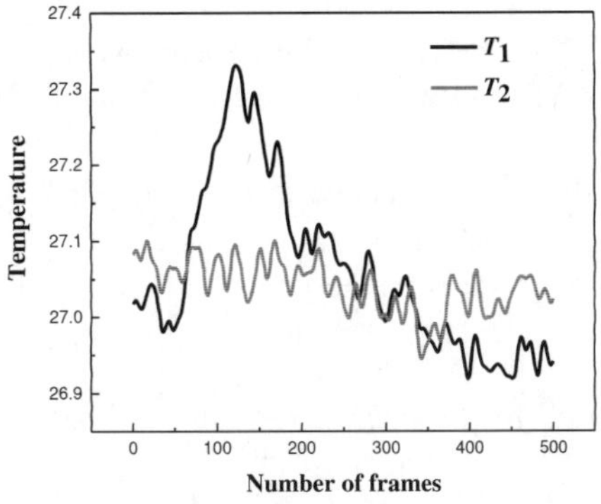

(a) Using no couplant

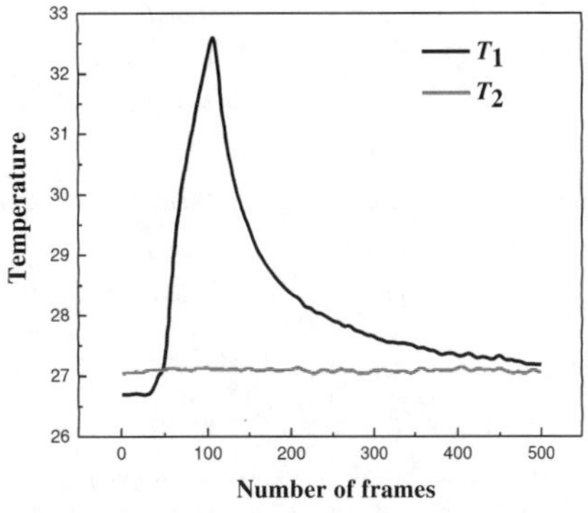

(b) Using sheet metal as couplant

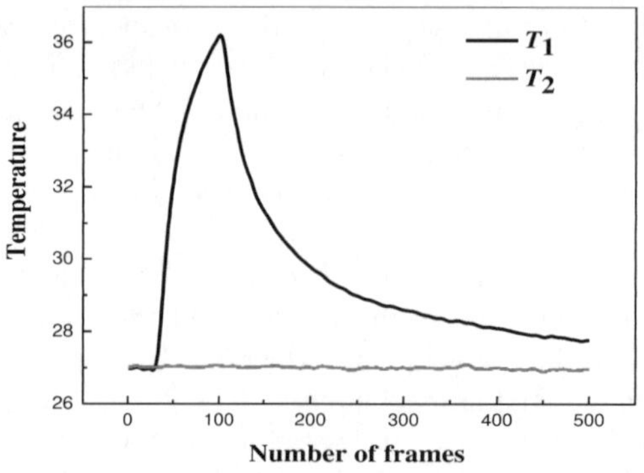

(c) Using medical adhesive plaster as couplant

Figure 6. Temperature vs. time curve in non-damage region and damage region.

In Figure 6 (a), after ultrasonic excitation, the temperature increased by 0.4℃ in the damage region and it was not obvious, while the temperature fluctuated basically at 27.05℃ without significant increase and decrease in the non-damage region. So there is no hot spot in the corresponding infrared image sequences. This shows that the ultrasonic wave cannot be effectively injected into the blade without couplant. In Figure 6 (b), after ultrasonic excitation, the temperature increased rapidly by 6℃ in the damage region, while the temperature fluctuated basically at about 27℃ without obvious change in the non-damage region. So there are obvious hot spots in the corresponding infrared image sequences. In Figure 6 (c), after ultrasonic excitation, the temperature increased rapidly by 9℃ in the damage region, while the temperature fluctuated basically at about 27℃ without obvious change in the non-damage region. So there are larger and brighter hot spots in the corresponding infrared image sequences.

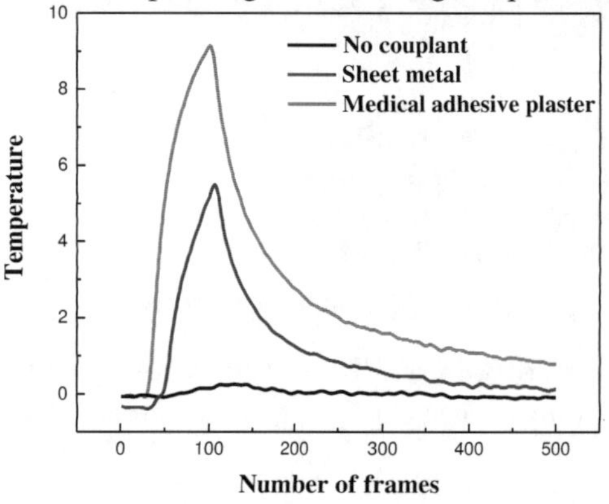

Figure 7. Temperature difference vs. time curve in non-damage region and damage region

It can be seen from Figure 7 that there is almost no temperature difference between the damage region and the non-damage region without couplant. When using sheet metal as couplant, the surface temperature difference reached a maximum of 5.5℃ in the 106th frame. When using medical adhesive plaster as couplant, the surface temperature difference reached a maximum of 9.1℃ in the 101th frame. Therefore, the defects of the blade cannot be detected without couplant and the medical adhesive plaster is a type of better couplant compared to the sheet metal.

4. Conclusion

In this work, ultrasonic infrared thermography was used to detect the crack in the blade, and three groups of comparative test were carried out respectively by using no couplant, sheet metal and medical adhesive plaster as couplant. The results obtained show that couplant is the key factor to the detection efficiency of ultrasonic infrared thermography, and the medical adhesive plaster is a type of better couplant compared to the sheet metal. Moreover, the ultrasonic infrared thermography can effectively detect the crack in the blade with the detection time of 1s, which has instructional significance for the rapid detection of various blades.

Acknowledgments
This work was supported by National Natural Science Foundation of China (Grant No. 51605481 and 51575516)

References

[1] JS Rao. Application of fracture mechanics in the failure analysis of a last stage steam turbine blade. Mechanism & Machine Theory, 1998, 33(5): 599-609.
[2] Qing-he Wang. Inspection method and influence factors of turbine blade surface cracks. China Plant Engineering, 2017, (6): 72-73(In Chinese).
[3] E Jasinien, R Raiutis, R Literis, et al. NDT of wind turbine blades using adapted ultrasonic and radiographic techniques. Insight: Non-Destructing Testing and Condition Monitoring, 2009, 51(9): 477-483.
[4] SY Zhang. Ultrasonic infrared thermography and its applications in nondestructive evaluation. Applied Acoustics, 2004, 23(5): 1-6.
[5] A. Mian, X. Han, S. Islam, et al. Fatigue damage detection in graphite/epoxy composites using sonic infrared imaging technique. Compos. Sci. Technol, 2004, 64(5): 657–666.
[6] B Yang, Y Huang, L Cheng. Defect detection and evaluation of ultrasonic infrared thermography for aerospace CFRP composites. Infrared Physics & Technology, 2013, 60(60): 166-173.
[7] L. Junyan, L. Liqiang, W. Yang. Experimental study on active infrared thermography as a NDI tool for carbon–carbon composites. Compos. B Eng, 2013, 45(1): 138–147.
[8] FZ Feng, CS Zhang, QX Min, et al. Identification and Reconstruction of Cracks in Ultrasonic Infrared Thermography. Applied Mechanics & Materials, 2012, 249-250(4): 46-50.
[9] Y Hong, PC Miao, ZN Zhang, et al. Installation and application of ultrasonic infrared thermography. Acoustical Science & Technology, 2004, 25(1): 77-80.
[10] E. G. Henneke, K. L. Reifsnider, W. W. Stinchcomb. Thermography An NDT Method for Damage Detection. Journal of Metals, 1979, 31(9): 11-15.
[11] ADAMS R D, CAWLEY P, PYE C J, et al. A vibration technique for non-destructively assessing the integrity of structures. Journal of Mechanical Engineering Science, 1978, 20(2): 93-100.
[12] L. D. Favro, X. Han, Z. Ouyang, et al. Infrared Imaging of Defects Heated by a Sonic Pulse. Review of Scientific Instruments, 2000, 71(6): 2418-2421.
[13] J. Lu, X. Han, G. Newaz, et a1. Study of The Effect of Crack Closure in Sonic Infrared Imaging. NDT&E, 2007, 22(2): 127-135.
[14] Jin Guofeng, Zhang Wei, Song Yuanjia, et al. Numerical Simulation for Ultrasonic Infrared Thermal Wave Detecting of Curvature Structural Crack. Science Technology and Engineering, 2013, 3(13): 776-779(In Chinese).
[15] Hui Liu. Research on ultrasonic infrared lock-in thermography for nondestructive testing [D], Harbin Institute of Technology, 2010 (In Chinese).

IOP Conf. Series: Materials Science and Engineering **389** (2018) 012024 doi:10.1088/1757-899X/389/1/012024

Effect of Temperature on Tensile Fatigue Life of Natural Rubber

J Wu[1*], L Chen[1], H H Li[1], B L Su[1*] and Y S Wang[1]

[1]Center for Rubber Composite Materials and Structures, Harbin Institute of Technology, Weihai, China

*Corresponding Author / E-mail: wujian@hitwh.edu.cn, benlongsu@hitwh.edu.cn

Abstract. Research on rubber fatigue properties is of great significance for improving the durability of rubber products. Natural rubber is widely used in rubber products, it is very important to study the effect of temperature on its fatigue properties. Here, tensile fatigue tests of natural rubber have been carried out under different temperatures and strains for studying the influence of temperature and strain on the fatigue life. Results indicate that fatigue life decreases when test temperature increases, and it decreases obviously when temperature is higher than 75 ℃; with increasement of temperature, tensile strength decreases, and the elongation at break increases; the fatigue life decreases with the increasement of strain, and the decreasement of fatigue life is more remarkable under small strain.

1. Introduction

Due to its elasticity and large deformation capacity, rubber is widely used in automotive, machinery, aerospace and other industries, these applications makes it ideal for vibration isolators, elastic elements, seals of spacecraft and other key structures [1]. Rubber properties have a direct impact on the life of products, especially mechanical properties. Fatigue is the primary failure mode of vibration damping rubber structural parts for aerospace. Fatigue properties of rubber material are affected by many environmental factors, especially the temperature. Due to the effects of ambient temperature and heat in rubber structures, the fatigue life decreases when the temperature of fatigue damage position increases, which has a great impact on the use of rubber structures. It is necessary to study the influence of temperature and strain on fatigue of natural rubber.

Rubber fatigue has been widely studied at home and abroad. When natural rubber material was studied by uniaxial tensile test with positive R, the critical region cracks prone to change into secondary cracks and cracks branching [2]. A uniaxial tensile and compression fatigue tests of carbon black filled natural rubber were carried out by Cam [3, 4], the fatigue crack morphology has been studied. Fatigue test of the natural rubber with different carbon black component was studied by Kim [5], which including N330, N650 and N990. Test results showed that different carbon black materials caused different fatigue life of vulcanized rubber. A uniaxial fatigue test of filled natural vulcanized rubber has been studied by Mars and Fatemi [6, 7]. When the initial load increased, the fatigue life increases. Then, results indicated that the fatigue cracks proned to produce and arise in a special plane whether proportional loading or nonproportional loading under multiaxial loading. The static loading test and the cyclic loading test of natural rubber between air and vacuum have been carried out by Gent and Hind [8]. Results indicated that oxygen promoted the growing of fatigue cracks.

IOP Conf. Series: Materials Science and Engineering **389** (2018) 012024 doi:10.1088/1757-899X/389/1/012024

In summary, current researches of rubber fatigue are mainly focused on the influence of material, however, influence of coupling of temperature and strain is also important. For this reason, uniaxial tensile fatigue tests of NR under different temperatures and strains have been carried out for studying the influence of temperature and strain on fatigue life, which provides the theoretical basis for the design of rubber damping products.

2. Experiment set up

2.1. Materials and equipments

A uniaxial tension test at different temperatures is performed by using natural rubber dumbbell specimen with a type 1 shape according to ISO Standard 37:2011, then the stress-strain curves for tension can be obtained. The test is performed by a high-low temperature tension machine. Fig. 1 shows the production process of rubber specimen.

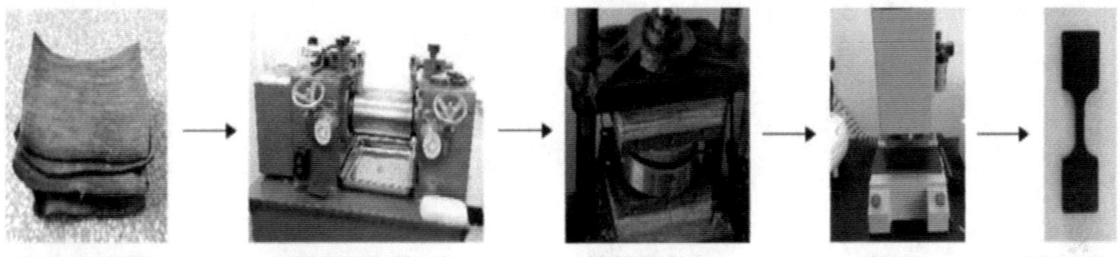

Figure 1. The production process of specimen.

The fatigue test at different temperatures has been carried out by using natural rubber dumbbell specimen. Fig. 2 shows the dimensions of dumbbell sample with a thickness of 2 mm. Fig. 3 shows the fatigue machine of integrated MUF-1050.

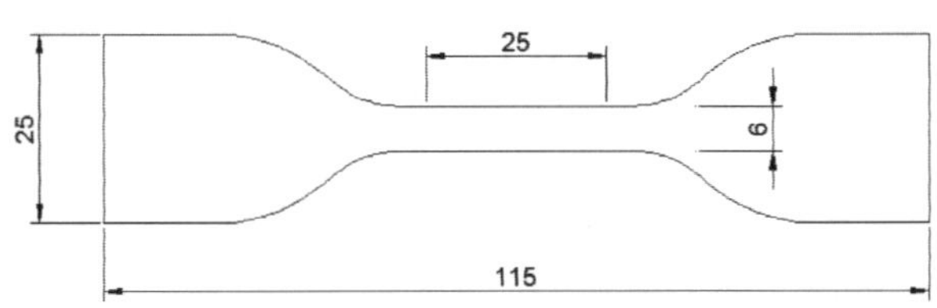

Figure 2. Schematic of dumbbell specimen.

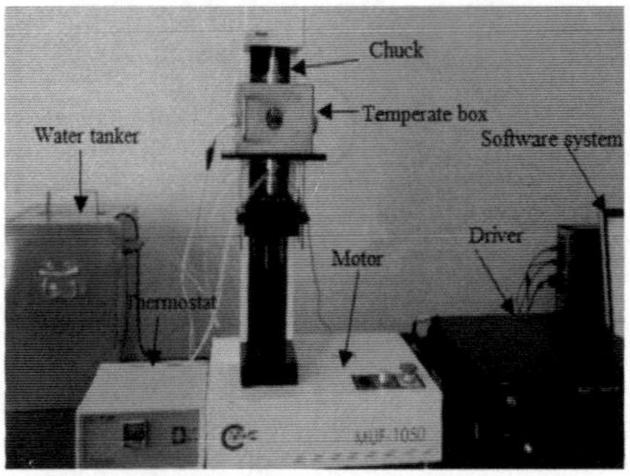

Figure 3. Fatigue machine.

2.2. Experiment methods

In this study, the stroke controlled method is used by a sine waveform of 5Hz, which can eliminate the effect of creep. Uniaxial tensile fatigue tests of rubber dumbbell specimen have been carried out under the temperature of 20℃-110℃. Every 15℃ for a gradient, each temperature is divided into ten conditions according to strain of 60%-150%. Five tests are carried out under each condition, and the mean of values is calculated as the experimental data. Experimental parameters are shown in table 1.

Table 1. Experimental parameters used in tests.

	Experimental parameters
Temperature (℃)	20,35,50,65,80,95,110
Strain	150%,140%,130%,120%,110%,100%,90%,80%,70%,60%

IOP Conf. Series: Materials Science and Engineering **389** (2018) 012024 doi:10.1088/1757-899X/389/1/012024

3. Results and discussion

The definition of fatigue crack nucleation life is the number of cycles when the cracks of a small size appear on the sample. In fatigue test, it can be observed that the cycles of fatigue crack nucleation period is far more than the cycles of crack expansion period. Therefore, the fatigue crack nucleation life can replace the fatigue failure, which is defined as the number of cycles when the maximum load decreases 30%.

3.1. Effect of temperature on Fatigue Performance

The mechanical properties at different temperatures can be obtained by uniaxial test data, tensile strength and elongation at break under different temperatures are shown in Fig. 4. Tensile strength of rubber decreases with increasement of temperature, and elongation at break increases. The increasement of ambient temperature exacerbates thermal motion, which changes the arrangement of rubber molecular chain. with the effect of external forces, rubber molecular chain is easier to move in the stress direction and harder to rehabilitate.

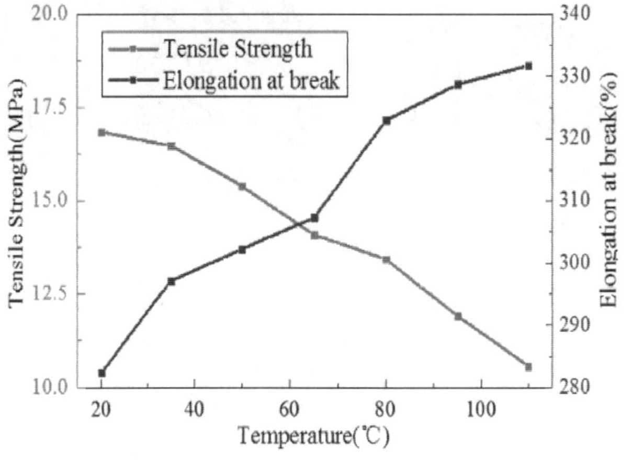

Figure 4. Tensile strength and elongation at break under different temperatures.

The fatigue life and the tensile strength are shown in Fig. 5. Fatigue life and tensile strength are directly proportional relationship, with the reduction of tensile strength, the fatigue resistance of rubber weakened.

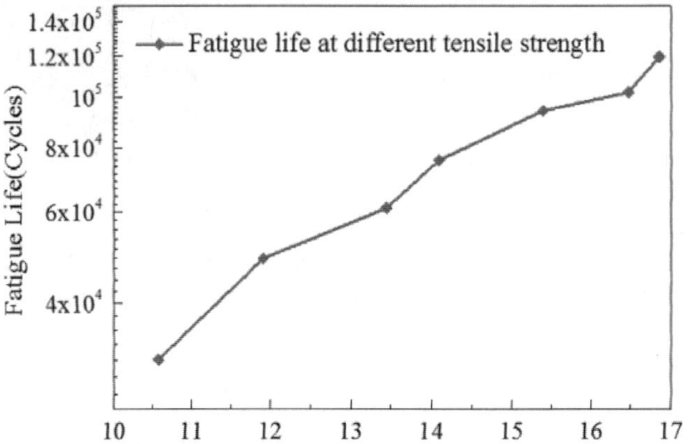

Figure 5. The variation of fatigue life with tensile strength.

3.2. *Effect of Strain on Fatigue Life*

Fig. 6 shows the maximum load with the fatigue life at 20℃ with different strains, the growth of fatigue cracks can be reflected by the variation of the maximum load to a certain degree.

The maximum load decreases according to the decreasement of strain, and the cycles requires to drop to a critical values with increasement of strain. A decrease of the maximum load is observed, which is known as Mullin's effect. The crack first produces and expands at a faster rate in the rubber specimen with larger strain, the lower cycles required to reach the critical value, results show that the maximum load decreases faster of rubber specimen with a larger strain, fatigue fracture appeared firstly and the corresponding maximum load sharply decreased to 0.

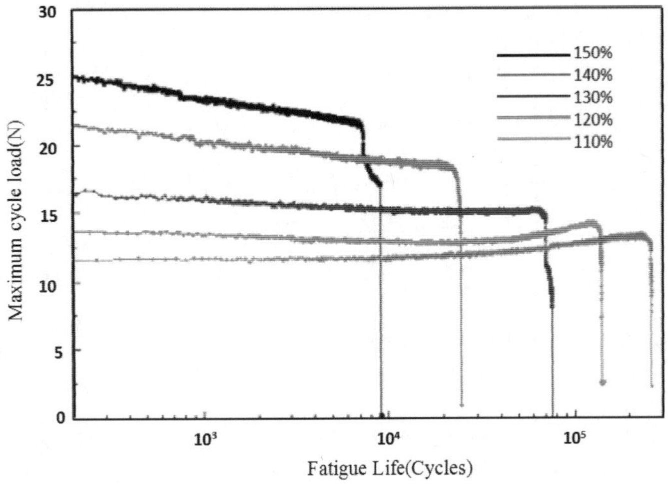

Figure 6. Cyclic maximum load during fatigue process change diagram.

The fatigue life of rubber specimens under different strains are shown in Fig. 7. Whether low or high temperature, the fatigue life decreases with increasement of strain. In the case of strain less than 90%, the fatigue life decreases sharply with increasement of strain, and tends to be gentle when the strain reaches 110%.

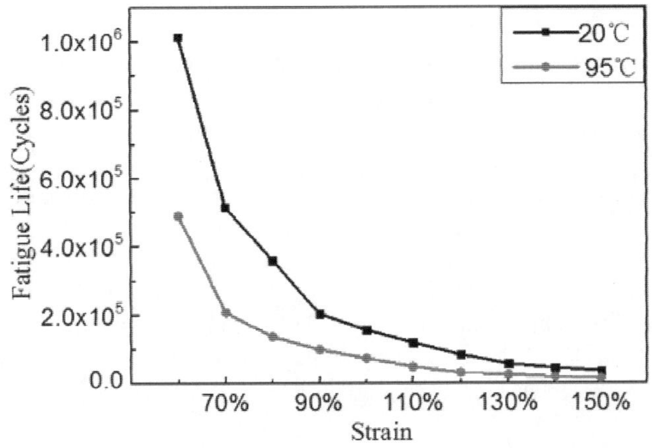

Figure 7. Fatigue life of rubber specimen under different strains

3.3. *Effect of temperature on Fatigue Life*

Fig. 8 shows the fatigue life with the change of temperature.

The fatigue life decreases with increasement of temperature under the same strain, which decreases obviously when the temperature is up to 95℃. Fatigue life under small strain conditions is more significantly affected by temperature, and the fatigue life of 110℃ is less than 25% of 20℃ under the strain 70%. Due to the increasement of temperature, which makes the rubber molecular chain more free and weakens the relationship between molecular, and makes the breakage between the molecular chains easier in the circular movement, the cycles required for fracture failure reduces in the experiment.

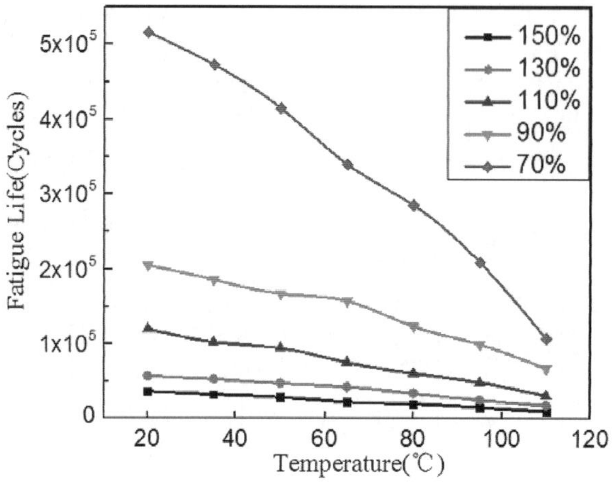

Figure 8. Fatigue life of rubber specimen at different temperatures

4. Conclusions

Natural rubber is widely used, fatigue is one of the main properties of rubber products, which has a significant impact on the life of rubber products, especially the damping rubber products. A uniaxial tension fatigue test under different temperatures and strains have been carried out by natural rubber dumbbell specimen. Then, the influence of temperature and strain on fatigue life of rubber have been studied. According to the analyses above, we have the following points:

(1)Fatigue life of rubber specimen decreases with increasement of strain. When the strain is less than 90%, the fatigue life decreases sharply with increasement of strain.

(2)The tensile strength of rubber materials decreases with increasement of temperature, and the elongation at break increases.

(3)Fatigue life decreases with increasement of temperature, which decreases obviously when the temperature is higher than 75 ℃. Fatigue life under small strain conditions is more significantly affected by temperature, and the fatigue life of 110℃ is only about 25% of that of 20℃ under the strain 70%.

(4)The influence laws of temperature and strain on the fatigue life have been proposed, which provides the theoretical basis for the design of rubber damping products.

Acknowledgments

This work is funded by major special project of Shandong province independent innovation achievements transformation (2014ZZCX03407) and Major Program of National Natural Science Foundation of China (51790502).

References

[1] Mars W V and Fatemi A 2002 A literature survey on fatigue analysis approaches for rubber *J. Fatigue* **24** pp 949–961
[2] Saintier N, Cailletaud G and Piques R 2011 Cyclic loadings and crystallization of natural rubber: An explanation of fatigue crack propagation reinforcement under a positive loading ratio *Mater. Sci. Eng., A* **528** pp 078–1086
[3] Cam J B L, Huneau B and Verron E 2008 Description of fatigue damage in carbon black filled

natural rubber *Fatigue Fract. Eng. Mater. Struct* **31** pp 1031–1038
[4] Cam J B L, Huneau B and Verron E 2013 Fatigue damage in carbon black filled natural rubber under uni- and multiaxial loading conditions *J. Fatigue* **52** pp 82–94
[5] Kim J H and Jeong H Y 2005 A study on the material properties and fatigue life of natural rubber with different carbon blacks *J. Fatigue 27* pp 263–272
[6] Mars W V and Fatemi A 2004 Observations of the constitutive response and characterization of filled natural rubber under monotonic and cyclic multiaxial stress states *ASME J. Eng. Mater. Technol.* **126** pp 19–28
[7] Mars W V and Fatemi A 2005 Multiaxial fatigue of rubberpart II: experimental observations and life predictions *Fatigue Fract. Eng. Mater. Struct* **28** pp 523–538
[8] Gent A N and Hindi M 1989 Effect of oxygen on the tear strength of elastomers *Rubber Chemistry and Technology*. **63** p 40

IOP Conf. Series: Materials Science and Engineering **389** (2018) 012025 doi:10.1088/1757-899X/389/1/012025

Degassing of Pure Copper Melt by Ultrasonic Cavitation

J W Li [1,*], X Ma [1], L L Yang [1], Y L Teng [1], F Li [2] and L H Meng[3]
[1]College of Materials Science & Engineering, Dalian Jiao Tong University, 794 Yellow River Road, Dalian, Liaoning 116028, CHINA
[2]School of Materials Science and Engineering, Huazhong University of Science and Technology, Wuhan, Hubei 430074, CHINA
[3]Ying Kou Vocational and Technical College, Yingkou, Liaoning, 115000, CHINA

*Corresponding author: joelee0527@163.com

Abstract. The effect of high intensity ultrasonic vibration (UV) on degassing of pure copper has been studied in this paper. The methods of reduced pressure test (RPT) and direct hydrogen and oxygen measurements are used for the evaluation of degassing efficiency on hydrogen and oxygen concentrations. The results showed that high intensity UV had a significant degassing effect for the commercially pure copper (CP-Cu) melt. With UV, the density index Di was reduced from 11.94 to 1.13%, and the product of hydrogen-oxygen concentration ($[\%H]\cdot[\%O]$) in the liquid Cu was decreased to 1.440×10^{-5} from 6.624×10^{-5}. It was also found that electric power and ultrasonic treatment time could influence the degassing efficiency.

1. Introduction

Usually, in copper foundries, several degassing techniques have been developed for the analysis of hydrogen in pure Cu [1]. These techniques can be divided into three categories: Vacuum Degassing is capital intensive, limited by the vacuum pressure that can be created over the melt, and can lead to the loss of desirable elements. In some cases, Argon or Nitrogen is used in a lance application for degassing. The purge gas collects hydrogen, and to a certain extent, precipitated oxide phases from the melt. In the case of Chemical Degassing, some of the reagent metals (Rare earths, Li and Ca) can be expensive, and the compounds of the undesirable elements may remain as inclusions in the metal. Therefore, there exists a need for a commercially viable, energy efficient, and environmentally sound molten metal degassing technique which overcomes the aforementioned disadvantages.

Ultrasonic degassing (USD), an environmentally clean and cheap technique, uses high intensity ultrasonic vibrations (UVs) to generate oscillating pressures in liquid metal, which is a possible way to overtake such drawbacks and to improve hydrogen removal [2]. USD is common in molten Al [3], but this method is not as of yet utilized in pure copper or copper alloy foundries to any great extent.

In this work, an experimental device has been built in Dalian Jiaotong University for the degassing of commercially pure copper (CP-Cu) using UV at a frequency of 20 kHz and vibration intensities up to 3000 W. Parametric studies have been carried out to investigate the efficacy of the USD of molten pure Cu under reduced pressure. This article mainly focuses on the effect of UV on hydrogen and oxygen concentrations and porosity formation for CP-Cu.

2. Experimental

2.1. Raw materials and USD Apparatus

The copper of 500 g was held in a graphite crucible with a 50 mm diameter and 75 mm height, then melted in a standard silicon carbide resistance furnace. Melt temperature was controlled with a PID temperature controller. When the copper were completely melted, it was standed 30 min for homogenization. Afterward, the crucible with the pure Cu melt was took out rapidly by crucible tongs from the furnace. Immediately USD was carried out under the various processing conditions, which included the electric power, degassing time, temperature of melt, and ambient humidity. The electric power was 1000, 2000 and 3000 W. The degassing time was 5, 10 and 15 s. The humidity was varied from 46, 56 to 66%. Four melt temperatures, 1210, 1230, and 1250 °C were tested, respectively.

In addition, USD combined with chemically degassing was also carried out in air. Chemical degassing was conducted by using dry nontoxic degasser tablets (NDT) containing nitrate and active carbon (1.0 wt. % of the molten Cu). For comparison, the single NDT and without USD were first carried out in the Cu melt. The experimental processing is shown in Figure 1.

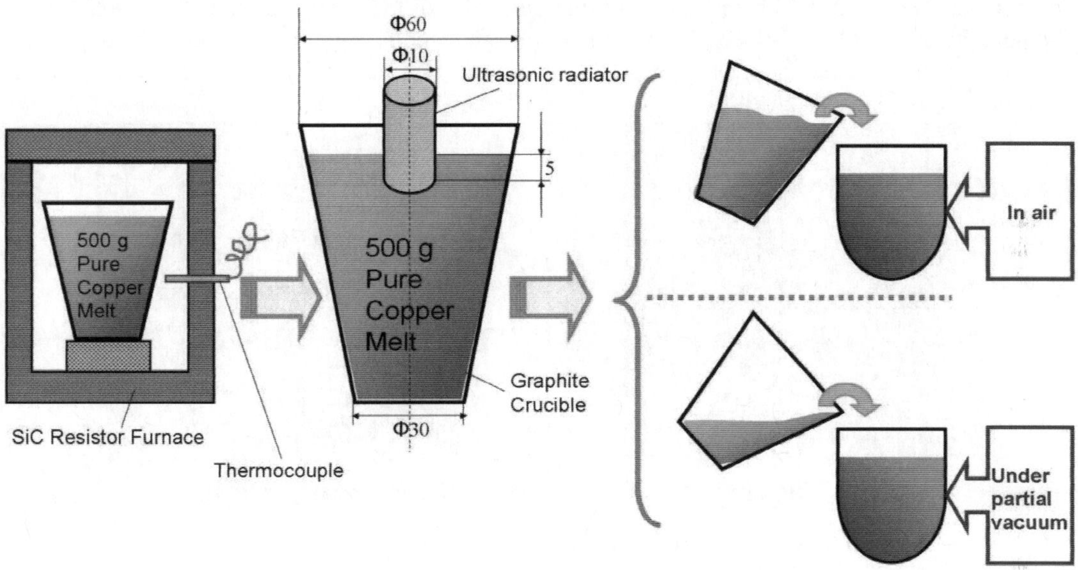

Figure 1. Schematic illustration of experimental processing.

2.2. Reduced pressure test

A Reduced Pressure Test (RPT) [4] was used to evaluate porosity levels in the pure Cu. Molten Cu (~200 g) was poured into a preheated clay-graphite cup and allowed to solidify under a reduced pressure of 73 mbar. At the same time, atmospherically solidified cup sample was also prepared from the same melts for comparison (see Figure 1).

The gases content (mainly H_2 and H_2O [1]) was then assessed by calculating the density index D_i [5] from the measured densities using a MP61001J balance:

$$D_i = \frac{D_a - D_v}{D_a} \quad (1)$$

where, D_i is the density index, D_a is the density of the sample solidified in air and D_v is the density of the sample solidified under partial vacuum. D_i can be used as a measure of the gases content. The smaller the D_i value, the lower the content of hydrogen and oxygen. After measuring the density, the RPT specimens were sectioned in the middle vertically and were polished to reveal the extent of the gas porosity.

2.3. Direct measurement of hydrogen, oxygen concentration

According to the Ostrom's studies [6], the molten copper mainly evolve water vapour with some hydrogen. Therefore, both the elements of hydrogen and oxygen were selected to test. To confirm the RPT results, the hydrogen and oxygen concentrations of samples solidified in air were melted in a high purity graphite crucible (the inner diameter of 12.7 mm and height of 24 mm) and directly measured using the LECO-TCH600 analyzer. The test precision is 0.001 ppm. For comparison, the hydrogen and oxygen concentrations in the non-treated melt were also directly measured at the same conditions.

3. Results

3.1. Samples solidified at atmospheric pressure and partial vacuum without UV

Figure 2 shows the porosity levels in the specimens without melt degassed in air (Figure 2a) and under partial vacuum (Figure 2b). The top surface of the sample solidified in air is a little of convex. There is a flat porosity about 10 mm in length near the top of the sample, and a round pores about 1 mm in diameter inside the sample (seen in Figure 2a). Such porosities are formed due to gaseous evolution during the final stage of solidification.

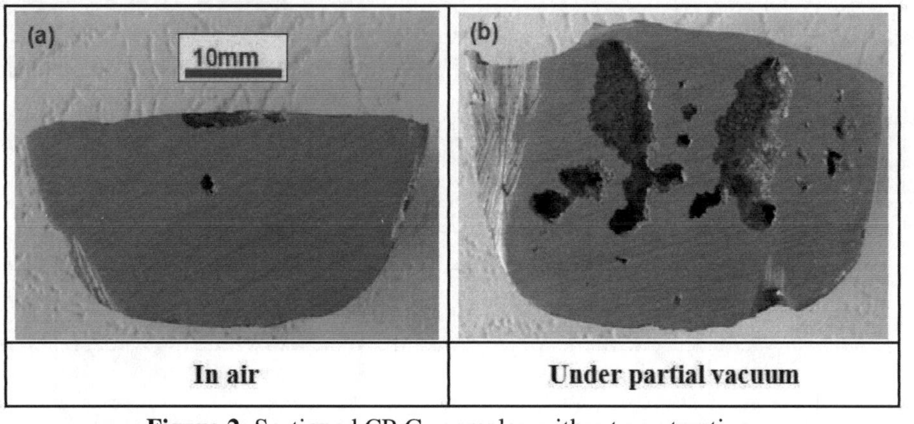

Figure 2. Sectioned CP-Cu samples without any treating.

Under partial vacuum, the gases content is high. The sample has a convex top surface, the morphology of porosities look like a honeycomb with size in order of centimetre, and is full of big and small cavities (Figure 2b). They are formed in liquid state during or shortly after columnar crystals coherency bridging. When shrinkage occurs, there is a negative pressure in the sample, more pores will be formed because of higher gases level, which can offset the solidification shrinkage. Therefore, there is an obvious convex top surface, as shown in Figure 2b. The measured densities D_a and D_v were 8.2953 and 7.3051 g cm^{-3}, respectively. The calculated density index D_i was 11.94%, which implies the gases level is high.

3.2. Degassing by UV

In the same conditions as the non-treated one, the pure Cu melt was treated by UV at 1210 ^{o}C and 3000 W for 15 s. The measured densities D_a and D_v were 8.5849 and 8.4875 g cm^{-3}, respectively; moreover, the calculated density index D_i was 1.13%, which indicates the gases level for the treated Cu melt is low. The comparison of the densities and the density indices between treated and non-treated samples is shown in Figure 3, which shows that the D_i value from 11.94% (without UV) to 1.13% (with UV). The result indicates that UV has a significant degassing effect on the CP-Cu.

Figure 4 shows the vertical cross-section of the treated samples solidified in air (Figure 4a) and under partial vacuum (Figure 4b). For the treated sample solidified in air, due to the lower gases level, it is more difficult to form porosity during solidification, and shrinkage cannot be so easily offset by gases porosity. Therefore, the concave top surface was formed, as shown in Figure 4a. Although USD can degas the pure Cu melt, there is still some hydrogen and oxygen in the molten Cu. Because there is less offsetting by the formation of pores, there will be more shrinkage. Consequently, more shrinkage porosity was formed. For the treated sample solidified under partial vacuum (Figure 4b), the pores are smaller and much less number, the top surface is more convex than the non-treated sample (Figure 2b). This is due to the reduction in hydrogen and oxygen concentrations by USD.

In order to confirm the degassing effect by USD, the hydrogen and oxygen concentrations in the Cu samples solidified in air before and after USD was also directly measured using the LECO analyzer. The results are shown in Figure 5. After USD, the hydrogen and oxygen concentration product $[\%H]\cdot[\%O]$ was reduced from 6.624×10^{-5} to 1.440×10^{-5}, which supports the results of RPT. Due to the difficulties of direct hydrogen and oxygen measurement on a laboratory scale, based on the coincidence of the results between the RPT and the direct hydrogen, oxygen measurement, therefore, only the RPT was used to evaluate the gases level in the melts processed under different conditions.

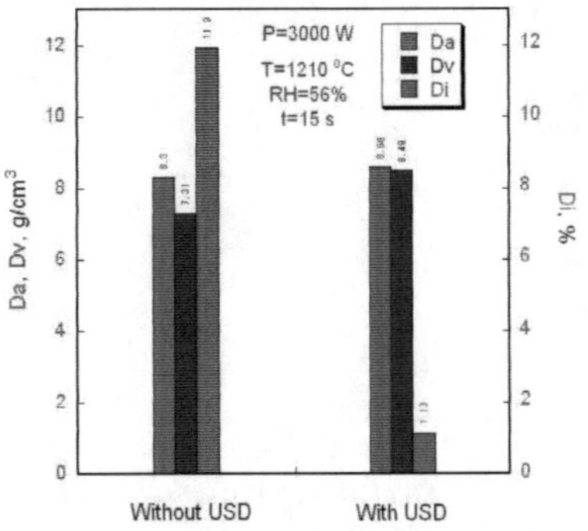

Figure 3. Comparison of densities and density indices of CP-Cu with and without USD.

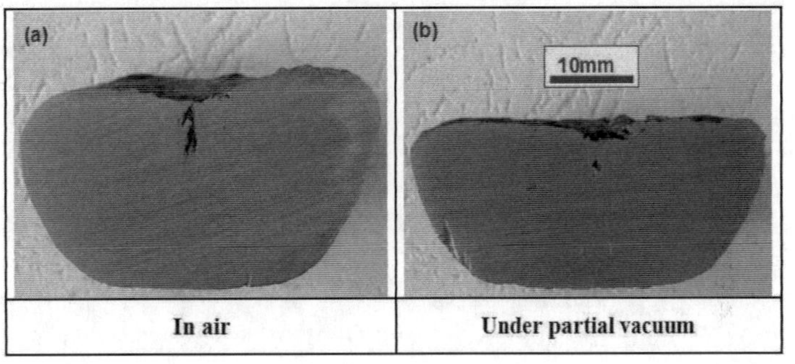

Figure 4. Sectioned CP-Cu samples solidified after UV 15 s.

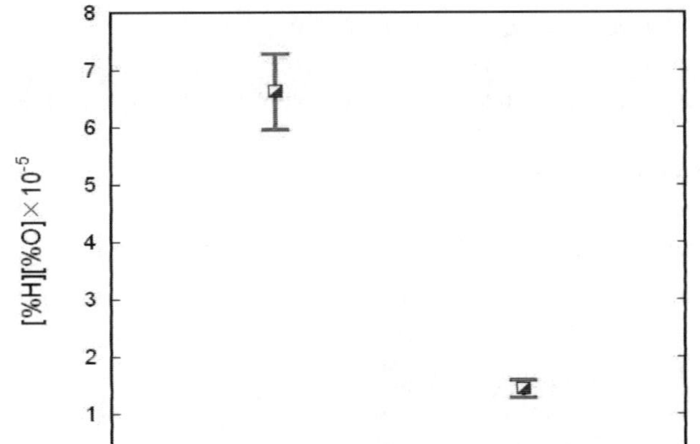

Figure 5. $[\%H]\cdot[\%O]$ product in CP-Cu melt with and without USD.

3.3. Evaluation of USD for processing parameters

It is necessary to study the effect of processing parameters on the degassing efficiency and to optimise the processing conditions. In the present work, ultrasonic electric power and ultrasonic treatment time were studied.

To evaluate the effect of ultrasonic electric power, the molten Cu was treated at 1210 ^{o}C for ultrasonic electric power of 0 (*i.e.* without USD), 1000, 2000 and 3000 W for 15 s. The values for D_a, D_v and D_i are shown in Figure 6. With the increase of electric power, D_i decreases rapidly from 11.94, 8.08, 3.34 and 1.13 %, respectively. It seems that the USD efficiency is electric power dependent, and that this has an evident influence on both the gases level and the Cu ingot density.

The effect of degassing time on degassing effect was investigated by USD the CP-Cu melt at 1210 °C for 5, 10 and 15 s. During degassing the ultrasonic electric power was kept at 3000 W and the ambient humidity was 56%. The results are shown in Figure 7. With increased USD time, D_i does not change much within the first 5 s and then rapidly reaches a minimum level, indicating that USD time has a significant degassing effect on CP-Cu, and the optimal degassing time is 15 s in this work.

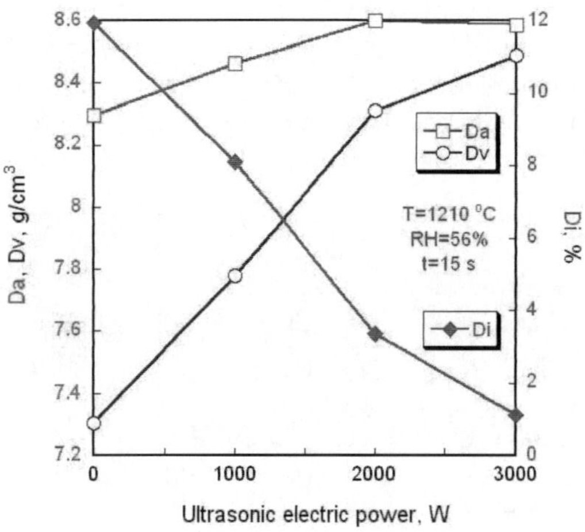

Figure 6. Variation in densities and D_i of CP-Cu as function of ultrasonic power.

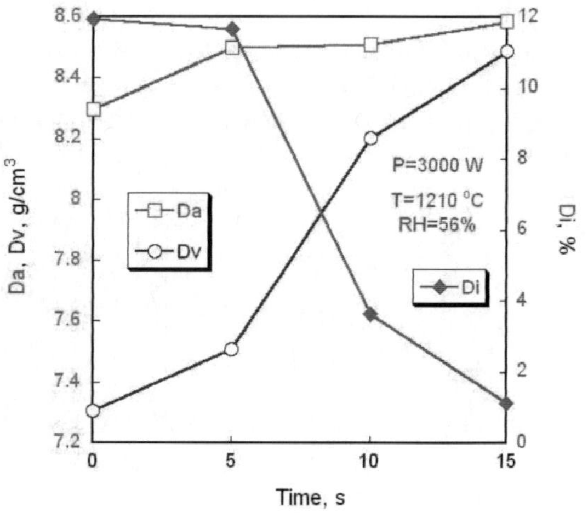

Figure 7. Variation in densities and D_i of CP-Cu as function of ultrasonic time.

4. Discussion

4.1. Degassing and Deoxidation by Ultrasonic Cavitation

Figure 5 illustrates the hydrogen and oxygen concentration in molten Cu by USD above liquidus. It seems, USD not only can degas the hydrogen but also can deoxidize the copper melt. The mechanism of ultrasonic degassing is closely related to the phenomenon of cavitation in the molten Cu. Figure 8 presents the schematic diagram of the degassing and deoxidation mechanism by ultrasonic cavitation. Generally, the mechanism of degassing and deoxidation above liquidus can be divided into four stages:

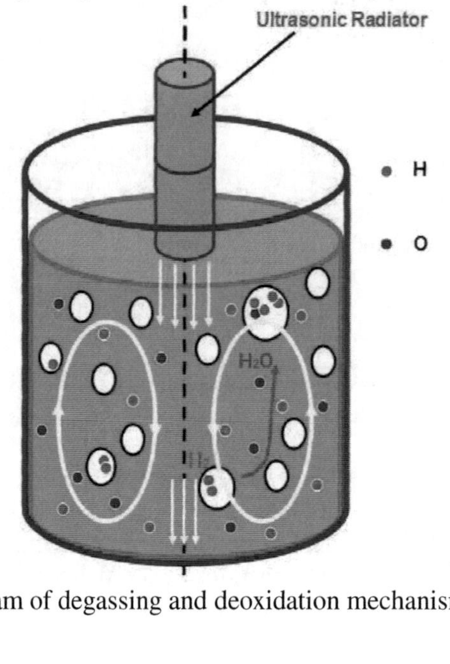

Figure 8. Schematic diagram of degassing and deoxidation mechanism by ultrasonic cavitation.

(1) Nucleation of cavitation bubbles on nuclei (usually nonwettable oxide particles containing cavities) and growth of the bubbles due to the diffusion of hydrogen atoms from the surrounding Cu melt to the bubbles. *i.e.*, for the elementary reaction can be expressed as [7]:

$$2H\ (melt) = H_2\ (g) \tag{2}$$

(2) In the present case of deoxidation of molten Cu, the reaction in equilibrium at the gas/melt interface is represented by [7]:

$$H_2\ (g) + Cu_2O\ (melt) = 2Cu\ (melt) + H_2O\ (g) \tag{3}$$

As a result, the reaction formed vapor, which act as nucleating sites further promote the formation of cavitation bubbles in liquid Cu and create a large number of cavitation bubbles which accumulate hydrogen, because vapor does not dissolve in molten Cu.

(3) Coalescence of bubbles to form large bubbles [3, 8].

(4) Float of large bubbles to the surface of the molten Cu and escape of the bubbles at the top melt surface [3].

5. Conclusions

High-intensity UV was introduced in the molten pure Cu. Both RPT and direct hydrogen and oxygen measurement were carried out to study the effect of UV on hydrogen, oxygen concentration and porosity formation. Based on the experimental results, the following conclusions can be drawn:

(1) USD has a significant degassing effect on molten pure Cu, which can reduce the hydrogen and oxygen concentration product $[\%H]\cdot[\%O]$ in pure Cu from 6.624×10^{-5} to 1.440×10^{-5} and the density index D_i from 11.94 to 1.13%.

(2) With the increase of ultrasonic electric power and ultrasonic treatment time, the density index D_i decreases rapidly and then reaches a minimum value. On the contrary, with the increase of melt temperature and ambient humidity, the density index D_i increases gradually and then reaches to a maximum value.

(3) The USD not only can degas the hydrogen but also can deoxidize the Cu melt.

Acknowledgements

The authors are gratefully acknowledge Dalian Jiaotong University (project no. 054078) and the NSF of Liaoning Province (project no. 20170540140), and to thank Shanghai Research Institute of Materials (SRIM) for providing LECO-TCH600 for directly measuring the hydrogen and oxygen concentrations.

References

[1] Campbell J. *Castings*. 2nd edn. Oxford: UK Butterworth Heinemann; 2003.
[2] Abramov O V. *Ultrasound in Liquid and Solid Metals*, London: CRC, 1994.
[3] Li J W, Momono T. Application of ultrasonic treating to degassing of metal ingots. *Mater Lett* 2008; 62(25): 4152-4.
[4] Zuo Y B, Jiang B, Zhang YJ, Fan Z. Degassing LM25 aluminium alloy by novel degassing technology with intensive melt shearing. *Int J Cast Met Res* 2013; 26: 16-21.
[5] Samuel A M, Samuel F H. The reduced pressure test as a measuring tool in the evaluation of porosity/hydrogen content in Al-7 wt pct Si-10 vol pct SiC(p) metal matrix composite. *Metall Trans A* 1993; 24: 1857-68.
[6] Trojan P K, Flinn R A, Ostrom T R. Melt control variables in copper-base alloys. *AFS Trans* 1982; 90: 729-44.
[7] Yamauchi C, Sano K. Kinetic studies on the deoxidatio of molten copper by hydrogen gas. *J Jpn Inst Metals* 1969; 33: 1249-55. (Japanese)
[8] Xu H, Jian X, Meek T, Han Q. Degassing of molten aluminum A356 alloy using ultrasonic vibration. *Mater Lett* 2004; 58: 3669-73.

IOP Conf. Series: Materials Science and Engineering **389** (2018) 012026 doi:10.1088/1757-899X/389/1/012026

Effect of the Surface Scratch Angle on the Mechanical Characteristics of Nickel-Based Alloy

Cui Yinghao[1], Xue He[1,*], Zhao Lingyan[2], Yang Hongliang[1], Zhang Jianlong[1] and Wang Weibing[1]

[1] School of Mechanical Engineering, Xi'an University of Science and Technology, Xi'an 710054, China
[2] School of Science, Xi'an University of Science and Technology, Xi'an 710054, China

*xue_he@hotmail.com

Abstract. Stress corrosion cracking (SCC) produced by surface scratches has great effect on the safety and reliability of nuclear power plant. To understand the effects of the angle of surface scratches on SCC behavior of nickel-based alloy, the stress and strain field nearby the surface scratches of nickel based alloy is simulated using ABAQUS. Results show that the high stress and strain region increases as the angle increases, which leads to an increase in SCC growth rate. The high stress and strain region offset toward the scratch angle, which indicates that the crack is more likely to crack and extend along the direction of scratch angle.

1. Introduction

Generally, SCC is a failure mechanism that is caused by environment, susceptible material and tensile strain of nuclear power structural materials under high temperature and high pressure aqueous environments [1]. Owing to the oxide film formed on its surface, the nickel base alloy is a strong anti-corrosion material, and is commonly used to fabricate transfer tubes of pressurized water. However, research shows that surface scratch tends to appear during manufacture, transport and installation of steam generator [2]. And many research result also shows that surface scratch is one of important origins of SCC in steam generators, which also seriously threaten the safety and reliability of nuclear power plant [3-4]. Micro-cracks induced by surface scratch will also greatly influence the SCC when the corrosion-enhanced dislocation emission and motion reach a certain condition [5-6]. Mechanical state around the surface scratch is one of the main mechanical factors determining SCC crack growth rate for nuclear pressure vessels and steam generators [7].

However, the effects of the variation of the angle of surface scratch on mechanical field of nickel-based alloy has not been investigated and understood. Based on a finite element method (FEM), the effects of scratch angle on the SCC stress field was also analyzed, which is of great significance for the study of SCC crack initiation and propagation behavior.

2. Finite element modeling

2.1. Specimen geometric model

Figure 1 shows the cross-section morphology of Alloy 690TT under SCC using scratch technique in high temperature oxygenated water at 325 ℃ after constant load experiment [2]. Mechanical cracks is

IOP Conf. Series: Materials Science and Engineering **389** (2018) 012026 doi:10.1088/1757-899X/389/1/012026

produced at the bottom of scratch groove during slow strain rate tensile stage. The oxides were also formed on scratch groove.

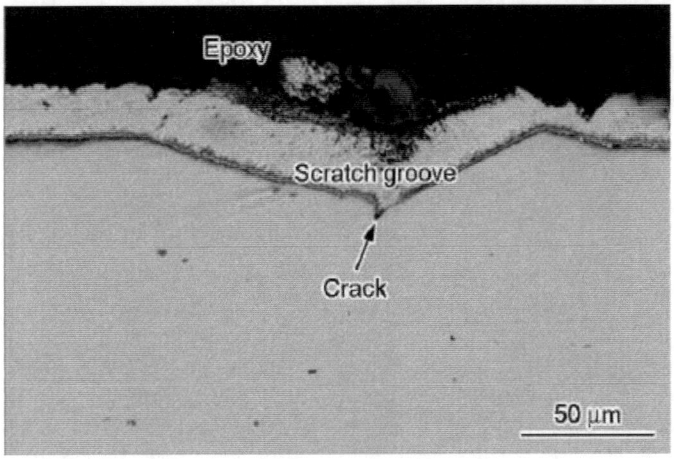

Figure 1. Scratch section of stress corrosion cracking

A finite wide plate, tensile specimen with a single edge crack is selected to simulate this process. The plate width W is 80 mm and length 2L is 160 mm, as shown in Fig.2(a). The α is the surface scratch angle and its values equals to 0°, 15° and 30° respectively in this study, which is shown in Fig.2(b).

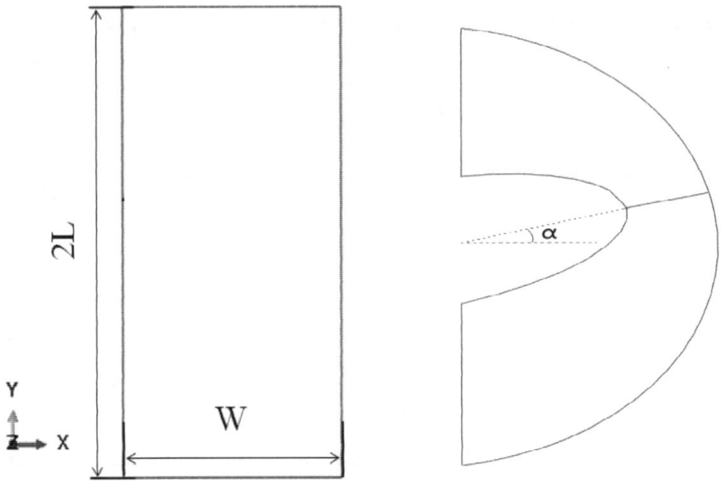

(a) Geometry size of wide plate (b) Morphology of the surface scratch

Figure 2. Geometry size of finite wide plate specimen

The finite element mesh of the finite wide plate global model and the local region around the surface scratch are shown in Figure 3(a) and Figure 3(b), CPE8R is used in simulation and total number is 3819. To obtain a more accurate crack tip stress-strain, the mesh is refined around the surface scratch and its number is 1375.

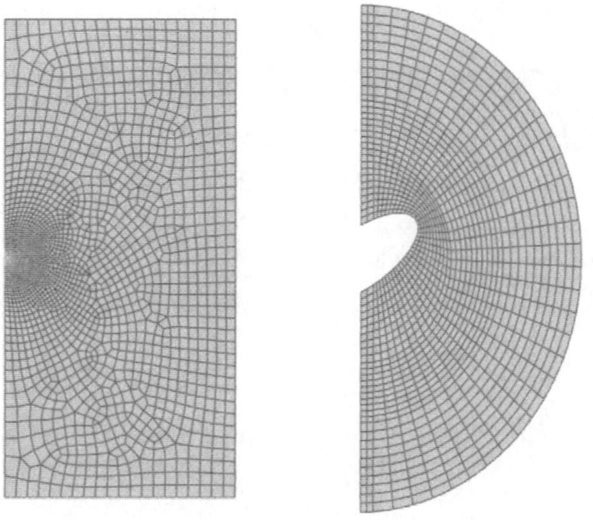

(a) Global mesh model (b) Mesh at surface scratch zone

Figure 3. Finite element mesh model of plate specimen

2.2. Material model

The Ni-based Alloy 600 is a power-hardened material whose constitutive relationship is usually characterized by the Ramberg-Osgood relationship [8], and its form is as follows:

$$\frac{\varepsilon}{\varepsilon_0} = \frac{\sigma}{\sigma_0} + \alpha \left(\frac{\sigma}{\sigma_0} \right)^n \tag{1}$$

Where ε is the strain, including elastic and plastic strain. σ is the total stress; ε_0 is the yield strain of the material, σ_0 is the yield stress of the material, and n is the strain hardening exponent of the material, α is the offset coefficient of the material.

Material mechanical parameters of nickel-based alloy 600 are: Young's modulus E is 190GPa, Poisson's ratio ν is 0.3, yield stress $\sigma_{0.2}$ is 436MPa, strain hardening exponent n is 5.29 and the offset coefficient α is 1 [9]. The main component of oxide film is Cr_2O_3, which is generally simplified as a linear elastic material, and its Young's modulus E is 19GPa, Poisson's ratio ν is 0.3 [10].

2.3. Loading and boundary conditions

In surface scratch period, external load including working load and residual stress has little mechanical effect on the SCC, so the load in this period is main the film induced stress. Because of Aiming at the different coefficients of thermal expansion of the base metal and the oxide film, the growth process of oxide was simulated by expansion. The thermal expansion coefficient β and the temperature difference ΔT was assumed to be 0.001 and 6 ℃, respectively [5-6].

3. Results and Discussion

3.1. Effect of surface scratch angle on Mises stress of nickel-based alloy

With different surface scratch angle under the same film induced stress, the Mises stress distribution in front of the crack front is shown in Figure 4. Figure 4 shows that the region of high Mises stress at the crack front is smallest when the angle α=0° and the high stress region increases as the angle increases. Meanwhile, the Mises stress becomes small and small when the distance ahead of the crack front is larger. It also shows that the high stress region offset toward the direction of the scratch angle, which indicates that the crack is more likely to crack and extend along the direction of surface scratch.

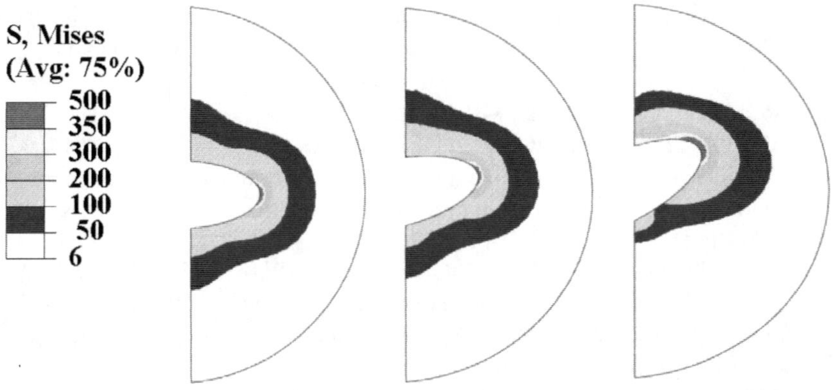

(a)α=0° (b)α=15° (c)α=30°

Figure 4. Distribution of Mises stress under different angle of surface scratch

3.2. Effect of surface scratch angle on PEEQ of nickel-based alloy

With different surface scratch angle under the same film induced stress, the PEEQ distribution in front of the crack front is shown in Figure 5. Figure 5 shows that the PEEQ at the crack front also increases as the angle increases. The larger the distance ahead of the crack front, the smaller the PEEQ is. It also shows that the high strain region offset along the scratch angle, which indicates that the crack is more likely to crack and extend along the direction of scratch angle.

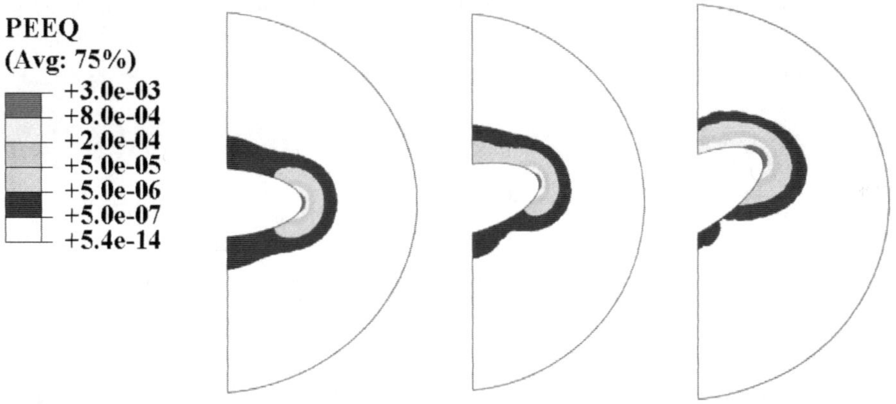

(a)α=0° (b)α=15° (c)α=30°

Figure 5. Distribution of Mises stress under different angle of surface scratch

4. Conclusions

(1) The high Mises stress region of nickel-based alloy increases as the surface scratch angle increases, and the larger the distance ahead of the crack front, the smaller the Mises stress is. The plastic strain has the same rules.

(2) The high stress and strain region offset toward the surface scratch angle, which indicates that the crack is more likely to crack and extend along the direction of scratch angle.

Acknowledgments

This work is financially supported by the National Natural Science Foundation of China (Grant nos. are 51475362, 11502195 and 2018QDJ009).

References

[1] Andresen P L, Karen G, Lawrence N J. Stress Corrosion Cracking of Sensitized Type 304 Stainless Steel in 288C Water: A Five Laboratory Round Robin[M]. John Wiley & Sons, Inc. 2013.

[2] Meng F J, Wang J Q, Han E H et al. MICROSTRUCTURE NEAR SCRATCH ON ALLOY 690TT AND STRESS CORROSION INDUCED BY SCRATCHING [J], 2011, 47(7): 83

[3] Staehle R W, Gorman J A. Quantitative Assessment of Submodes of Stress Corrosion Cracking on the Secondary Side of Steam Generator Tubing in Pressurized Water Reactors: Part 3[J]. Corrosion -Houston Tx-, 2012, 60(2):115-180.

[4] Ding X S. IGA/IGSCC AND PROTECTION ON SECONDARY SIDE OF STEAM GENERATOR TUBGES IN NUCLEAR POWER PLANTS[J], 2002, 23(10): 441

[5] Li J X, Wang Y B, Qiao L J et al. Film-Stress induced stress corrosion cracking of austenitic stainless steel[J], 2002, 38(8): 861 (in Chinese)

[6] Li J X, Gao K W, Su Y J et al. Study on Stress Corrosion of Stainless Steel Caused by Passivation Film Stress [J], 2003, 33(9): 796

[7] He X, Ogawa K, Shoji T. Effect of welded mechanical heterogeneity on local stress and strain ahead of stationary and growing crack tips[J]. Nuclear Engineering & Design, 2009, 239(4):628-640.

[8] Ueda Y, Shi Y, Sun S. Effect of crack depth and strength mis-matching on the relation between J-integral and CTOD for welded tensile specimens(mechanics, strength & structure design) [J]. Transaction of JWRI, 1997, 26(1): 133-140.

[9] He X, Li Y. Micro-mechanical State at Tip of Environmentally Assisted Cracking in Nickel-based Alloy[J]. Rare Metal Materials & Engineering, 2016, 45(3):537-541.

[10] Xue H, Xue X, Tang W. Analysis on Mechanical Property of Oxide Film in Stress Corrosion Cracking Tip of Nickel-based Alloys[J]. Rare Metal Materials & Engineering, 2011, 40(7):1188-1191.

IOP Conf. Series: Materials Science and Engineering **389** (2018) 012027 doi:10.1088/1757-899X/389/1/012027

Three-Dimensional Analysis of the Effect of Out-Of-Plane Constraint on the Local Plastic Strain around Crack Front

Zhao Lingyan[1, *], Cui Yinghao[2] and Yang Fuqiang[1]

[1]School of Science, Xi'an University of Science and Technology, Xi'an 710054, China

[2]School of Mechanical Engineering, Xi'an University of Science and Technology, Xi'an 710054, China

*gloomy2@foxmail.com

Abstract. In the last few decades, constraint effects and plastic strain ahead of crack front have played important roles in stress corrosion cracking (SCC) behaviour of reactor pressure vessels (RPV) and pipes. At the same average J-integral, the local three-dimensional stress-strain fields around crack fronts were simulated, the effects of out-of-plane constraint (specimen thickness) on plastic strain, plastic zone and SCC driving force along crack fronts were analyzed. The results show that the strains of thinner specimens near the free surface are more sensitive to out-of-plane constraint effect. The strain gradients of thinner specimens are more sensitive to out-of-plane constraint, especially near the free surface. The results may be used in incorporating out-of-plane constraint effects in structures for improving structural integrity assessments and analyses of RPVs and pipes.

1. Introduction

Both fracture behaviour and constraint effects play important roles in structural integrity assessment and design, especially for integrity analyses of RPVs and pipes. As the resistance of structures against the crack-tip plastic deformation, constraints have attracted much attention in the research of material fracture behaviour. Constraints can be divided into in-plane constraint and out-of-plane constraint. The in-plane constraint, such as the crack length, relates to the specimen dimension in the direction of growing crack. While, the out-of-plane constraint, such as the specimen thickness, relates to the specimen dimension parallel to crack front. In the last few decades, many constraint parameters has been investigated to establish a unified correlation of in-plane and out-of-plane constraints with fracture toughness for improving structural integrity accuracy [1-3]. However, a great deal of evidence and experiments indicated that the material plasticity and residual plastic strain ahead of crack front also have great effect SCC behaviours of RPVs and pipes [4,5], which has been paid more attention in the nuclear power industry. In order to increase the accuracy of structural integrity assessment, the constraint effects on the plastic strain around crack fronts and the crack driving force must be considered.

At the same average J-integral, the local stress-strain fields ahead of crack fronts were calculated by using elastic-plastic finite element method (EPFEM). Moreover, the effect of out-of-plane constraint on plastic strain, plastic zone and SCC driving force along the specimen crack fronts were analyzed. The SCC crack growth rates of C(T) specimen with different thickness were also discussed in this paper.

2. Finite element modelling

2.1. Specimen model

A compact tension (CT) specimen with a constant load was used to measure SCC growth rate in high temperature water environment [6]. Five types of C(T) specimens (specimen width W=50 mm) with thickness-to-width ratio B/W=0.0625, 0.125, 0.25, 0.5 and 1 were chosen, and their relative crack length a/W=0.5 after pre-cracking. The geometry and size configurations of a C(T)50 specimen is shown in figure 1. It can be considered that the out-of-plane constraints were changed by the changing specimen thickness B, and the specimen sizes are shown in table 1.

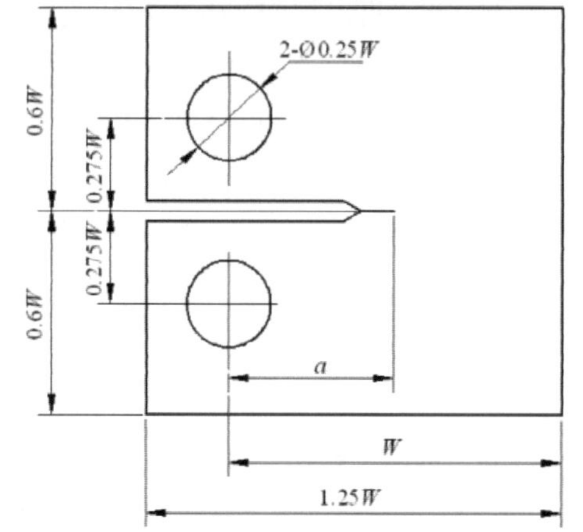

Figure1. Geometry of C(T)50 specimen (W=50 mm)

The calculated results of fracture toughness are also listed in table 1. The stress-strain field distributions along crack fronts were calculated at the same average J-integral value of J_{ave}=20 kJ/m^2 along the specimen crack fronts with different out-of-plane constraints, and the maximum values of J-internal J_{max} at the crack fronts mid-point were also listed. The values of fracture toughness K_{max} in table 1 were calculated from J_{max} by equation 1.

$$K_{max} = \sqrt{J_{max} \frac{E}{1-v^2}} \tag{1}$$

Where, E is the Young's modulus and v is the Poisson's ratio.

Table 1. C(T) specimen sizes and the calculated results of fracture toughness

Specimen	a/W	B (mm)	B/W	P (kN)	K_{max} (MPa·m$^{1/2}$)	J_{max} (kJ·m^{-2})
C(T)50	0.5	3.125	0.0625	1.83	74.78	23.67
		6.25	0.125	3.7	74.56	23.53
		12.5	0.25	7.57	74.27	23.35
		25	0.5	15.6	74.05	23.21
		50	1	32.4	74.02	23.19

2.2. Material model

The pressure vessel steel 22NiMoCr3-7 is investigated in this paper, and its non-linear relationship between stress and strain beyond yield can be described by Ramberg-Osgood equation [7]:

$$\frac{\varepsilon}{\varepsilon_0} = \frac{\sigma}{\sigma_0} + \alpha\left(\frac{\sigma}{\sigma_0}\right)^n \tag{2}$$

Where ε is the strain, including elastic and plastic strain. σ is the total stress; ε_0 is the yield strain of the material, σ_0 is the yield stress of the material, and n is the strain hardening exponent of the material, α is the material offset coefficient.

The mechanical properties data and the Ramberg-Osgood constitutive parameters of 22NiMoCr3-7 steel at -60℃ are given in table 2.

Table 2. Mechanical properties of 22NiMoCr3-7 steel at -60℃[8]

Material	Young's modulus E (MPa)	Poison ratio ν	Yield stress σ_0 (MPa)	Hardening exponent n	Hardening parameter α
22NiMoCr3-7	215 000	0.3	517	3.7	6.2

2.3. FE Model

Five 3D finite element models in table 1 were built by the commercial finite element code ABAQUS [9]. The eight-node isoperimetric elements with reduced integration (C3D8R) were used for all models. To reduce the computation time, considering the loading and geometry symmetry, only 1/4th of the specimen was modeled. The symmetry boundary conditions were applied on the un-cracked ligaments. The finite element model meshes of C(T)50 specimen with B=25 mm and a/W=0.5 are shown in figure 2.

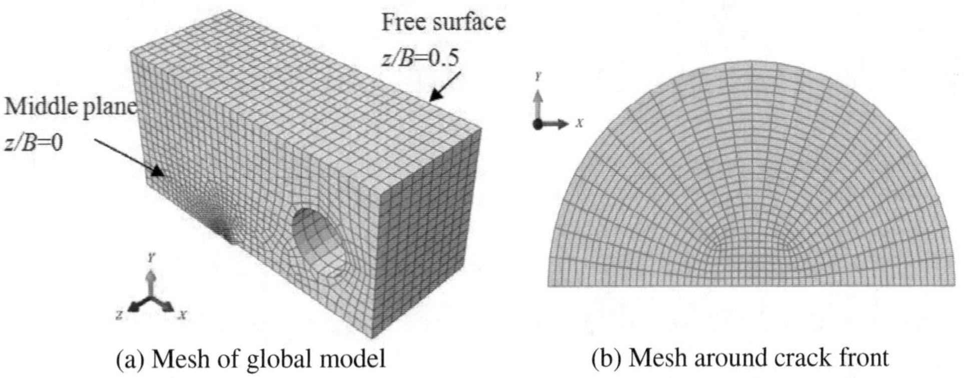

Figure 2. The global model mesh and mesh around crack front of C(T)50 specimen with B=25 mm and a/W=0.5

3. Results and Discussion

Mechanical affecting factors of crack driving force, such as plastic zone, plastic strain and plastic strain rate ahead of crack tips are crucial to quantitatively predict SCC growth rate. Therefore, to analyze the effect of out-of-plane constraint on the plastic strain along crack front, the 5 specimens with different out-of-plane constraints in table 1 were modelled and the local stress-strain fields around crack fronts with different specimen thickness were calculated by FEM. The loading condition is at the same average J-integral value of J_{ave}=20 kJ/m^2 along crack fronts (specimen thickness) for the C(T)50 specimens with different out-of-plane constraint in table 1.

3.1. Local strain zones around crack fronts

The equivalent plastic strain distributions around crack fronts are shown in figure 3, the radius of the researched zone is 0.5 mm. From figure 3(a) to (e), the high strain zones have similar distributions and the low strain zones decrease with the increasing specimen thickness B, which implies that the strains of the low strain zones are more sensitive to the out-of-plane constraints in the centre region.

IOP Conf. Series: Materials Science and Engineering **389** (2018) 012027 doi:10.1088/1757-899X/389/1/012027

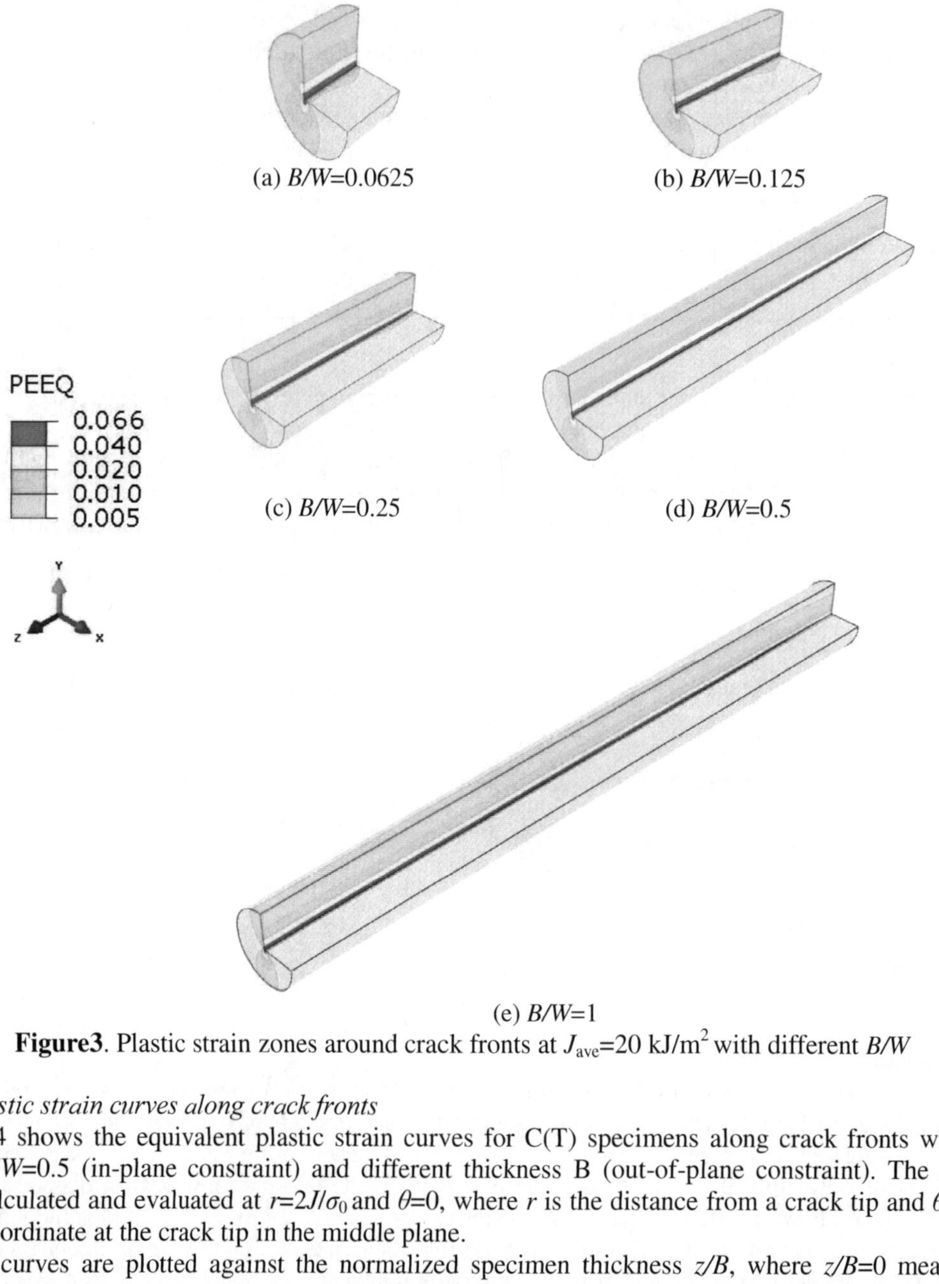

Figure3. Plastic strain zones around crack fronts at J_{ave}=20 kJ/m^2 with different *B/W*

3.2. Plastic strain curves along crack fronts

Figure 4 shows the equivalent plastic strain curves for C(T) specimens along crack fronts with the same *a*/*W*=0.5 (in-plane constraint) and different thickness B (out-of-plane constraint). The strains were calculated and evaluated at $r=2J/\sigma_0$ and $\theta=0$, where *r* is the distance from a crack tip and θ is the polar coordinate at the crack tip in the middle plane.

The curves are plotted against the normalized specimen thickness *z/B*, where *z/B*=0 means the middle plane and *z/B*=0.5 means the free surface of the specimens. It can be seen from figure 4 that the centre region (from *z/B*=0 to 0.4) of all specimens has lower strain, and the surface region (near *z/B* = 0.5) has higher strain. The stains along crack fronts of thicker specimens in the centre region (from *z/B*=0 to 0.3) have small variation while the stains of thinner specimens near the free surface (from *z/B*=0.3 to 0.5) have great variation with the increasing thickness B. This indicates that the strains of thinner specimens near the free surface are more sensitive to out-of-plane constraint effect induced by thickness B.

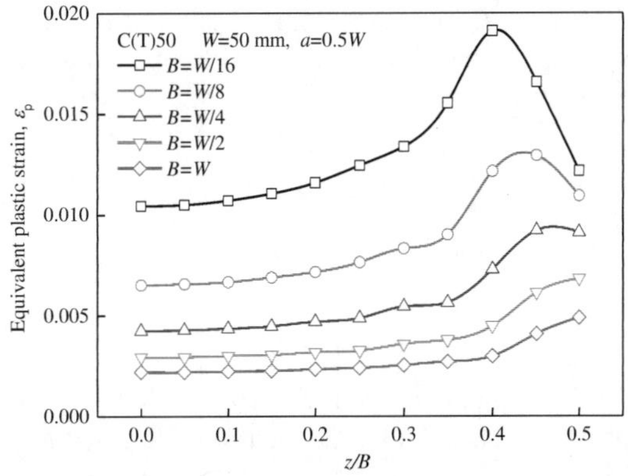

Figure 4. Equivalent plastic strain along crack fronts at J_{ave}=20 kJ/m^2

3.3. Plastic strain rate curves along crack fronts

In the film slip/dissolution oxidation model and FRI model of estimating SCC growth rate, the equivalent plastic strain rate can be used as the basic factor to quantitatively estimate SCC growth at flaws of actual LWR components in a specific high temperature oxygenated aqueous system [10]. Combined with the elastic-plastic FEM, the equivalent plastic strain gradient of a regular electrochemical environment where the crack fronts exposed can be calculated and the result curves are shown in figure 5.

The strain gradients along crack fronts in the centre region (from *z/B*=0 to 0.3) have small variation with the increasing thickness B, while the strain gradients of thinner specimens near the free surface (from *z/B*=0.3 to 0.5) are much higher than those of thicker specimens. It indicates that the SCC driving force decreases with the increasing thickness B, and the thinner specimens have much higher SCC crack growth rates than the thicker specimens at the same average J-integral. Figure 5 shows that the strain gradients of thinner specimens are more sensitive to out-of-plane constraint, especially near the free surface.

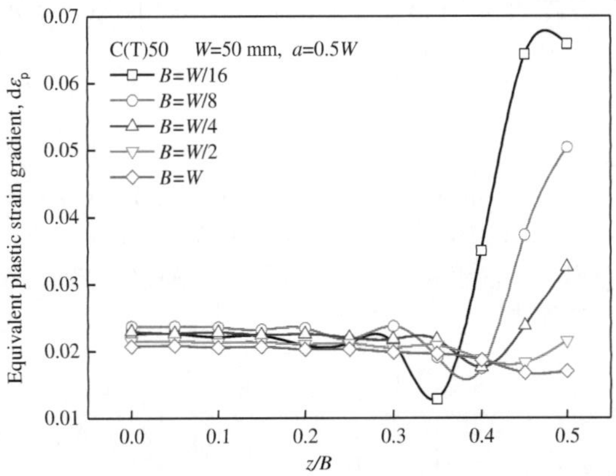

Figure 5. Equivalent plastic strain gradient along crack fronts at J_{ave}=20 kJ/m^2

4. Conclusion

Three-dimensional finite numerical simulations are carried out to analyze the stress-strain fields and plastic strain gradients along crack fronts with different out-of-plane constraints. The effect of out-of-plane constraint on crack driving force is also considered. Conclusions are as follows:

(1) The stains along crack fronts of thicker specimens in the centre region (from z/B=0 to 0.3) have small variation while the stains of thinner specimens near the free surface (from z/B=0.3 to 0.5) have great variation with the increasing thickness B. The strains of thinner specimens near the free surface are more sensitive to out-of-plane constraint effect induced by thickness B.

(2) SCC driving force decreases with the increasing thickness B, and the thinner specimens have much higher SCC crack growth rates than the thicker specimens at the same average J-integral. The strain gradients of thinner specimens are more sensitive to out-of-plane constraint, especially near the free surface.

Acknowledgements
This work is financially supported by the National Natural Science Foundation of China (Grant nos. are 11502195 and 51475362).

References

[1] Mu M Y, Wang G Z, Tu S T and XUAN F Z. (2016) Three-dimensional analyses of in-plane and out-of-plane crack-tip constraint characterization for fracture specimens. *Fatigue. Fract. Eng. M.*, 39, pp1461-1476
[2] Yang J, Wang G Z, Xuan F Z and Tu S T. (2014) A unified correlation of in-plane and out-of-plane constraint with fracture resistance of a dissimilar metal welded joint. *Eng. Fract. Mech.*, 115, pp296-307
[3] Mu M Y, Wang G Z, Xuan F Z and Tu S T. (2015) Unified correlation of in-plane and out-of-plane constraints with cleavage fracture toughness. *Theor. Appl. Fract. Mec.*, 80, pp121-132
[4] Li G F and Congleton J. (2000) Stress corrosion cracking of a low alloy steel to stainless steel transition weld in PWR primary waters at 292℃. *Corros. Sci.*, 42, pp1005-1021
[5] Li G F, Li G J, Fang K and Peng J. (2011) Stress corrosion cracking behavior of dissimilar metal weld A508/52M/316L in high temperature water environment. *Acta. Metall. Sin.*, 47, pp797-803
[6] Kim Y J, Kim J S, Cho S M and Kim Y J. (2004) 3-D constraint effects on J, testing and crack tip constraint in M(T), SE(B), SE(T) and C(T) specimens: numerical study. *Eng. Fract. Mech.*, 71, pp1203-1218
[7] Zhao L Y, Xue H, Yang F Q, Suo Y H. (2014) Numerical investigation on stress corrosion cracking behavior of dissimilar weld joints in pressurized water reactor plants. *Frat. Integrita Strutr.*, 29, pp410-418
[8] Hebel, J., Hohe, J., Friedmann, V. and Siegele, D. (2007) Experimental and numerical analysis of in-plane and out-of-plane crack tip constraint characterization by secondary fracture parameters. *Int. J. Fract.*, 146, pp173-188
[9] ABAQUS. (2015) ABAQUS/Standard User's Manual Version 6.14. *ABAQUS Inc*
[10] Xue H and Shoji T. (2007) Quantitative Prediction of EAC Crack Growth Rate of Sensitized Type 304 Stainless Steel in Boiling Water Reactor Environments Based on EPFEM. *J. Press. Vess.-T. ASME*, 129, pp460-467

IOP Conf. Series: Materials Science and Engineering **389** (2018) 012028 doi:10.1088/1757-899X/389/1/012028

Study on the Influence of Hemispherical Dome Thickness on Aerodynamic Thermal Radiation

Yin Hang[1,] *, Yang Hongyu[2], Ju lin[3] and Zhang yu[4]

[1.]School of mechanical and electrical engineering, Northeast Forestry University, Harbin
[2.]Luoyang electro-optical equipment research institute, Henan, Luoyang
[3.] Research center of space optical engineering, Harbin Institute of Technology, Harbin
[4] Internal and external decoration research institute, Great Wall motor co. LTD., Bao ding

*784441228@qq.com

Abstract. Aspheric optical lens is a key optical component used in infrared optical system for transmission, refraction and incident light. The application prospect in the field of modern national defense science and technology and civil industrial products is promising. At the same time, a large number of aspheric infrared optical lenses are used. The processing precision and machining materials of aspheric infrared optical lens are proposed. The surface quality and precision are higher and higher. To ensure the accuracy of the surface profile of the aspheric infrared optical lens. In the grinding process, the grinding wheel center should be aligned as far as possible with the axisymmetric turning center of the rotary surface part. There are many cutting edges on the surface of the grinding wheel in the grinding process. It is often difficult to determine the precise position of the grinding point for a knife. The inaccuracy of the knife will result in the error of the knife. It is a difficult problem to make the precision of the knife in the aspheric grinding process. To improve the accuracy of the knife, in this paper, the influence of X-axis on the surface profile accuracy of aspheric surface infrared optical lens is analysed. This paper proposes a method to compensate the error of the knife caused by the axis of the X-axis. The accuracy of surface profile of aspheric surface infrared optical lens is improved. The results show that the error of X-axis is corrected by polka dot method. The surface profile accuracy of the aspheric surface infrared optical lens was 382nm before the adjustment. The surface profile accuracy of the aspheric surface infrared optical lens is 145nm after the adjustment of the knife. It has obviously improved.

1. Introduction

In recent years, the demand for aspheric infrared optical lenses is increasing. At the same time, the processing precision and machining materials of aspheric infrared optical lens are presented with new requirements. In this context, this paper starts with the application of industrial application. The monocrystalline silicon with a diameter of 20mm and a thickness of 10mm was studied. The aspheric surface infrared optical lens is experimentally studied with grinding technology. To ensure the accuracy of the surface profile of the aspheric infrared optical lens. In the grinding process, the grinding wheel center should be aligned as far as possible with the axisymmetric turning center of the rotary surface part. In this paper, the error analysis of the knife during the grinding process of aspheric surface infrared optical lens is analyzed and the compensation is made to improve the grinding process

IOP Conf. Series: Materials Science and Engineering **389** (2018) 012028 doi:10.1088/1757-899X/389/1/012028

[1]. At present, the knife method is mainly used for the acoustic emission method, the main shaft power method, the test grinding method and the trigger method. There are many cutting edges on the surface of grinding wheel during grinding. It is often difficult to determine the precise position of the grinding point for a knife. The inaccuracy of the knife will result in the error of the knife, how to be more accurate in the grinding process of the knife is a difficult problem.

Using the vertical grinding method, the aspheric infrared optical lens is grinded. In the case of a knife error. The axis of the circular arc of the end section of the grinding wheel is combined with the axis z axis of the aspheric surface. The circular arc formed by the grinding point of grinding wheel is tangent to the aspheric curve. Because of the existence of the knife error. In the xz plane, the center axis of the grinding wheel often deviates from the axis of symmetry of the aspheric surface. That's the error in the x direction. Deviation distance for Δx as shown in figure 1.

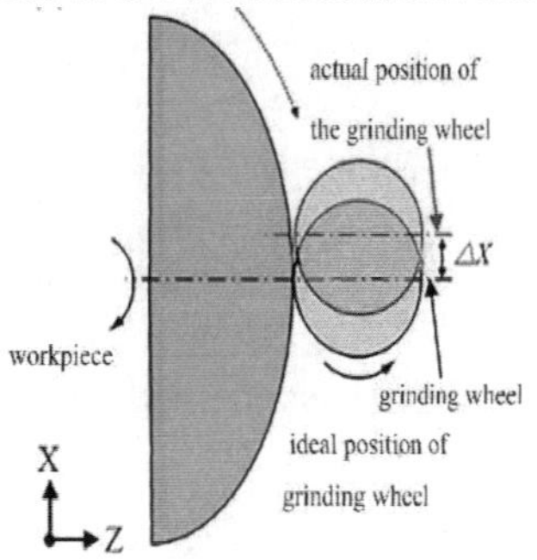

Figure 1. Grinding wheel to knife error

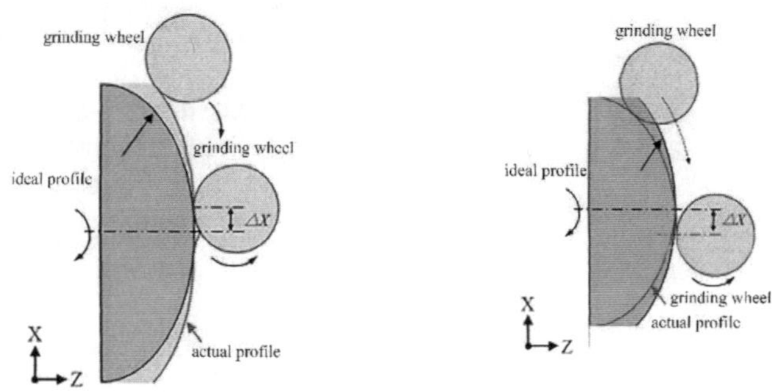

(a) (b)

Figure 2. Influence of x direction on the precision of machining shape

2. The X-axis`s mathematical model of knife error establish in grinding process

Set of cutter error value of the x axis Δ x, when Δ x > 0 for grinding wheel movement had a lens center axis; Δ x < 0 for grinding wheel movement through the lens center axis. The contour trajectory of the aspheric infrared optical lens in the xz plane is designed to be a circle of radius R. The diameter of the aspheric infrared optical lens is D and radius of the grinding wheel is Rs. The curve of contour error is the error curve obtained by subtracting the actual aspheric curve from the ideal aspheric curve. Take the convex lens as an example, the ideal track of the aspheric infrared optical lens, the grinding wheel is grinding into the center by the right end of the lens. When there is no X-axis to the knife error, the center axis of the grinding wheel coincides with the axis z of the lens center. The ideal contour trajectory of the xz plane can be expressed as:

$$y=\sqrt{R^2-x^2} \quad (-\frac{1}{2}D<x<\frac{1}{2}D) \tag{1}$$

In the processing of aspheric infrared optical lens, If the x axis to the error Δ x knife, when Δ x > 0, called a centerline of cutter error, Distances from the center of the lens in Δ x axis location stop grinding wheel. Aspheric center area near Δ x radius is not normal grinding. However, the grinding wheel has a radius Rs, and the cutting depth is also produced in the area that is not normally grinded, and the cutting depth is related to the grinding wheel radius Rs. The grinding trajectory of the edge of the lens and the center axis of the lens will be deviated from the ideal trajectory. The track and depth of the grinding wheel will change. Grinding wheel of cutter offset Δ x distance, namely the actual grinding trajectory arc corresponds to circle the center offset Δ x distance, lens diameter D is the same, actual grinding trajectories have changed in whole. The influence of the knife error on the surface profile accuracy of the lens is shown in figure 2 (a). When Δ x < 0 grinding wheel movement is called a lens center shaft, actual infrared optical aspheric lens grinding path corresponds to circle the center a central axis z axis Δ x distance. Grinding wheel movement has been near axis cause lens center axis Δ

IOP Conf. Series: Materials Science and Engineering **389** (2018) 012028 doi:10.1088/1757-899X/389/1/012028

radius are two grinding x scope. When the grinding track of the lens changes, the grinding depth will increase. The influence of the center axis on the precision of the surface profile is shown in figure 2 (b).

2.1. The X-axis does not cross the central axis of the lens mathematical model of error curve is established

Due to the symmetry of the aspheric infrared optical lens, the mathematical model of the actual trajectory of the grinding process can be programmed only on the x axis. The X-axis is a negative semi-axis grinding track, and the axisymmetric curve can be drawn with x=0 as the axis of symmetry. The actual grinding trajectory of x axis is divided into two parts. Curve of the highest point x = 0 to x = x Δ part and X = Δ x to lens edges, grinding the actual track y is：

$$\begin{cases} y = \sqrt{R^2 - (x - \Delta x)^2} \\ y = \sqrt{R_s^{\,2} - (x - \Delta x)^2} \end{cases} \qquad \begin{cases} (0 < x < \Delta x) \\ (x > \Delta x) \end{cases} \tag{2}$$

When D=2R, the grinding track image is shown in figure 3.

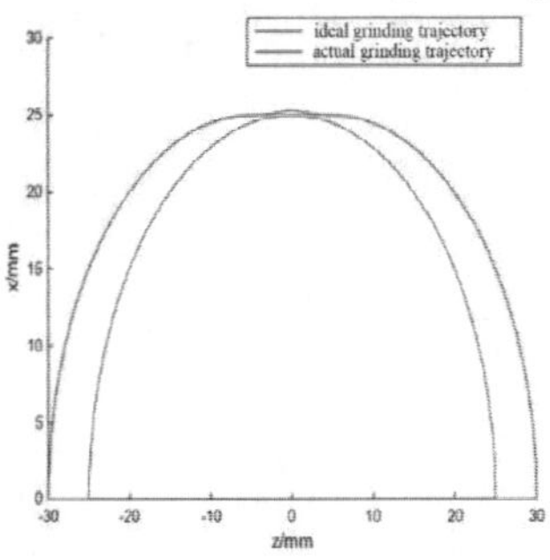

Figure 3. Grinding track of the grinding wheel without the central axis of the lens

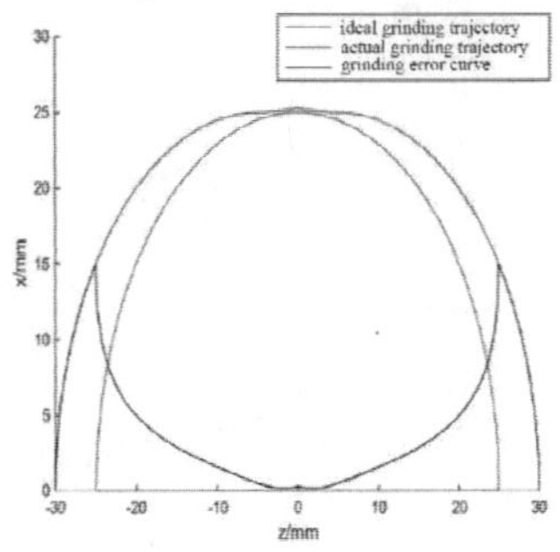

Figure 4. Grinding error curve of grinding wheel without the center axis of the lens

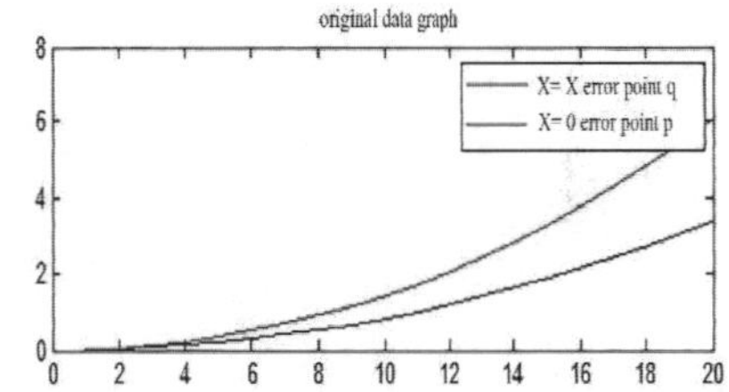

Figure 5. p and q change withΔ x curve at two points

The curve of contour error is the error curve obtained by subtracting the actual aspheric curve from the ideal aspheric curve. The error curve was obtained by using MATLAB software, as shown in figure 4.

When -30<x<-25, the error value increases sharply from 0 to the maximum of the error. The effective grinding range of workpiece selection cannot be too large. It must avoid the area of maximum error. The smaller the effective part, the smaller the error. Because of the actual machining, the X-axis has very little error. The scope of the sharp increase in error is small and negligible.

When the x axis of cutter error Δ x > 0, grinding wheel movement has not passed the central axis of the lens. With the change of the error Δ x corresponding value against the knife, the actual grinding trajectory and error curve will change accordingly. In order to find out the variation of the error curve in the grinding process and the influence of the knife error on the surface profile accuracy of the aspheric optical lens. The two special points on the error curve are observation points as shown in figure 5. X=0 error curve corresponding points p and x=Δ x error curve corresponding points q. Using MATLAB to calculate different error against the knife Δ x, the value of p, q two corresponding, observe change as Δ x p, q value of the two axes in the changes, can be reflected in the workpiece diameter D phase. The Δ x knife error increase, the greater the value of p and q, the surface profile accuracy of workpiece is lower. The smaller the p and q, the higher the surface profile accuracy. The accuracy of the surface profile of the aspheric surface infrared optical lens can be improved by reducing the X-axis

2.2. The X-axis cross the central axis of the lens mathematical model of error curve is established

The motion trajectories of aspheric infrared optical lens and grinding method are symmetrical. Therefore, in establishing the mathematical model, we also need to analyze the x positive semi-axis trajectory, and the X-axis is the symmetric curve with the z-axis as the axis of symmetry. When the knife error is over the central axis, the actual trajectory equation of grinding is y. When D=2R, the trajectory diagram is shown in figure 6.

$$y=\sqrt{R^2-(x+\Delta x)^2} \qquad (x>0) \tag{3}$$

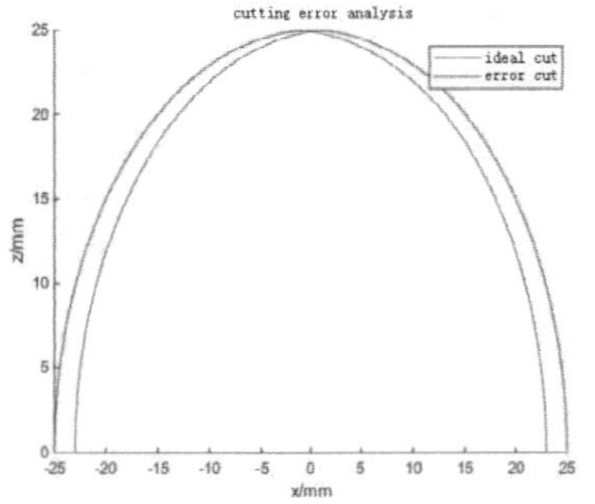

Figure 6. Actual grinding trajectory of the grinding wheel through the central axis

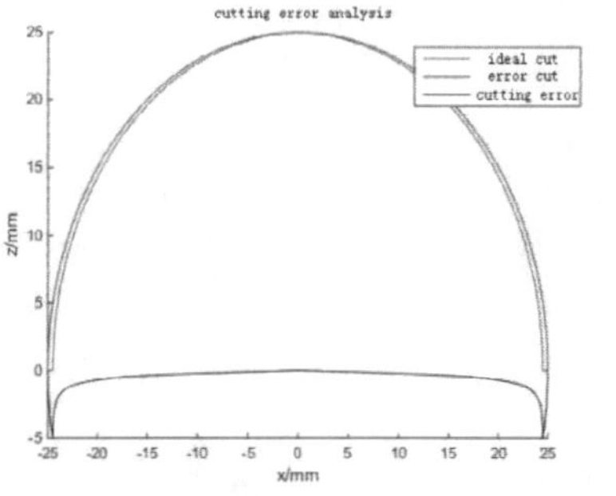

Figure 7. Grinding error curve of grinding wheel over lens center

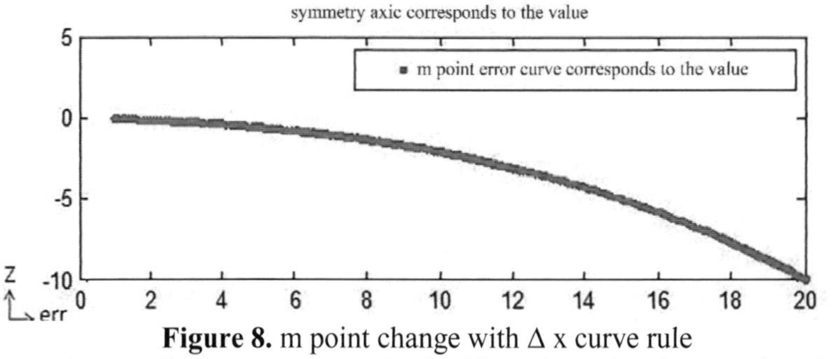

Figure 8. m point change with Δ x curve rule

The curve of contour error is the error curve obtained by subtracting the actual aspheric curve from the ideal aspheric curve. The error curve was obtained by using MATLAB software, as shown in figure 7.

When the deviation on the x axis knife $\Delta x < 0$ through the lens center axis, With the change of the corresponding value of cutter deviation Δx, actual grinding trajectory and error curve will change. In order to find out the law of the error curve with the change of error value. We take the error curve Δx error of the corresponding values for the observation point m, as shown in figure 8, Using MATLAB to calculate change with error Δx point m, take x=0, the error curve corresponds to m. In the case that the diameter D of the workpiece is constant, the error value of the knife increases with the X-axis over the center axis, absolute value of m is getting bigger, the surface profile accuracy of workpiece is lower, conversely, the smaller the absolute value of m points, the higher the accuracy of the surface profile of the work piece [2].

The axial positive deviation of the grinding wheel is the same as that of the axial negative deviation. That is, when the grinding wheel passes the center axis of the lens and the center axis of the lens is not shifted. The absolute value of the residual error of the aspheric surface is the same, but the positive and negative signs are opposite. The residual error of the convex aspheric surface is positive when the grinding wheel has a center axis of the lens. It means that the aspheric infrared optical lens after grinding and the machining allowance can be compensated. When the grinding wheel has the offset of the center axis of the lens. The residual error of the convex aspheric optical lens is negative, which means that the grinding amount exceeds the preset value. Although it can also be compensated for processing, it requires more time to correct and difficult to correct. Therefore, when machining convex aspheric surface, it is necessary to avoid the error of the center axis of the grinding wheel. When the grinding wheel has not been shifted by the center axis of the lens, it can be compensated for the knife error to obtain the workpiece with higher surface profile precision [3].

3. The center method determines the study of knife error

3.1. Theoretical analysis

The grinding of aspheric surface is characterized by symmetry. We need to analyse the x positive half axis. The workpiece has only one center point, and the actual grinding track produces two circles with a symmetry of x=0. Due to error Δ x knife, result in artifacts area center Δ x radius is not grinding, Analysis of the semi-spherical grinding trajectory of the workpiece with x - positive half axis. Is equal to the ideal trajectory translational Δ x distance to the right, we need to find the two centers of the right and left hemisphere trajectories. Can draw the value of Δ x and compensate. Using MATLAB, the data obtained from the measurement is drawn into a function curve. As a result of the numerical control extra bed for knife error within 0.5 mm, and Δ x location near the center axis inevitably. Therefore, the position of the z axis at the edge of the aspheric infrared optical lens is selected, take the points as dense as possible, remove the point from the highest point of the z-axis at 0.5mm. The circle is fitted to the point and the center of the left and right arc is found, center distance center axis of symmetry is Δ x.

In the actual machining process, it is necessary to pre-process the nc machine tool. The surface profile of the pre-processed workpiece can be detected by on-line detection and the surface location data point and the actual contour curve of the workpiece can be obtained. Use MATLAB to draw into the actual machining path curve function, and the function analysis of the data found grinding wheel for Δ x knife error value. Due to the possibility of eccentric error in the installation of grinding wheel, it is impossible for the center of the circle to be absolutely symmetrical according to the data fitting circle. In order to reduce the data processing error, use MATLAB to find the peak value of the z-axis corresponding to the x value before taking the point. This x value is the central axis of symmetry of the actual trajectory. Use of symmetry axis x numerical correction from the above two center point, accurate grinding wheel of cutter error value of Δ x as shown in figure 9 [4].

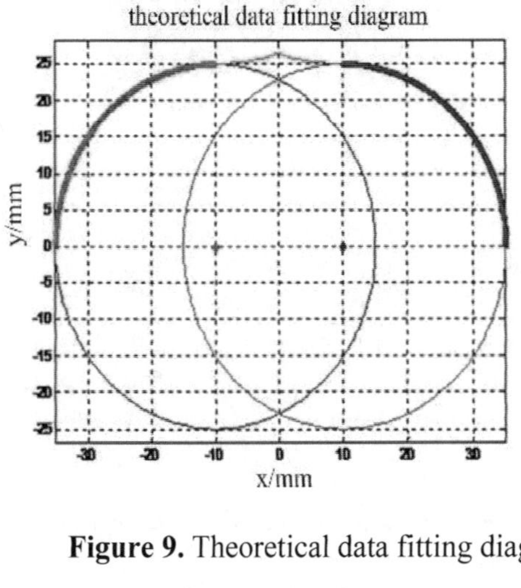

Figure 9. Theoretical data fitting diagram

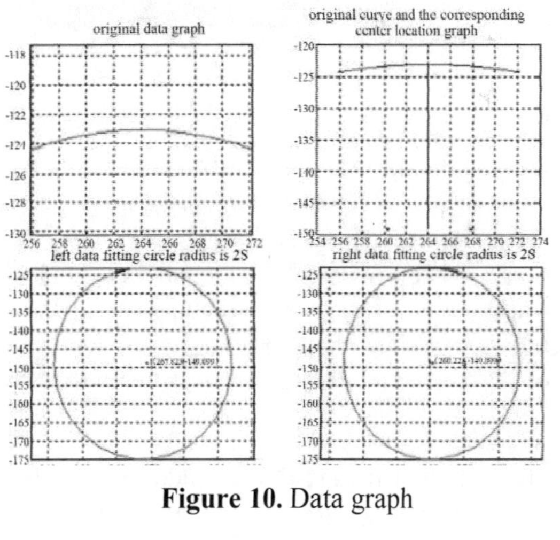

Figure 10. Data graph

3.2. Error compensation experiment analysis results

Before the grinding of aspheric surface infrared optical lens, we need to fix the X-axis for the knife error. It is necessary to use specimens for grinding experiments to obtain spherical optical lenses, determine the x axis to the knife error and make compensation. The experiment used single crystal silicon with a diameter of 200mm and 10mm as the material for grinding. The precision of the surface profile of the aspheric surface infrared optical lens is measured by contact measuring instrument.

Output x axis per 0.001mm unit, the coordinates of the non-spherical optical lens corresponding to the Y-axis, and the data analysis of each point coordinate on the aspheric surface is shown in figure 10.Using MATLAB software to analyse the data to obtain the error point.

Table 1. Grinding parameters

feed speed F(mm/min)	grinding depth ap(μm)	grinding wheel radius r (mm)	grinding wheel speed Ns(r/min)	work piece speed Nw(r/min)
3	2	40	6000	160

After sorting out the data points, the data obtained by the center method were 267.823mm and 260.223mm respectively. According to the theoretical model, the error value of the two symmetric center distance is 0.300mm. According to the analysis of the error value in the numerical control add bed to adjust. After the adjustment, the same parameter is used to grind the optical lens, and the accuracy of the contour is measured and the accuracy of the contour is compared with that of the pre-adjusted contour, as shown in figure 11. The accuracy of the surface profile of the aspheric surface infrared optical lens was 235nm before the adjustment of the knife, and the accuracy of the surface profile of the aspheric infrared optical lens after the adjustment was 145nm. The surface profile accuracy of the aspheric surface infrared optical lens has been improved obviously before and after adjustment, and the method is feasible to compensate the error [5].

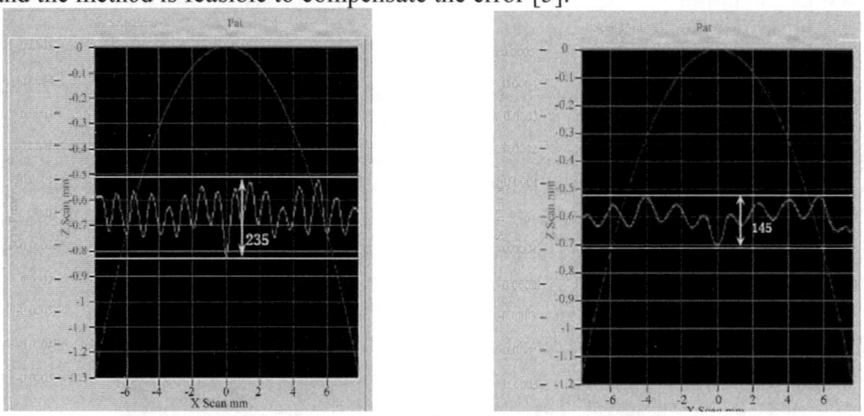

Figure 11. Measurement results of surface profile accuracy

4. Conclusion

Using MATLAB software for grinding trajectory simulation, the effect of the X-axis y axis on the surface profile accuracy of the aspheric surface infrared optical lens is analysed. The center axis of the X-axis is compensated by the center axis and accuracy of surface profile of aspheric surface infrared optical lens is improved. Combined with the grinding experiment, it is concluded that the center method can effectively compensate the X-axis for the error of the knife, so that the surface profile of the aspheric infrared optical lens is more accurate.

References

[1] Xu Jinkai and Yu Huadong *Axisymmetric aspheric surface parallel grinding Y axis compensates for knife error.* Journal of changchun university of technology. 2012,03:68-71

[2] Han Chengshun and Dong shen *Study on geometric model of axisymmetric aspheric surface ultra-precision grinding.* Optical technology. 2003,29(3):329-333

[3] Chen Mingjun,Zhang Feihu and Dongshen *Study on ultra-precision grinding of optical non - spherical surface devices.* Optical technology. 2001,27(6):512-515

[4] Chen Mingjun,Wang Lisong,Liang Yingchun and Lu Zesheng. *Ultra-precision machining method of plastic domain of brittle material.* Aviation precision manufacturing technology.2001,02:10-12.

IOP Conf. Series: Materials Science and Engineering **389** (2018) 012028 doi:10.1088/1757-899X/389/1/012028

[5] Ding Qiang. Research on compensation technology for precision grinding and polishing of aspheric surface. *Changchun polytechnic university*.2014.

Influence of Cold Working on the Mechanical Field around Creep Crack Tip of Stress Corrosion Cracking

Cui Yinghao[1], Zhao Lingyan[2,*] and Wang Weibing[1]

[1] School of Mechanical Engineering, Xi'an University of Science and Technology, Xi'an 710054, China
[2] School of Science, Xi'an University of Science and Technology, Xi'an 710054, China

*gloomy2@foxmail.com

Abstract. To understand the effect of cold working on the mechanical properties of stress corrosion cracking in 304 stainless steel, the influence of different cold working on the stress strain field near the crack front of 304SS was analyzed using ABAQUS. Results show that Mises stress will increase with increasing of the CW before creep. As the creep process is going on, the effect of cold working on the Mises stress around crack front will become weaker; Cold working will cause the decrease of the low-plasticity strain region at the crack front, but has little effect on the high plastic strain; Cold working will also cause the crack tip creep strain increase to a certain extent, indicating that the cold working will accelerate stress corrosion cracking to some extent.

1. Introduction

Stress corrosion cracking(SCC) is a failure mechanism that is caused by environment, susceptible material and tensile stress of nuclear power structural materials under high temperature and high pressure water environments [1]. During the manufacturing and assembly process under the nuclear power equipment, a certain degree of cold working will occur [2]. The amount of cold working will affect the stress corrosion rate and the crack tip mechanical state of the material in the nuclear environment. SCC caused by crack tip creep in high temperature water environment is one of the important factors lead to the failure of structures [3]. Therefore, it is very important to obtain accurate creep crack propagation rate for quantitative prediction nuclear structure life [4]. Many research shows that cold working has great effect on the creep mechanical field around crack tip [5-6].

Finite element model is established under different cold working using compact tension specimen using 304 stainless steel. The mechanical properties of crack tip region are studied.

2. Finite element modelling

2.1. Specimen model

The finite element analysis of the 1T-CT specimen was performed using the above material properties. The geometrical dimensions and the experimental procedure of the specimen confirmed to the ASTM399 standard [7], as shown in Figure 1.

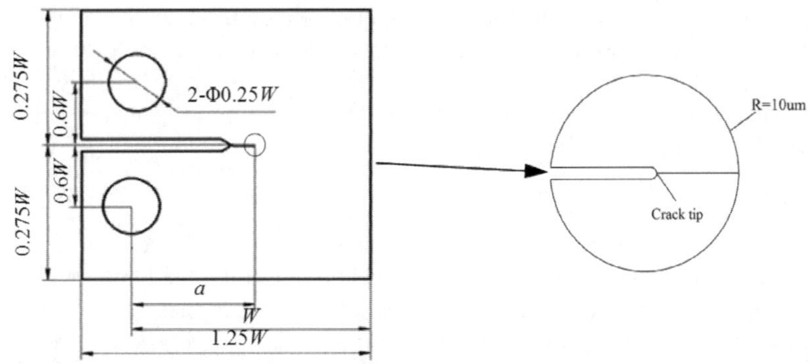

Figure 1. Geometry and dimensions of 1T-CT model

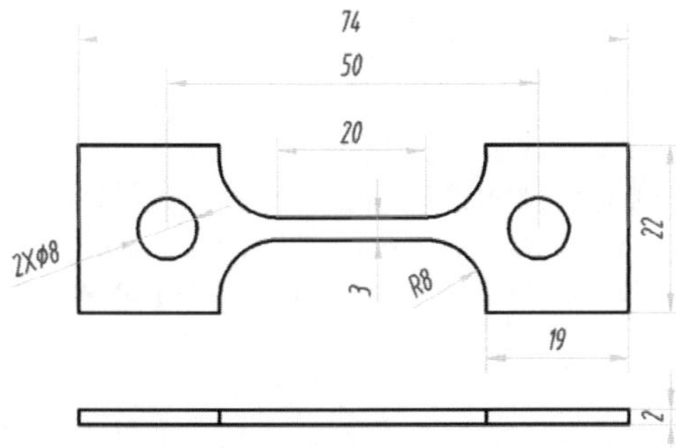

Figure 2. Plate tensile specimen of 304SS

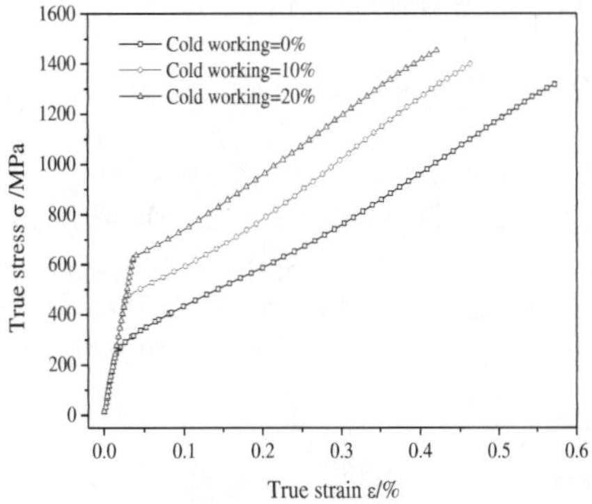

Figure 3. True stress-strain curves under different CW

2.2. Material model

The plate tensile specimen used in this study is 304 stainless steel and its geometry and dimensions are shown in Figure 2. Its chemical chemical composition includes C-0.043%, Si-0.4%, Mn-1.18%, P-0.033%, S-0.002%, Ni-8.04%, Cr-18.115%, N -0.046% etc. Tensile experiment was carried out in fatigue stretcher machine under different cold working rates, engineering stress and strain curve was derived and converted to true stress-strain curves. The material true mechanical curves was shown in Figure 3 under the amount of cold working is 0%, 10% and 20%, respectively.

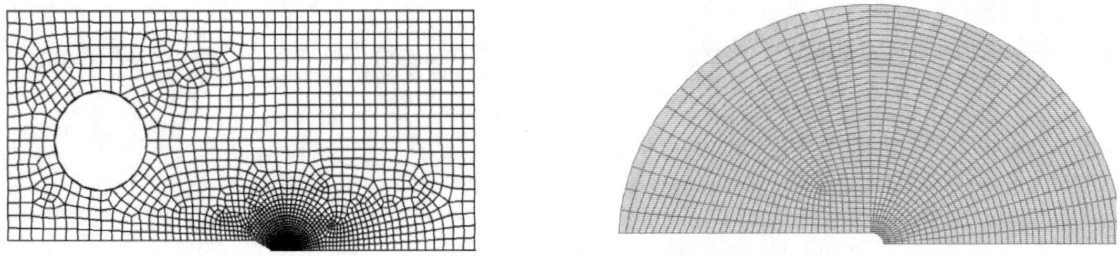

(a) Global mesh model (b) Mesh around the crack tip

Figure 4. Finite element mesh model of half model of 1T-CT

The creep rate is calculated using the power law model [8-9] in the FEM calculation:

$$\dot{\varepsilon}_{cr} = A\sigma^n t^m \tag{1}$$

Where $\dot{\varepsilon}_{cr}$ is the creep strain rate. σ is the applied stress; A is the power law multiplier, the value is 1.24×10^{-23}, n is the creep exponent, and the value is 7.5.

To eliminate the stress singularity generated by the crack tip [10], the crack tip radius was taken as 0.5μm in the finite element simulation. Because of the symmetry of the sample structure and load, 1/2 model was established in my study. The CPE8R was adopted in mesh type, as shown in Figure 4a. To obtain more detailed and accurate crack tip stress and strain, the mesh around crack tip was refined and its number was 15027, as shown in Figure 4b.

2.3. Loading and boundary conditions

Due to symmetry of CT specimen, only the half specimen was modelled. Boundary condition is applied to the symmetry plane of the model. Loading is applied to the load hole by using the multiple point constraints facility in ABAQUS, which joins the hole centre to the nodes of the whole surface. The concentrated force is applied on the reference point to ensure the crack tip stress intensity factor K_I=10 MPa·m$^{1/2}$ [9]. Two analysis steps are set. The first one is the loading process, which produces elastic-plastic deformation on the sample; The other is the process of constant load, setting the creep equivalent time to 1000 hours.

3. Results and Discussion

3.1. Effect of cold working on Mises stress around crack tip

The Mises stress distribution around the crack tip under different CW before and after creep is shown in Figure 5 and Figure 6, respectively. Figure 5 shows that Mises stress decreases with increasing of the CW before creep. Figure 6 shows that Mises stress tends to be the same under different CW after creep. Creep has great effect on the Mises stress.

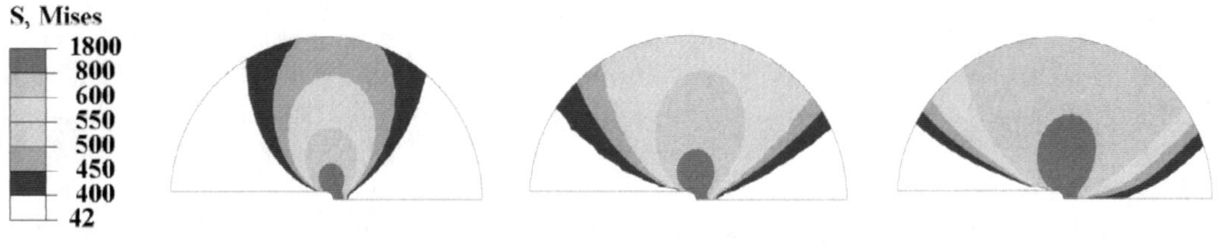

(a) Cold working=0% (b) Cold working=10% (c) Cold working=20%

Figure 5. Comparison of the Mises stress distribution around crack tip before creep (MPa)

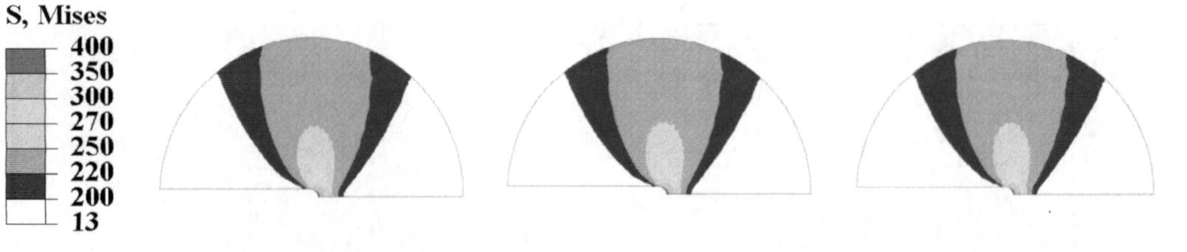

(a) Cold working=0% (b) Cold working=10% (c) Cold working=20%

Figure 6. Comparison of the Mises stress distribution around crack tip after creep (MPa)

3.2. Effect of cold working on PEEQ around crack tip

The PEEQ distribution around the crack tip under different CW before and after creep is shown in Figure 7 and Figure 8, respectively. Figure 7 shows that cold working has little effect on the high PEEQ zone near crack front, but can cause low PEEQ zone decreases with increasing of the CW. Compared Figure 7 with Figure 8, CW has little effect on the PEEQ.

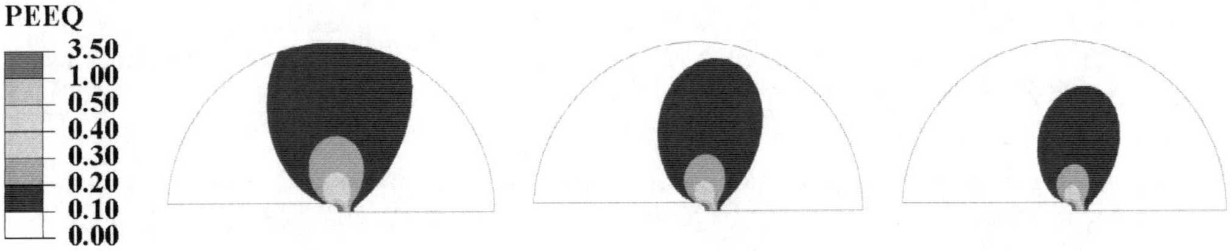

(a) Cold working=0% (b) Cold working=10% (c) Cold working=20%

Figure 7 Comparison of the creep strain distribution around crack tip before creep

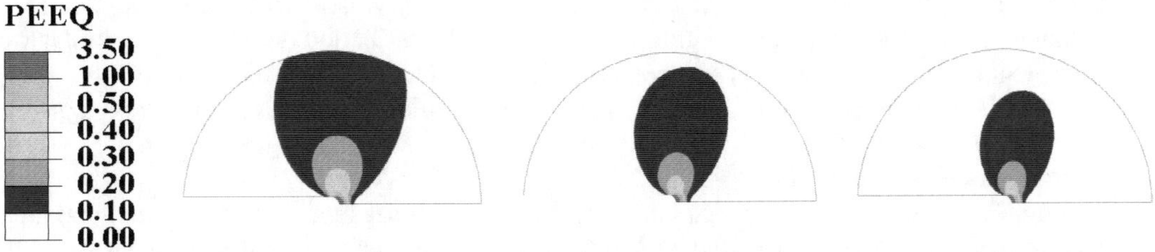

(a) Cold working=0% (b) Cold working=10% (c) Cold working=20%

Figure 8. Comparison of the creep strain distribution around crack tip after creep

3.3. Effect of specimen thickness on SCC crack tip creep strain

The creep strain distribution around the crack tip under different CW is shown in Figure 9. which shows that creep strain increases with the increasing of CW, indicating that the cold working will accelerate stress corrosion cracking to some extent.

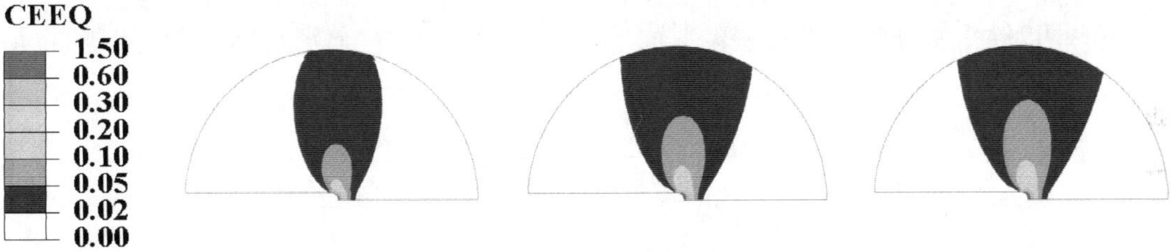

(a) Cold working=0% (b) Cold working=10% (c) Cold working=20%

Figure 9. Comparison of the creep strain distribution around crack tip

4. Conclusions

(1) CW will increase the Mises stress at the crack tip before creep, and the stress around crack tip will release because of creep. As the creep process going on, the effect of CW on the Mises stress around crack front will become weaker;

(2) CW will cause the decrease of the low-plasticity strain region at the crack front, but has little effect on the high plastic strain; CW will also cause the crack tip creep strain increase to a certain extent, indicating that the CW will accelerate stress corrosion cracking to some extent.

Acknowledgements

This work is financially supported by the National Natural Science Foundation of China (Grant Nos. are 11502195 and 2018QDJ009).

IOP Conf. Series: Materials Science and Engineering **389** (2018) 012029 doi:10.1088/1757-899X/389/1/012029

References

[1] Andresen P L, Karen G, Lawrence N J. Stress Corrosion Cracking of Sensitized Type 304 Stainless Steel in 288C Water: A Five Laboratory Round Robin[M]. *John Wiley & Sons, Inc.* 2013.

[2] Hou Juan, Peng Qunjia, Shoji Testuo et al. Study of microstructure and stress corrosion cracking behaviour in welding transition zone of Ni-based alloys[J]. *ACTA METALLURGICA SINICA*, 2010, 46(10): 1258-1266.

[3] Zhang L, Chen K, Du D, et al. Characterizing the effect of creep on stress corrosion cracking of cold worked Alloy 690 in supercritical water environment[J]. *Journal of Nuclear Materials*, 2017, (492): 32-40.

[4] Peng Qunjia, Xue He, Hou Juan. Role of water chemistry and microstructure in stress corrosion cracking in the fusion boundary region of an Alloy 182-A533B low alloy steel dissimilar weld joint in high temperature water[J]. *Corrosion Science*, 2011, 53(12): 4309-4317.

[5] Donghai Du, Kai Chen, Hui Lu et al. Effects of chloride and oxygen on stress corrosion cracking of cold worked 316/316L austenitic stainless steel in high temperature water[J]. *Corrosion Science*, 2016, (110): 134-142.

[6] J. Hou, T. Shoji, Z. P. Lu et al. Residual strain measurement and grain boundary characterization in the heat affected zone of a weld joint between Alloy 690TT and Alloy 52[J]. *Journal of Nuclear Materials*, 2010, 397(1): 109-115.

[7] ASTM Standard E399-90. Annual book of ASTM standards[M]. *USA: ASTM International*, 2002.

[8] Quintero H, Mehmanparast A. Prediction of creep crack initiation behaviour in 316H stainless steel using stress dependent creep ductility[J]. *International Journal of Solids & Structures*, 2016, 97–98(10): 101-115.

[9] Xue He, Guo Rui, Chen Zheng, et al. Effect of oxide film thickness on creep at tip of stress corrosion cracking[J]. *China Sciencepaper*, 2016, 22(11): 2619-2622.

[10] Cui Yinghao, Xue He, Zhao Lingyan. Effect of creep property mismatch on mechanical field at crack tip of stress corrosion cracking[J]. *Hot working technology*, 2017, 46(12): 81-85.

IOP Conf. Series: Materials Science and Engineering **389** (2018) 012030 doi:10.1088/1757-899X/389/1/012030

Gold Nanoparticles with Immobilized β-cyclodextrin-capsaicin Inclusion Complex for Prolonged Capsaicin Release

Y S Pestovsky and A Martínez-Antonio

Biological Engineering Laboratory, Genetic Engineering Department, Center for Research and Advanced Studies of the National Polytechnic Institute (Cinvestav), Campus Irapuato. Km. 9.6 Libramiento Norte Carretera Irapuato-León, Irapuato, Guanajuato, 36824, México

E-mail: iurii.pestovskii@cinvestav.mx; agustino.martinez@cinvestav.mx

Abstract. Inclusion complexes of capsaicin with β-cyclodextrin (prepared at three different concentrations of β-cyclodextrin) were immobilized onto the surface of concentrated gold nanoparticles synthesized using the Frens method. Capsaicin release for 2 days into pure water and acetonitrile was recorded. The nanoparticles were characterized with absorbance spectra and transmission electron microscopy images. An inclusion complex without gold nanoparticles also displayed capsaicin release with a lag period of 1.5 days that suggested capsaicin release from gold nanoparticles was due to decomposition of the inclusion complex. Neither the inclusion complex nor the free capsaicin formed nanoparticles themselves. It was shown that physical immobilization of capsaicin onto the surface of gold nanoparticles without the aid of β-cyclodextrin did not lead to prolonged release. For comparison, inclusion complex encapsulation into chitosan nanoparticles was also carried out, but the release kinetics was less pronounced compared to that of gold nanoparticles.

1. Introduction

Capsaicin, a main pungent component in 'hot' chili peppers [1-3], can be used for treatment of obesity [2], osteoarthritis [3], diabetic neuropathy [3], post-therapy and postherpetic neuralgia [3]. Treatment with capsaicin causes sensory chemodenervation and thus improves myocardial function [4]. It was found to be a potent inhibitor of platelet aggregation [5], to induce apoptosis in cancer cells [6] and to have an analgesic effect [3, 7]. But its practical applications are limited by its very low water solubility [8-10] with unstable solutions [11] and strong gastrointestinal irritation [8]. Water solubility can be improved by the formation of an inclusion complex with β-cyclodextrin derivatives [8-10], even though the cavity diameter of β-cyclodextrin (0.6-0.65 nm) [12] is smaller than a capsaicin molecule (1.142 nm). This inclusion complex solution still contains free capsaicin at high concentrations [10] and only partially solves the problems mentioned above.

The aim of this study is to create nanoparticulated forms of capsaicin by means of the immobilization of its β-cyclodextrin inclusion complex onto the surface of gold nanoparticles. Nanostructured drug delivery architectures are promising candidates that will enable efficient and targeted delivery of drug compounds [13]. The advantages of gold nanoparticles include their ability to penetrate inside living cells, the possibility to synthesize stable particles with predefined size [14], and the possibility to produce multifunctional nanoparticles by binding chemically distinct moieties to the same nanoparticle [15]. For example, it would also be possible to immobilize antibodies on the nanoparticle for targeted drug delivery [16]. To our knowledge, there is no report to date dealing with immobilization of capsaicin onto the surface of gold nanoparticles.

2. Materials and methods

2.1. Materials

Capsaicin, chitosan, glutaraldehyde, phosphotungstic acid hydrate, sodium citrate, sodium tripolyphosphate and tetrachloroauric acid ($HAuCl_4$) were purchased from Sigma-Aldrich (USA). Cyclodextrin was a gift from Wacker Chemie AG (Germany). Glacial acetic acid was purchased from Karal, S. A. de C. V. (Mexico). Ethanol was purchased from Reasol (Mexico). Water used in all the experiments was distilled and subsequently purified by the Thermo Barnstead EASYpure II system, model No. D8611.

2.2. Preparation of β-cyclodextrin-capsaicin inclusion complex

β-cyclodextrin solution in water was mixed with 100 μl of 15 mg/ml capsaicin solution in ethanol and with an additional amount of ethanol. The mixture was shaken at 250 rpm (BIOSAN PSU-20i, EU) at 4°C for 2 days in the dark.

2.3. Synthesis of gold nanoparticles using the Frens method

1 ml of 1% sodium citrate was added to 50 ml of boiling 10^{-2}% tetrachloroauric acid solution. The solution was boiled for 5 min [17]. The colorless solution soon turned light blue, then light red and finally dark red. Then the heater was turned off, and the solution was allowed to cool down to room temperature (about 27°C). The solution was stored at 4°C in the dark.

2.4. Concentrating gold nanoparticles by centrifugation

The solution of gold nanoparticles was transferred to 8 plastic test tubes (1.5 ml for each tube) and centrifuged at 6000 rpm for 20 min using centrifuge 5430 with rotor FA-45-30-11 (Eppendorf, Germany). The precipitate was resuspended in fresh portions of solution by briefly vortexing at a maximal rate (Vortex-Genie 2 G-560, Scientific Industries, Inc., USA) and then by orbital shaking at 250 rpm for 10 minutes. Then 4 test tubes were centrifuged, and the precipitate was resuspended in solution from 4 remaining tubes. Two tubes were centrifuged again, and the precipitate was resuspended in a solution from the 2 remaining tubes. After each centrifugation step the supernatant was discarded [18].

2.5. Immobilization of the inclusion complex onto the surface of gold nanoparticles for capsaicin release in water

Concentrated gold nanoparticles were centrifuged at 6000 rpm for 20 min. The supernatant was discarded, but a small amount remained (about 50 μl). The precipitate was resuspended in this solution by briefly vortexing, and then 1.5 ml of the solution of the inclusion complex prepared with 900 μl of 6.3 mg/ml β-cyclodextrin and with 700 μl of ethanol was added. The solution was shaken at 250 rpm for 10 min in the dark. Then it was centrifuged at 8000 rpm for 20 min. After discarding the supernatant, the residual capsaicin concentration in the supernatant was determined by HPLC. The precipitate was resuspended in 1.5 ml of water by orbital shaking at 250 rpm for 10 min in the dark. In a control experiment, concentrated gold nanoparticles as prepared were resuspended in 900 μl of water, then 700 μl of ethanol was added. 100 μl of 15 mg/ml capsaicin in ethanol was added after briefly vortexing, followed by orbital shaking at 250 rpm for 10 minutes in the dark. All of the experiments involving capsaicin were carried out at 4°C as recommended [11] to enhance capsaicin stability.

2.6. The investigation of storage stability of gold nanoparticles with immobilized inclusion complex at different temperatures

500 μl of the solution of gold nanoparticles with immobilized β-cyclodextrin-capsaicin inclusion complex in water were stored for 6 days in temperature regulated rooms at 4°C, 28°C and 37°C, respectively. After that, the samples were analyzed by means of scanning electron microscopy (SEM).

2.7. Immobilization of inclusion complexes onto the surface of gold nanoparticles for capsaicin release in acetonitrile

Concentrated gold nanoparticles were centrifuged at 6000 rpm for 20 min. The supernatant was discarded, but a small amount remained (about 50 μl). The precipitate was resuspended in this solution by briefly vortexing, and then 1.5 ml of the solutions of inclusion complexes prepared with 900 μl of 3.1, 6.3 or 9.6 mg/ml β-cyclodextrin and with 700 μl of ethanol were added. The solution was shaken at 250 rpm for 10 min in the dark and stored overnight at 4°C. Then it was centrifuged at 8000 rpm for 20 min. The supernatant was discarded, and the precipitate was resuspended in 1.5 ml of acetonitrile by orbital shaking at 250 rpm for 10 min in the dark. All of the experiments involving capsaicin were carried out at 4°C.

2.8. Preparation of chitosan particles crosslinked with tripolyphosphate

The inclusion complex solution prepared with 1100 μl of 3.93 mg/ml β-cyclodextrin and with 400 μl of ethanol was added to 107 μl of 0.5 mg/ml chitosan in 0.2% CH_3COOH followed by vortexing for 10 min. Then 400 μl of 0.5 mg/ml sodium tripolyphosphate was added, and after vortexing for 10 min precipitate appeared. The solution was centrifuged at 7500 rpm for 5 min. The supernatant was carefully removed leaving the precipitate stuck to the wall of the tube. It was resuspended in 2 ml of water by vortexing for 10 min. All of the experiment was carried out at 4°C.

2.9. Preparation of chitosan particles crosslinked with tripolyphosphate and glutaraldehyde

The inclusion complex solution prepared with 1000 μl of 5.8 mg/ml β-cyclodextrin and with 500 μl of ethanol was added to 107 μl of 0.5 mg/ml chitosan in 0.2% CH_3COOH. After vortexing for 10 min 300 μl of 0.5 mg/ml sodium tripolyphosphate was added followed by vortexing for 10 min. 1500 μl of the solution was centrifuged at 14000 rpm for 5 min. The precipitate was resuspended in 1.5 ml of 50% glutaraldehyde by vortexing for 10 min. Precipitate appeared after overnight shaking at 250 rpm in the dark. It was separated from glutaraldehyde solution by centrifugation at 5000 rpm for 5 min and resuspended into 2 ml of water by vortexing for 10 min. All of the experiment was carried out at 4°C.

2.10. Characterization of nanoparticles

Absorbance spectra were acquired by means of a microplate reader BioTek Synergy 2 (USA). Transmission electron microscopy (TEM) images were acquired by means of a Philips FEI Morgagni 268D microscope (USA). 7 μl of the solutions were dropped onto 200 mesh formvar/carbon-coated copper grids (Ted Pella, Inc., USA), and the remaining liquid was collected by filter paper after 20 min of incubation under a Petri dish. In the case of chitosan nanoparticles, 7 μl of 1% phosphotungstic acid was dropped onto the grids thereafter, and the remaining liquid was collected by filter paper after 15 min of incubation under a Petri dish. SEM images were acquired by means of a JSM-7800F microscope (JEOL, Japan). The solutions were dropped onto silicon and allowed to dry. The images were processed using ImageJ software bundled with Java 1.8.0_112 operating under Linux. At first, the background was subtracted at a rolling ball radius of 50 pixels, and then the threshold was set using the default algorithm. The nanoparticles were approximated by ellipses, and the histograms of the minor axes length distribution of these ellipses were plotted.

2.11. The investigation of kinetics of capsaicin release

Samples having a volume of 350 μl were taken from the solutions of nanoparticles with immobilized inclusion complexes at different time points, centrifuged at 14800 rpm for 30 min, and then the upper 300 μl of supernatant was used for HPLC analysis. All of the experiment was carried out at 4°C in the dark. Before HPLC analysis, the solutions were stored at -20°C in the dark.

IOP Conf. Series: Materials Science and Engineering **389** (2018) 012030 doi:10.1088/1757-899X/389/1/012030

2.12. HPLC analysis
An Agilent Technologies 1200 Series chromatograph (USA) equipped with column Zorbax Eclipse XDB-C18 (4.6 × 150 mm, particle size 5 μm) and Diode Array Detector operating at 280 nm was used for HPLC analysis. Capsaicin was eluted with 60% acetonitrile in water at 25°C using isocratic pump at 1 ml/min.

3. Results and discussion
Blank gold nanoparticles have an absorbance maximum at 518 nm. After the inclusion complex immobilization, the absorbance maximum shifts to 520.5 nm. Plasmon resonance shift to longer wavelengths corresponds to the presence of immobilized molecules [19]. The nanoparticles have a shell giving them an irregular form. This shell can be ascribed to the immobilized inclusion complex. The size of single nanoparticles has somewhat increased after inclusion complex immobilization (figure 1). The control experiment has not revealed any nanoparticles in the capsaicin solution and in the solution of the inclusion complex.

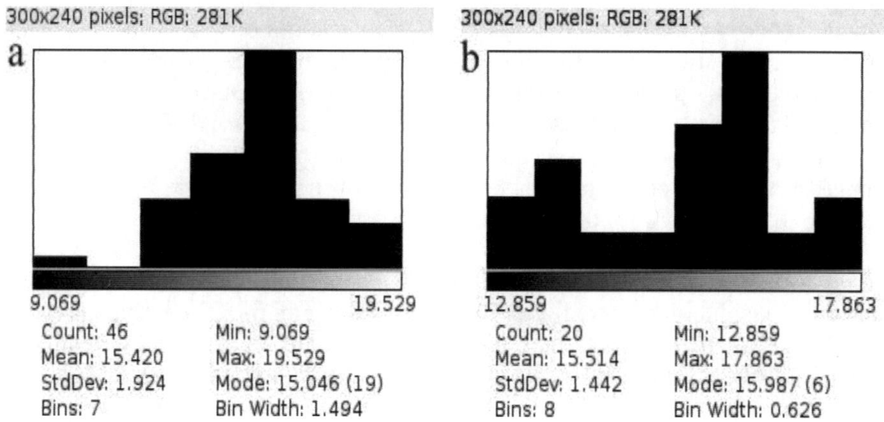

Figure 1. The histograms of length distribution of minor axes of ellipses of the gold nanoparticles before (a) and after (b) β-cyclodextrin-capsaicin inclusion complex immobilization.

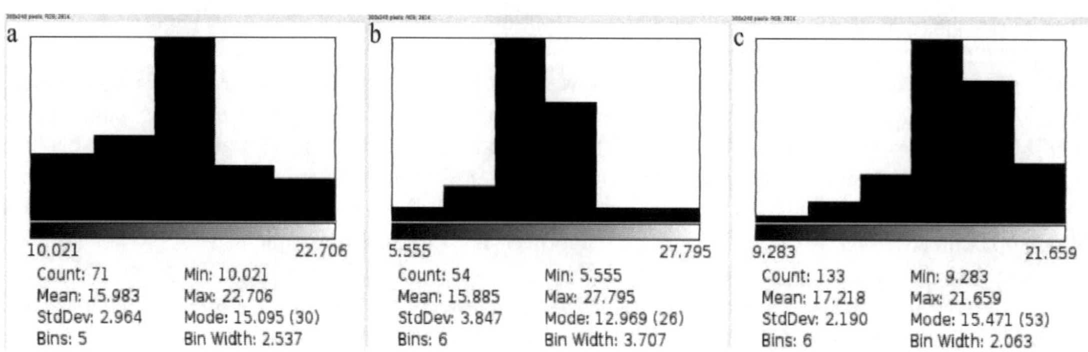

Figure 2. The histograms of length distribution of minor axes of ellipses of the gold nanoparticles with immobilized β-cyclodextrin-capsaicin inclusion complex after 6 days of storage at 4°C (a), 28°C (b) and 37°C (c). The Y axis represents frequency.

The investigation of the storage stability of the solution of gold nanoparticles with immobilized β-cyclodextrin-capsaicin inclusion complex in water at different temperatures has shown that the data calculated from SEM images for 4°C and 28°C (figure 2) is not significantly different from each other and from those calculated from TEM images (figure 1(b)). However, at 37°C the size of the nanostructures is somewhat larger (figure 2(c)) because of an increased aggregation rate. Moreover, a temperature rise must destabilize capsaicin [11]. Therefore, the storage of the solutions of gold nanoparticles with immobilized β-cyclodextrin-capsaicin inclusion complexes at 4°C is preferred.

As can be seen from the release curves (figure 3), encapsulated capsaicin is liberated for at least 2 days. When glutathione was released from sodium tripolyphosphate-crosslinked chitosan nanoparticles with encapsulated α-cyclodextrin-glutathione inclusion complex into simulated gastric and intestinal media for 1 hour, encapsulation efficiency was 6% [20]. Methotrexate and calcium folinate encapsulated in sodium tripolyphosphate-crosslinked chitosan nanoparticles containing their inclusion complexes with β-cyclodextrin were released in phosphate buffered saline for 24 hours with a burst release for 30 min [21]. One of the best results was obtained for the release of cisplatin from nanoparticles consisting of crosslinked α-cyclodextrin — 28 hours of release in phosphate buffered saline with an encapsulation efficiency of 37.5% [22]. The nanoparticles with immobilized capsaicin (without β-cyclodextrin) have not shown any capsaicin release kinetics at all. An inclusion complex without nanoparticles has shown some capsaicin release with a lag period of 1.5 days. Therefore, the release of capsaicin from gold nanoparticles with an immobilized inclusion complex may be explained by decomposition of the complex. The encapsulation yield calculated by dividing the difference between residual capsaicin mass in the supernatant (see section 2.5) and initial capsaicin mass by initial capsaicin mass was 81%. In contrast, the 'encapsulation yield' calculated by dividing the highest capsaicin concentration registered in the solutions of gold nanoparticles (for those with an immobilized inclusion complex prepared at 6.3 mg/ml β-cyclodextrin and resuspended in acetonitrile) by capsaicin concentration registered in the solution of the inclusion complex after 2 days is 5%. This apparent inconsistency may be explained by significant amounts of capsaicin still being bound with nanoparticles even after 2 days of release. The results of the present work are comparable or even better than those reported for capsaicin immobilization in polymeric nanoparticles where release time into simulated gastric fluid and phosphate buffered saline in a dialysis bag was 60 hours with a burst release for 12 hours and an encapsulation efficiency of 81.5% [23]. The current results were also better than those reported for solid lipid nanoparticles with 14 hours of capsaicin release in a 1:1 ethanol-citrate-phosphate buffer in a diffusion cell with cellophane membrane [24, 25] and an encapsulation efficiency of 90.89% [24] or over 60% [25]. This superiority may be explained by dual immobilization: capsaicin is encapsulated in β-cyclodextrin, and the inclusion complex is immobilized onto the surface of gold nanoparticles giving additional steric hindrance due to the statistical orientation of immobilized inclusion complexes. For comparison, chitosan nanoparticles with a diameter about 50 nm and an encapsulated β-cyclodextrin-capsaicin inclusion complex were also prepared. As can be seen from figure 3, gold nanoparticles displayed much better capsaicin release kinetics that supports the conclusion above.

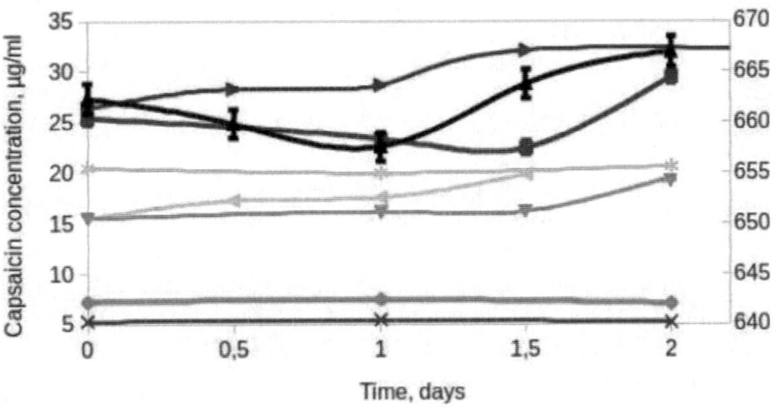

Figure 3. The dependence of capsaicin concentration on time for gold nanoparticles with an inclusion complex in water (triangles), for gold nanoparticles with capsaicin in water (rhombi), for gold nanoparticles with an inclusion complex at 3.1 (inverted triangles), 6.3 (triangles turned right) and 9.6 mg/ml β-cyclodextrin (triangles turned left) in acetonitrile; for chitosan particles crosslinked with tripolyphosphate (asterisks; the values are reduced by a factor of 4); for chitosan particles crosslinked with tripolyphosphate and glutaraldehyde (crosses); for an inclusion complex (squares, additional Y axis).

4. Conclusions

β-cyclodextrin-capsaicin inclusion complex can be physically adsorbed onto the surface of gold nanoparticles prepared by the Frens method. The nanoparticles show capsaicin release into water or acetonitrile for at least 2 days. This is better or comparable with the results reported previously for the nanoparticles releasing capsaicin or other low molecular weight drugs. This capsaicin release is caused by inclusion complex decomposition. The solutions of nanoparticles should not be stored at high temperatures, and 4°C is recommended. Chitosan nanoparticles crosslinked with tripolyphosphate or glutaraldehyde and tripolyphosphate with an encapsulated β-cyclodextrin-capsaicin inclusion complex show a less pronounced capsaicin release when compared to gold nanoparticles.

Acknowledgments

The authors thank María Christian Albor Cortés (Centro de Investigaciones en Óptica) for acquisition of SEM images and Carolyn Smith (Peace Corps Response) for copy editing the manuscript.

References

[1] Caterina M J, Schumacher M A, Tominaga M, Rosen T A, Levine J D and Julius D 1997 *Nature* **389** 816–24
[2] Szallasi A and Blumberg P 1999 *Pharmacological Reviews* **51(2)** 159–211
[3] Fusco B M and Giacovazzo M 1997 *Drugs* **53(6)** 909–14
[4] Bencsik P, Kupai K, Giricz Z, Görbe A, Huliák I, Fürst S, Dux L, Csont T, Janscó G and Ferdinandy P 2008 *British Journal of Pharmacology* **153** 488–96
[5] Wang J-P, Hsu M-F and Teng C-M 1984 *Thrombosis Research* **36(6)** 497–507
[6] Lu H-F, Chen Y-L, Yang J-S, Yang Y-Y, Liu J-Y, Hsu S-C, Lai K-C and Chung J-G 2010 *Journal of Agricultural and Food Chemistry* **58** 12999–13005
[7] Menéndez L, Lastra A, Hidalgo A and Baamonde A 2004 *Life Sciences* **74** 3235–44
[8] Zhao Y, Sun C, Shi F, Firempong C K, Yu J, Xu X and Zhang W 2016 *Pharmaceutical Biology* **54(1)** 130–8
[9] Chen X, Sun X, Ren K, Zhang X, Zhang Z and Gong T 2010 *Arzneimittelforschung* **60(9)** 571-4
[10] Zi P, Yang X, Kuang H, Yang Y and Yu L 2008 *International Journal of Pharmaceutics* **358(1-2)** 151–8
[11] Kopec S E, DeBellis R J and Irwin R S 2002 *Pulmonary Pharmacology & Therapeutics* **15(6)** 529–34
[12] Szejtli J 1998 *Chemical Reviews* **98** 1743–53
[13] Hughes G A 2005 *Nanomedicine: Nanotechnology, Biology and Medicine* **1(1)** 22–30
[14] Lévy R, Shaheen U, Cesbron Y and Sée V 2010 *Nano Reviews* **1** 4889
[15] Paciotti G F, Kingston D G I and Tamarkin L 2006 *Drug Development Research* **67(1)** 47–54
[16] Glazer E S, Zhu C, Massey K L, Thompson C S, Kaluarachchi W D, Hamir A N and Curley S A 2010 *Clinical Cancer Research* **16(23)** 5712–21
[17] Frens G 1973 *Nature Physical Science* **241** 20–2
[18] Pestovsky Y S 2016 *International journal of applied and fundamental research* **4(6)** 1077–82 (In Russian)

IOP Conf. Series: Materials Science and Engineering **389** (2018) 012030 doi:10.1088/1757-899X/389/1/012030

[19] Cao Z, Gong F-C, Tu M, Zeng M-X, Huang X-X, Zhang L, Tan S-Z, Sun L-X and Gu N 2009 *Analytical Letters* **42(17)** 2820–37
[20] Trapani A, Lopedota A, Franco M, Cioffi N, Ieva E, Garcia-Fuentes M and Alonso M J 2010 *European Journal of Pharmaceutics and Biopharmaceutics* **75** 26–32
[21] Ji J, Hao S, Liu W, Wu D, Wang T and Xu Y 2011 *Colloids Surf.,* B **83** 103–7
[22] Zhu W, Zhang K, Chen Y and Xi F 2013 *Langmuir* **29** 5939–43
[23] Peng W, Jiang X, Zhu Y, Omari-Siaw E, Deng W, Yu J, Xu X and Zhang W 2015 *Acta Pharmacologica Sinica* **36(1)** 139–48
[24] Sharma A, Jindal M, Aggarwal G and Jain S 2010 *Research Journal of Pharmaceutical, Biological and Chemical Sciences* **1(2)** 265–74
[25] Sharma A and Arora S 2011 *Indian Journal of Pharmaceutics* **2(1)** 29–41

IOP Conf. Series: Materials Science and Engineering **389** (2018) 012031 doi:10.1088/1757-899X/389/1/012031

Preparation and Luminescence Properties of $Ba_3Y(PO_4)_3:RE^{3+}$ (RE=Eu, Dy) Phosphors

KONG Shijin[1], ZHANG Zhanhui[a,1], HUANG Zhiliang[b,1], PENG Wei kang[1],QI Tonggang[1]

[1]School of Materials Science and Engineering, Wuhan Institute of Technology, Wuhan 430074, People's Republic of China

* correspondence: [a]zzhlzu@163.com; [b]hzl6455@126.com

Abstract: In order to investigate the luminescence properties of $Ba_3Y(PO_4)_3$ doped with rare earth ions, a series of Eu^{3+} and Dy^{3+} single doped $Ba_3Y(PO_4)_3$ phosphor powders were prepared by high temperature solid phase method. The relevant performance of samples were characterized by X ray diffraction (XRD) and fluorescence spectra. The results showed that the prepared phosphor was a cubic phase of bismuth ore and the space group is I-43d. Under the excitation of near violet light, the narrow band emission of $Ba_3Y(PO_4)_3$:0.08Eu^{3+} was nearly located at 594nm, 613nm and 702nm, which showed orange red light emission. The $Ba_3Y(PO_4)_3$:0.08Dy^{3+} narrow band emission was located at 485nm and 575nm respectively, which showed white light emission. The matrix material was doped with a single active ion, and the white light emission was achieved through its coordination in different light regions. The characteristic emission peak of Eu^{3+} in the red light region (580~650nm, $^5D_0$-^7F_J) could compensate for the lack of red components in the Dy^{3+} white light emission spectrum, thus the quality of white light of the single doped Dy^{3+}phosphor could be improved, and it was expected to be applied to light-emitting diode (LED) field.

1.Introduction

YAG:Ce blue plus yellow LED has been widely applied in many parts of the world, but its color saturation is relatively poor due to the lack of red light. In recent years, the research of fluorescent materials has been progressing rapidly[1-3]. Three main aspects are summarized: the first LED based on "blue and yellow" is based on the improvement of the color saturation of the phosphor by improving the red phosphor excited with the blue chip. Second, the excitation light source is ultraviolet and near ultraviolet chip, which is suitable for the white light produced by the excitation of RGB tricolor phosphor. The third way is to combine white light with UV and near ultraviolet chips to achieve white light. Therefore, the research of high efficiency single substrate white phosphor has great significance to further promote the industrialization of white LED. The study showed that different yellow and blue light intensity (Y/B) could be obtained by changing the type and doping amount of alkali metal ions. If Y/B was appropriate, the color coordinates were located in the white light area and warm white light, which was key to obtaining the single matrix and high quality phosphor. However, in many substrates, the Y/B values usually deviated from the ideal value so that it can not get warm white light. because the different doping concentration of the inorganic matrix Dy^{3+} will affect the Y/B value, the chromaticity stability of the phosphor prepared by doping Dy^{3+} is poor [4-5]. It had been studied that many Dy^{3+} activated luminescent materials (based on rare earth phosphate) had attracted much attention because of their low synthetic temperature, high brightness, good physical and chemical stability [6]. In addition, Eu^{3+} usually sends red light, and its characteristic emission peak (580~650nm, $^5D_0$-^7F_J) in the red light region can compensate for the lack of red components in the Dy^{3+} white light emission spectrum, thus improving the white light quality of the single doped Dy^{3+} phosphor. A single active ion is doped in the matrix, and white light emission is achieved through its coordination in different light regions. In this experiment, some different light emitting phosphors were compounded by doping rare earth Eu^{3+}, Dy^{3+},and they can be applied to the field of LED by blending and realizing white light.

2. Experiment

$BaCO_3$, Y_2O_3, $(NH_4)_2HPO_4$, Dy_2O_3 and Eu_2O_3 according to the stoichiometric ratio of $Ba_3Y(PO4)_3$(x=0,0.01,0.02,0.04,0.06,0.08,0.10) were accurately measured and placed in ceramic mortar. First, the medicine was ground with a proper amount of anhydrous ethanol, until the powder was mixed evenly.The samples were dried about 15min in the drying oven. The dry samples were precooked to 600°C in the crucible furnace with heat preservation for 1h.The grinding again samples with 10min was burned to 1150°C in the high temperature furnace and kept for 4h, and the samples were naturally cooled to room temperature.

X-ray (D8 Advance, Germany) diffractometer was used to measure the crystal phases of phosphors. Emission spectra of samples were measured by a fluorescence spectrophotometer (F-7000, Japan) with a 150W Xe lamp.

3. Results and Discussion

X ray diffraction pattern of sample $Ba_3Y(PO_4)_3$ are shown in Fig.1. As seen in Fig.1, the XRD patterns of phosphors after heating at 1100 and 1150°C were well agreed the standard card44-0318, but the crystal phase formed at 1150°C was better.$Ba_3Y(PO_4)_3$ are in cubic lattice constant of a=b=c=10.46Å, V=1146.25Å^3 space group I-43d. In the matrix of $Ba_3Y(PO_4)_3$, the radius of r=0.89 is r=0.89, while the radius of r=1.12 doped Eu^{3+}, Dy^{3+} radius and r=1.17 radius are similar to those of Y^{3+}. Therefore, $Eu^{3+}$and $Dy^{3+}$are more inclined to replace Y^{3+}, which is beneficial for the reduction of system energy and stability.

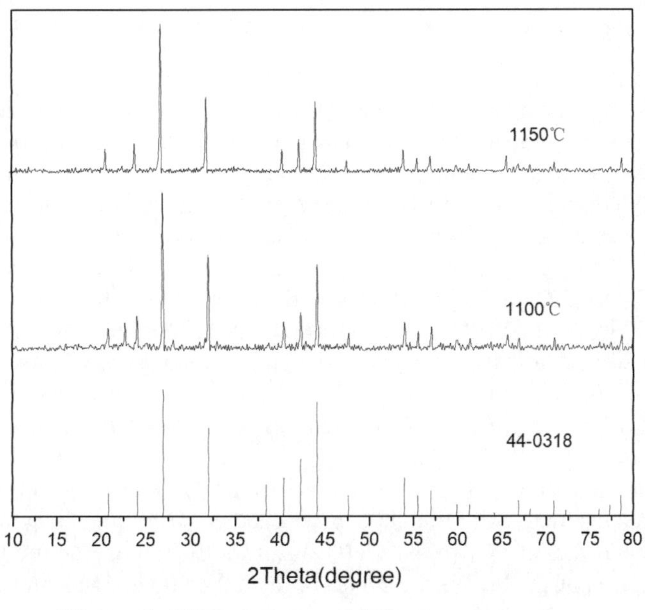

Figure 1. XRD patterns of different temperatures

As shown in Fig.2, the emission spectra of $Ba_3Y(PO_4)_3$ with different Eu^{3+} doping concentrations will be directly affected by the doping concentration of Eu^{3+}, and the luminous intensity [7] of $Ba_3Y(PO_4)_3$ will directly affect the emission spectrum. It can be seen from the diagram that the overall trend of luminescence intensity is increased with the increase of Eu^{3+} doping concentration, and the intensity decreases when the concentration increases to a certain extent. When the Eu^{3+} concentration is less than 2%, the luminescence intensity is weak because of the sample in less active ion. When the Eu^{3+} concentration is less than 8%, the emission intensity increased gradually with the increase of Eu^{3+} concentration. The light of 607nm and 691nm are rapid increased due to the $^5D_0$ to $^7F_2$ transition, and the light of 595nm also was increased by $^5D_0$ to $^7F_1$transition.But the change is relatively small compared with the former. When the concentration of $Eu^{3+}$is more than 8%, the emission strength decreases with the increase of the concentration of Eu^{3+}, which indicates that the high concentration of rare earth ions has a concentration quenching effect on the phosphor, thus reducing the relative intensity of the luminescence.

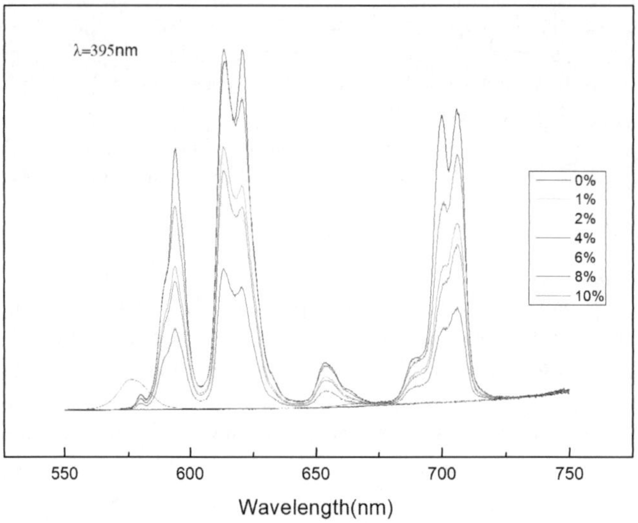

Figure 2. Emission spectra of $Ba_3Y(PO_4)_3$ with different Eu^{3+} doped contents.

Fig. 3 (A) is the excitation spectrum of $Ba_3Y(PO_4)_3$:$0.08Eu^{3+}$ emitted by 576nm, which are located at four different peaks of 350-475nm. The excitation peak of 395nm is the largest, and the most intense reason is the $^7F_0$to $^5L_6$transition of Eu^{3+}. The peak of 370nm is the $^7F_0$ to $^5H_3$ transition of Eu^{3+}, the peak of 383nm is the $^7F_0$ to $^5F_8$ transition of Eu^{3+}, and the peak of 463nm is the $^7F_0$ to $^5D_2$ transition of Eu^{3+}.The located at 389nm may be a f to f high-energy level transition absorption, and the $^7F_0$to $^5L_6$ of $Eu^{3+}$the transition occurs. As shown in Fig.3 (B), there are 6 emission peaks due to the transition emission of Eu^{3+}under 395nm excitation,and the emission peak at 672nm is the strongest. The peak of 590nm, 615nm and 621nm, 702nm, 712nm are respectively the $^5D_0$ to $^7F_1$, $^5D_0$ to $^7F_2$, $^5D_0$ to $^7F_3$, $^5D_0$ to $^7F_4$transition. The Eu^{3+} doped $Ba_3Y(PO_4)_3$ material is red because of the emission range of 620-760nm is the red light region ,which show the intensity of the emission peaks of 672nm, 702nm and 712nm is higher. The wavelengths of $^5D_0$ transition to ^7F_J (J=0-5) and $^5D_1$ transition to ^7F_J (J=0-5) are different with the addition of Eu^{3+} rare earth ions. The transition mechanism is the f-f transition[2], which is almost unaffected by the crystal field. The emission spectrum of $Ba_3Y(PO_4)_3$:$0.08Eu^{3+}$ is linear, and Eu^{3+}in the matrix lattice determines its emission wavelength. On the one hand, Eu^{3+} occupy the non inversion symmetry in the lattice. The red light is produced because of the 5d configuration mixed into the $^4F_6$ configuration, thus the uneven crystal field causes the F-F transition not to be maintained. On the other hand, the location of the Eu^{3+} is the center of the lattice inversion symmetry, and its emission spectrum is almost all in the $^5D_0$ to $^7F_1$ orange emission area.

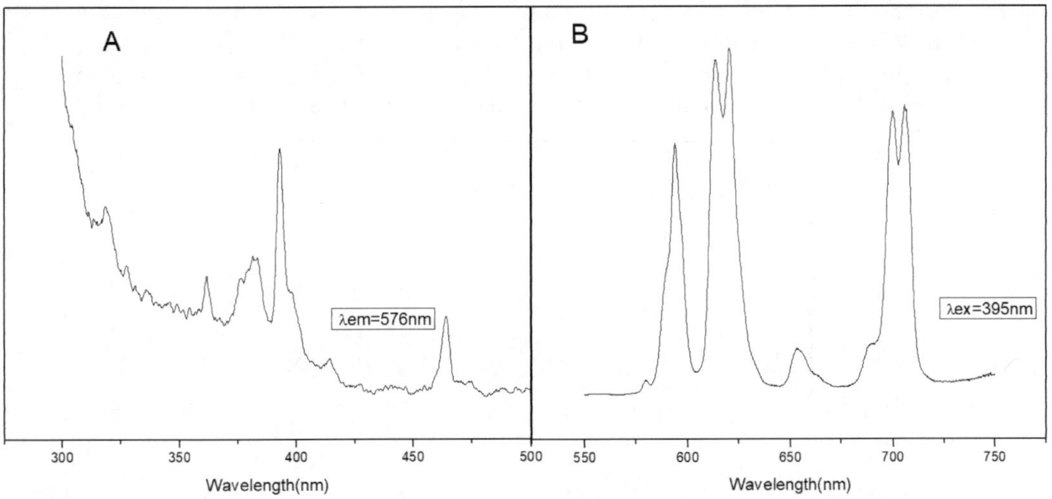

Figure 3. (A)The excitation spectra and the emission spectra(B) of $Ba_3Y(PO_4)_3$:0.08Eu^{3+}

Fig. 4 is the emission spectrum of $Ba_3Y(PO_4)_3$:0.08Dy^{3+}. There are three peaks with different emission intensities. The emission peak of 484nm corresponds to the blue light emission. It may be the $^4F_{9/2}$ to $^6H_{15/2}$ transition of Dy^{3+}. The emission peak of 575nm is yellow light emission, which may be produced by the $^4F_{9/2}$ to $^6H_{13/2}$ transition of Dy^{3+}. The emission peak of 671nm is corresponding to the weaker red emission relative to the former. The transition is produced because of the three emission peaks are located in the blue, yellow and red regions,thus they can be combined to get better white light.

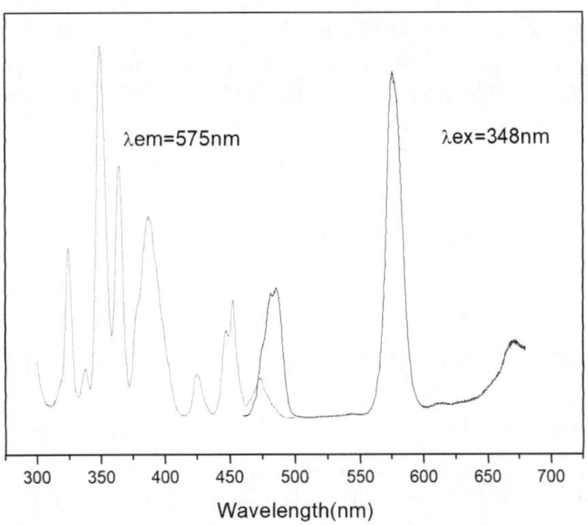

Figure 4 The excitation spectrogram and emission spectrogram of $Ba_3Y(PO_4)_3$:0.08Dy^{3+}

It can be seen from the excitation peak that the excitation peak located at 324nm may be caused by $^6H_{15/2}$ to $^6P_{3/2}$. The excitation peak of 340nm, 354nm, 425nm, 454nm and 474nm is respectively the $^6H_{15/2}$ to $^6P_{7/2}$, $^6H_{15/2}$ to $^6P_{5/2}$, $^6H_{15/2}$ to $^4M_{21/2}$, $^6H_{15/2}$ to $^4G_{11/2}$ and $^6H_{15/2}$ to $^4I_{15/2}$. The strongest excitation peak is the excitation peak at 354nm.As shown in Fig.5, it can be seen that the intensity of luminescence first increases with the increase of doping concentration, and then decreases with the increase of doping concentration. When the doping concentration of Dy^{3+} is less than or equal to 8%, the luminescence intensity of $Ba_3Y(PO_4)_3$:xDy^{3+} increases with the increase of Dy^{3+} doping concentration. When the Dy^{3+} concentration is equal to 8%, the luminescence intensity of the sample is the largest. When the doping concentration of Dy^{3+} increases to 10%, the luminescence intensity of the sample becomes weaker, which is both concentration quenching. The doping concentration of Dy^{3+} rises to a certain extent, and there will be no radiative energy transfer, resulting in the weakening of luminous intensity. The concentration quenching of inorganic nonconductive photoluminescent materials is caused by the electric multipole between ions -- Dexter theory [8].

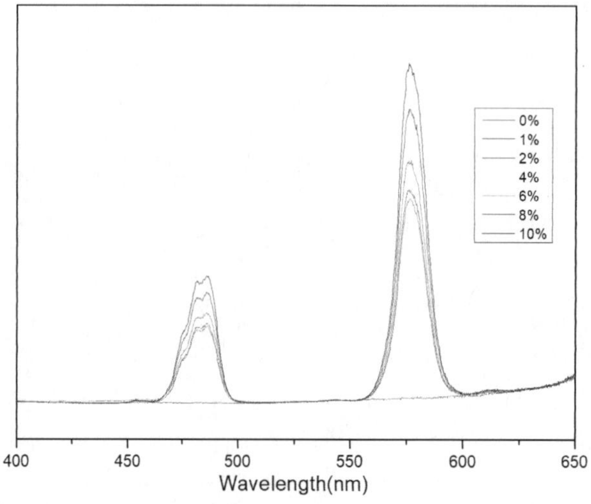

Figure 5. The Emission spectra of $Ba_3Y(PO_4)_3$ with different Dy^{3+} doped contents.

4. Conclusions

A series of Dy^{3+} and Eu^{3+} single doped $Ba_3Y(PO_4)_3$ phosphors were synthesized by the traditional high-temperature solid-state method. Under the excitation of near ultraviolet light, the phosphors of this series have great luminescence properties.$Ba_3Y(PO_4)_3$:Dy^{3+} phosphor white light emission and $Ba_3Y(PO_4)_3$:Eu^{3+} phosphor red light emission. Because doping Eu^{3+} ions compensate for the lack of red light components in Dy^{3+} cold white light in $Ba_3Y(PO_4)_3$:Dy^{3+}phosphor, the Dy^{3+} is changed by changing Dy^{3+}.The concentration and relative doping ratio of Eu^{3+} ions can achieve the luminescence color of phosphor, from cold white light to warm white light, and then to orange red light. When the concentration of doped ions is too high, the concentration quenching effect will happen. $Ba_3Y(PO_4)_3$: Dy^{3+} and $Ba_3Y(PO_4)_3$:Eu^{3+} have potential applications in near ultraviolet excited white light diode (NUV-WLED) devices.

Acknowledgments

This work was supported by the National Natural Science Foundation of China (51374155), the Hubei Province Key Technology R&D Program (2014BCB034) , the National Natural Science Foundation of Hubei province of China (2014CFB796) , the Science & Technology Pillar Program of Hubei Province (grant no. 2015BAA105), and National Innovation and Entrepreneurship Training Project of Local University Students of P.R. China (grant no.201310490008, 20141049047 and 20141049043) for the financial support.

References

[1] Guo N, Huang Y, You H, et al. 2010. $Ca_9Lu(PO_4)_7$:Eu^{2+},Mn^{2+}: A Potential Single-phased White-light-emitting Phosphor Suitable for White-light-emitting Diodes[J]. Inorganic Chemistry,49(23) : 10907-10913.

[2] Shang M,Li G,Kang X, et al. 2011. Tunable Luminescence and Energy Transfer Properties of Sr_3AlO_4F:RE^{3+} (RE=Tm/Tb,Eu,Ce) Phosphors[J]. Acs Applied Materials & Interfaces, 3(7) : 2738-2746.

[3] Bandi VR, Grandhe B K,Jang K,et al. 2012. Citric Based Sol gel Synthesis and Luminescence Characteristics of $CaLa_2ZnO_5$:Eu^{3+} Phosphors for Blue LED Excited White LEDs[J]. Journal of

Alloys and Compounds, 512(1) : 264-269.
[4] Han B, Yang G, Zhang G, et al.2008. Enhanced luminescence of $Ba_3La(PO_4)_3:Dy^{3+}$ by codoping Gd^{3+} ions and energy transfer between Gd^{3+} and Dy^{3+}[J]. Josa B, 25(12): 2057-2063.
[5] Liu X, Xiang W, Chen F, et al. 2013. Synthesis and photoluminescence characteristics of Dy^{3+},doped $NaY(WO_4)_2$, phosphors[J].Materials Research Bulletin, 48(2):281-285.
[6] Nagpure I M, Saha S, Dhoble S J. 2009. Photoluminescence and thermoluminescence characterization of Eu^{3+} and Dy^{3+}-activated $Ca_3(PO_4)_2$, phosphor[J]. Journal of Luminescence, 129(9):898-905.
[7] Wu Z C, Shi J X, Wang J, et al.2006. A novel blue-emitting phosphor $LiSrPO_4:Eu^{2+}$, for white LEDs[J]. Cheminform, 37(46): 2356-2360.
[8] Dexter D L, Schulman J H.1954. Theory of Concentration Quenching in Inorganic Phosphors[J]. Journal of Chemical Physics, 22(6):1063-1070.

Preparation of Flowerlike Porous $Mg_5(CO_3)_4(OH)_2 \cdot 4H_2O$ by CO_2 Bubble Templating

LEI Hong[1], WEI Yu[1], HUANG Zhiliang*[1], KONG Shijin[1] and YAO Donghui[1]

[1]School of Materials Science and Engineering, Wuhan Institute of Technology, Wuhan 430074, People's Republic of China

* correspondence: hzl6455@126.com

Abstract: Flowerlike porous basic magnesium carbonate ($Mg_5(CO_3)_4(OH)_2 \cdot 4H_2O$) were obtained successfully using magnesium chloride hexahydrate and ammonium bicarbonate as raw materials. We investigated the effect of reaction time and reaction temperature on the $Mg_5(CO_3)_4(OH)_2 \cdot 4H_2O$ preparation. In the first three hours' reaction, Magnesium carbonate trihydrate ($MgCO_3 \cdot 3H_2O$) whiskers generated, and transferred to the flowerlike porous $Mg_5(CO_3)_4(OH)_2 \cdot 4H_2O$ as the reaction time prolonged. During the transformation, the flowerlike $Mg_5(CO_3)_4(OH)_2 \cdot 4H_2O$ crystals firstly appeared at the head of the $MgCO_3 \cdot 3H_2O$ whisker.

1. Introduction

The morphology and properties of particles are closely related. To fabricate materials with certain morphology to achieve better performance, there has been increasing interests in controlled synthesis of inorganic microstructure with desirable shape and size[1-2].

Because of the special structure, spherical porous materials have advantages in easily moving, closely packing and occupying the available vacancies, which implies potential applications[3]. In recent years, in particular, spherical porous materials have attracted much attention and many studies have been carried out on the superiority of spherical porous materials. For example, Wang et al. have synthesized spherical porous $CaCO_3$ microparticle which could be applied as relatively safe drug vehicles. Yu *et al* have prepared spherical porous $LiFePO_4/C$ particles which provide better contact with electrolyte and are easier to bind than isolated $LiFePO_4$ particles[4-5].

This paper presents the formation process of flowerlike porous $Mg_5(CO_3)_4(OH)_2.4H_2O$. By investigating particles under various reaction times, the shape evolution of these flowerlike porous particles is discussed in detail. CO_2 bubble template was introduced to simulate the formation of flowerlike porous $Mg_5(CO_3)_4(OH)_2.4H_2O$, and the equation of reaction temperature and the pore size was given by the derivation of Laplace and Clapyeron equations.

2. Experimental

Magnesium chloride hexahydrate ($MgCl_2 \cdot 6H_2O$) and ammonium bicarbonate (NH_4HCO_3) were used and were of analytical grade. In this experiment, 0.4 mol/L of $MgCl_2 \cdot 6H_2O$ and NH_4HCO_3 (1/2 molar ratio of $MgCl_2 \cdot 6H_2O$) were mixed well and transferred into automatic Control Reaction Kettle. Then, the homogeneous reaction solution was heated at a constant pressure of 4 atm. The reaction temperature was maintained at 80 °C for a given reaction time. The reaction time was set as 3~6 h. After reaction finished, the special autoclave was taken out for ageing for 3 h. Afterwards, a filtration was implemented by vacuum suction filter method and washed with distilled water as well as ethanol. The washed residue was dried in vacuum oven for 2 h.

The crystal phases of the products were determined by XD-5A-type powder X-ray diffraction (XRD) with Ni-filtered Cu Ka radiation. The morphology of the product was observed using a JSM-5510LV scanning electron microscope (SEM).

3. Results and discussion

Figure. 1 shows XRD patterns of the mixtures produced under 80 °C for different duration times. The XRD pattern of the mixture prepared by reacting 3 h displays only the peaks of $MgCO_3 \cdot 3H_2O$. The XRD pattern of the mixture obtained after reacting 4 h are dominated by the peaks of $MgCO_3 \cdot 3H_2O$, with the main characteristic peaks of $Mg_5(CO_3)_4(OH)_2 \cdot 4H_2O$ just starting to show up in the XRD. 5 h later, $Mg_5(CO_3)_4(OH)_2 \cdot 4H_2O$ becomes the dominant phase in the mixture. $Mg_5(CO_3)_4(OH)_2 \cdot 4H_2O$ is the only phase present in the XRD pattern of the mixture when reaction time was increased to 6 h.

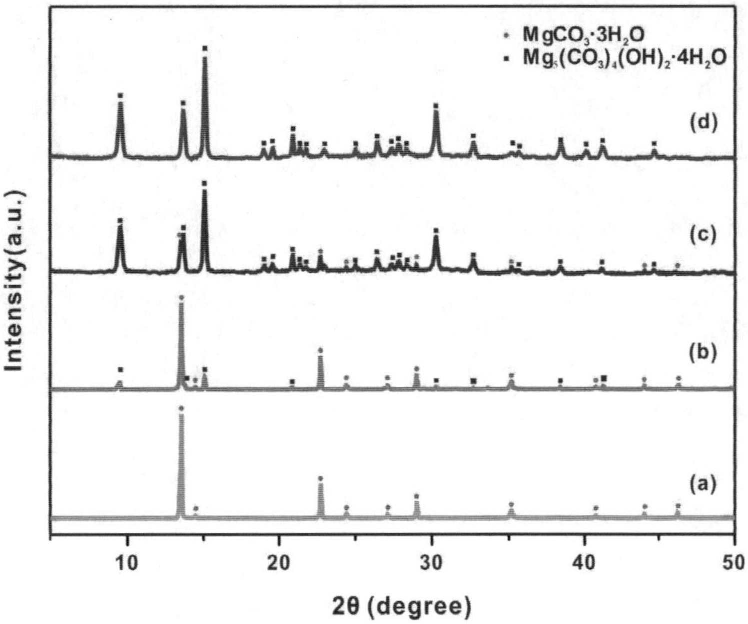

Figure 1. XRD patterns of the heated mixtures at 80 °C for different duration times (a: 3 h; b: 4 h; c: 5 h; d: 6h)

Figure. 2 shows SEM images of the particles synthesized at different reaction time, which indicates an obvious growth process occurring from whisker to flowerlike microsphere with porous structure. As shown in Fig. 2a, a large number of whiskers were obtained by reacting 3 h. Through the phase analysis, the product was determined as $MgCO_3 \cdot 3H_2O$ (Fig. 1a). The structure and surface of $MgCO_3 \cdot 3H_2O$ whiskers were unbroken and smooth. In addition, the length was about 140 μm and the aspect ratio was about 25. At the time of reacting 4 h, the sample was still whisker with reduced aspect ratio, but the surface became rough and a small number of leaf-like crystals appeared on the surface of the whisker. When the reaction time was increased to 5 h, the sample not only contained whisker product, but also began to appear flowerlike spherical particles. From a magnifying SEM image, it seemed that "connected flakes" gathered on the head of the whisker (Fig. 2c). When the samples were produced for reacting 6 h, the original whisker shape disappeared completely. All samples turned into the flowerlike porous $Mg_5(CO_3)_4(OH)_2 \cdot 4H_2O$ that are assembled by lamella (Fig. 2d).

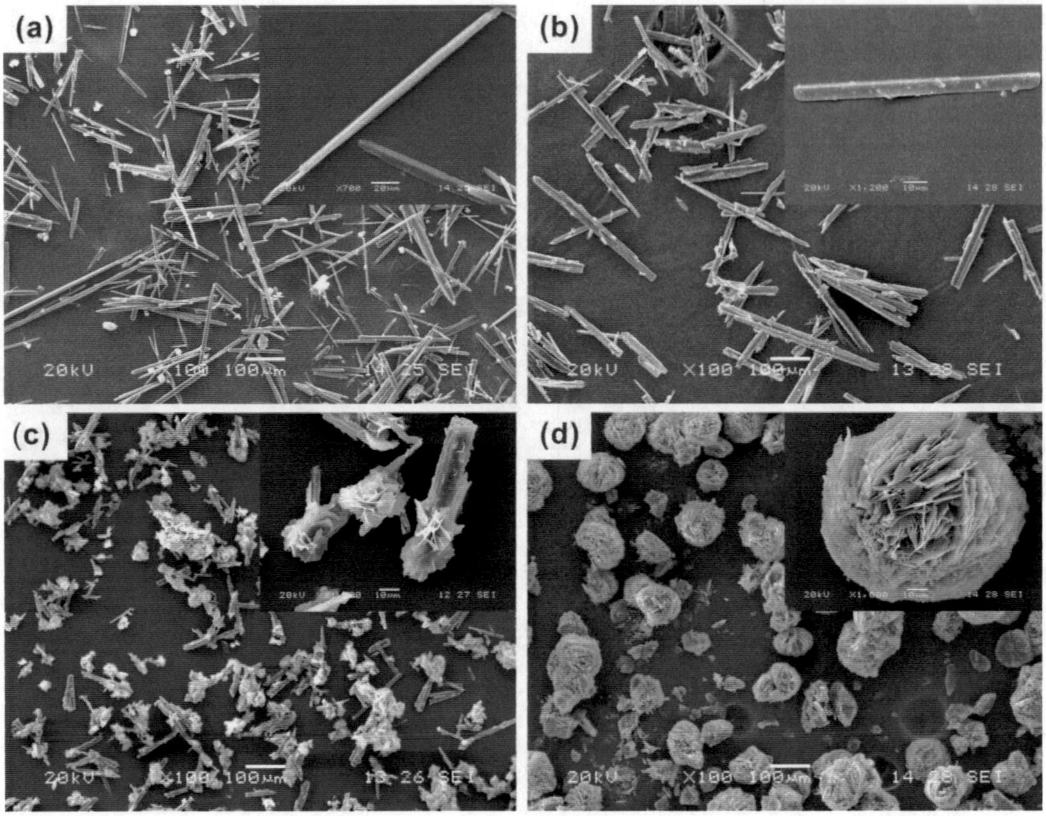

Figure 2. SEM images of the particles synthesized at different reaction time (a: 3 h; b:4 h; c: 5 h; d: 6 h)

The conversion reaction giving rise to the $Mg_5(CO_3)_4(OH)_2{\cdot}4H_2O$ crystal occurs via the formation of metastable $MgCO_3{\cdot}3H_2O$ in the synthetic process[6-8]. The reaction time has a significant impact on the phase transition process. Reactions of the transition process occur as follows:

$$MgCl_2 \cdot 6H_2O + 2NH_4HCO_3 \rightarrow MgCO_3 \cdot 3H_2O + 2NH_4Cl + CO_2 \uparrow + 4H_2O \quad (1)$$

$$5MgCO_3 \cdot 3H_2O \rightarrow Mg_5(CO_3)_4(OH)_2 \cdot 4H_2O \downarrow + CO_2 \uparrow + 10H_2O \quad (2)$$

As shown in Figure. 2d, the surfaces of the flowerlike $Mg_5(CO_3)_4(OH)_2{\cdot}4H_2O$ are assembled from a number of lamellae that interconnected to form an open porous structure. Generally, gas bubbles, such as CO_2 bubbles, can form open porous structure [9]. With the increase of reaction time, the NH_4HCO_3 slowly releases CO_2 bubbles. At the same time, in order to obtain lower surface tension to achieve a relatively stable state, small CO_2 bubbles gather into big spheres of micrometer size and preferentially adhere to the head of whisker (see Figure. 3). This process can be explained by Gibbs function:

$$dG = -SdT + Vdp + \sum_{\alpha} \sum_{B} \mu_{B(\alpha)} dn_{B(\alpha)} + \gamma dA_s \quad (3)$$

where G, S and V are Gibbs function, entropy and volume of system, separately; μ_B and n_B are chemical potential and amount of substance of component B in phase A, respectively; A_S is the interfacial area; γ is the interfacial tension. Since the temperature and pressure are constant, equation 3 can be simplified as follows:

$$dG = \gamma dA_s \quad (4)$$

If γ remains constant, upon integration equation 4 gives

$$G^{s} = \gamma A_{s} \tag{5}$$

Further differential equation can be obtained:

$$dG^{s} = \gamma dA_{s} + A_{s} d\gamma \tag{6}$$

Under the condition of constant temperature and constant pressure, the decrease of Gibbs function of interface system is a spontaneous process from the thermodynamic view. According to the equation 6, In order to reduce the interfacial Gibbs function to achieve relatively stable state, Many small CO_2 bubbles reduce its surface area by gathering into big CO_2 bubbles , and the head of whisker reduce the surface tension by dissolving and adsorpting CO_2 bubbles.

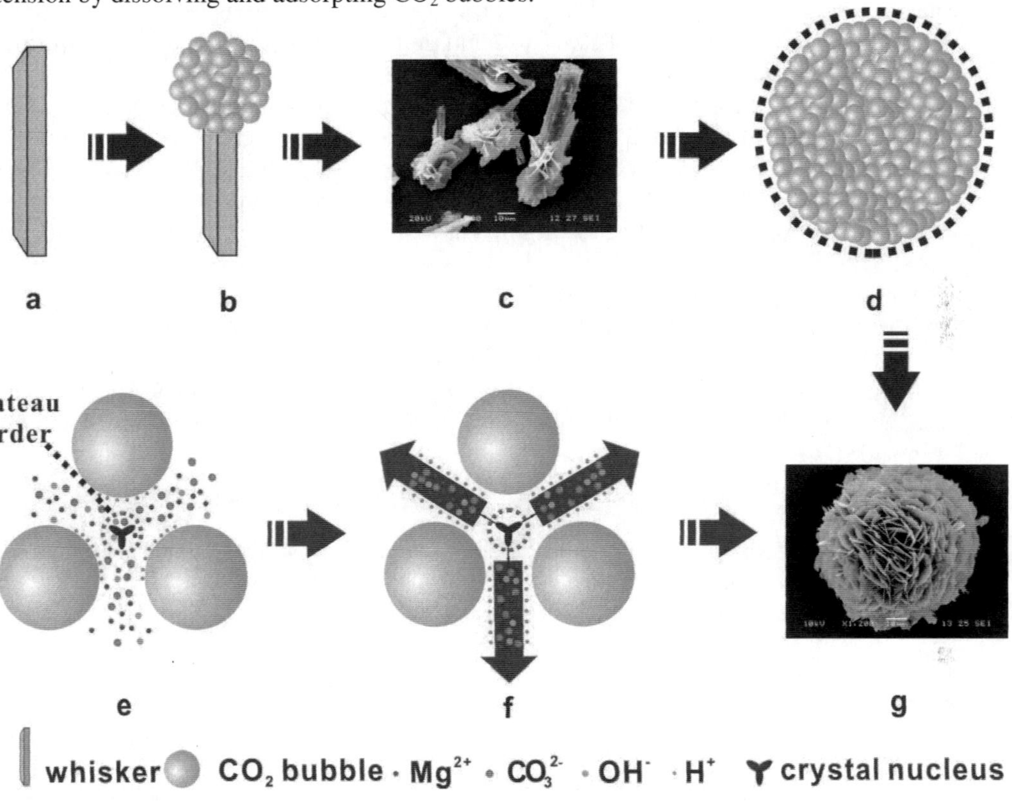

Figure 3. Illustration of the growth process of $Mg_5(CO_3)_4(OH)_2{\cdot}4H_2O$

4. Conclusions

Flowerlike porous $Mg_5(CO_3)_4(OH)_2{\cdot}4H_2O$ have been synthesized successfully using $MgCl_2{\cdot}6H_2O$ and NH_4HCO_3 as raw materials. The formation of porous structure and the control of system temperatures over the pore size of porous $Mg_5(CO_3)_4(OH)_2{\cdot}4H_2O$ were studied.

(1) The products synthesized at 3 h were $MgCO_3{\cdot}3H_2O$ whiskers. After a series of phase transition process, it turned out that the final products for 6 h are flowerlike $Mg_5(CO_3)_4(OH)_2{\cdot}4H_2O$ with porous structure. Moreover, flowerlike $Mg_5(CO_3)_4(OH)_2{\cdot}4H_2O$ crystals firstly appear at the head of the $MgCO_3{\cdot}3H_2O$ whiskers.

(2) The formation of porous structure of flowerlike $Mg_5(CO_3)_4(OH)_2{\cdot}4H_2O$ could be explained. The growth and gathering of the CO_2 bubbles induced the growth of the $Mg_5(CO_3)_4(OH)_2{\cdot}4H_2O$ flakes and formation of the porous structure.

Acknowledgments

This work was supported by the National Natural Science Foundation of China (51374155), the Hubei

IOP Conf. Series: Materials Science and Engineering **389** (2018) 012032 doi:10.1088/1757-899X/389/1/012032

Province Key Technology R&D Program (2014BCB034) and the National Natural Science Foundation of Hubei province of China (2014CFB796).

References

[1] Li G, Zhou M, Ma W W, et al. 2010. Controlled Size and Density Distribution of Nanoparticles by Thermal Ammonia Etching Method[J]. Journal of Wuhan University of Technology-Mater. Sci. Ed. 25(1): 108-111.
[2] Paul W, Sharma C P. 1999. Development of Porous Spherical Hydroxyapatite Granules: Application towards Protein Delivery [J]. Journal of Materials Science: Materials in Medicine,10(7): 383-388.
[3] Wang G, Liu H, Liu J, et al. 2010. ChemInform Abstract: Mesoporous $LiFePO_4/C$ Nanocomposite Cathode Materials for High Power Lithium Ion Batteries with Superior Performance [J]. Advanced Materials, 22(44): 4944-4948.
[4] Yu F, Zhang J, Yang Y, et al. 2010. Porous Micro-Spherical Aggregates of $LiFePO_4/C$ Nanocomposites: A novel and Simple Template-Free Concept and Synthesis Via Sol–Gel-Spray Drying Method [J]. Journal of Power Sources, 195(19): 6873-6878.
[5] Hong M H, Son J S, Kim K M, et al. 2011. Drug-Loaded Porous Spherical Hydroxyapatite Granules for Bone Regeneration [J]. Journal of Materials Science: Materials in Medicine, 22(2): 349-355.
[6] Hopkinson L, Rutt K, Cressey G. 2008. The Transformation of Nesquehonite to Hydromagnesite in the System CaO-MgO-H2O-CO2CaO-MgO-H2O-CO2: An Experimental Spectroscopic Study [J]. The Journal of Geology, 116(4): 387-400.
[7] Rautaray D, Sinha K, Shankar S S, et al. 2004. Aqueous Foams as Templates for the Synthesis of Calcite Crystal Assemblies of Spherical Morphology [J]. Chemistry of Materials, 16(7): 1356-1361.
[8] Kim T K, Yoon J J, Lee D S, et al. 2006. Gas Foamed Open Porous Biodegradable Polymeric Microspheres [J]. Biomaterials, 27(2): 152-161.
[9] Cheng X, Huang Z, Li J, et al. 2013, Self-Assembled Growth and Pore Size Control of the Bubble-Template Porous Carbonated Hydroxyapatite Microsphere [J]. Crystal Growth & Design, 10(3): 1180-1188.

IOP Conf. Series: Materials Science and Engineering **389** (2018) 012033 doi:10.1088/1757-899X/389/1/012033

Study on Dispersion Stability of SiC Slurry and Its Membrane Properties

MA Dong[1], CHEN Changlian*[1], HUANG Zhiliang*[1], HUANG Xiaoyu[1], LIANG Xin[1] and ZHOU Shicong[1]

[1]School of Materials and Engineering, Wuhan Institute of Technology, Wuhan 430205, People's Republic of China

*correspondence:cnsdqdccl@hotmail.com; hzl6455@126.com

Abstract: Silicon carbide (SiC) slurry with a low concentration was prepared, which was milled for 3 hours by mixing SiC powder with a certain amount of polyethylene glycol (PEG), tetramethyl ammonium hydroxide (TMAH), carboxymethyl cellulose sodium (CMC-Na), distilled water, and ethonal. The influence of the pH value on the Zeta potential and the additional components on the dispersion stability of the slurry were analyzed. The influence of various factors on slurry was studied by orthogonal experiment. The results showed that solid content had the greatest influence on viscosity and stability of slurry. After the slurry was coated on the surface of porous SiC ceramics and sintered at high temperature to form a SiC membrane, the micro-structure of SiC membrane and its pore size were characterized. The result showed that the pore size distribution of SiC membrane was narrow, and the average pore diameter was about 1μm. Meanwhile, the resultant grains tend to be round, uniform, single phase. Based on the above, we could know that the slurry has a good membrane-forming properties.

1. Introduction

With the increase of environmental protection in China, membrane materials are widely used in filtration and purification of from industrial wastewater and waste gas [1-4]. Silicon carbide materials and their products have strong covalent bonds. They are widely concerned with the excellent characteristics of high temperature resistance and corrosion resistance. They are widely used in the purification of water for food, fermentation, pharmaceutical, biological products, plant extraction, chemical industry, textile and other fields, the concentration of chemical products, the recycling treatment of the production process and the treatment of waste water. The effect of different dispersants and pH values on the rheological property and stability of SiC slurry was studied. The optimum mixture ratio of the slurry used in low concentration SiC membrane was obtained. Finally, the SiC ceramic membrane with narrow pore size distribution was prepared, and the phase, structure and properties of the membrane were also characterization [5-8].

2. Experimental

A certain amount of deionized water, different content of polyethylene glycol (PEG), four methyl ammonium hydroxide (TMAH), sodium carboxymethyl cellulose (CMC-Na) and silicon carbide (SiC) powder are mixed to the ball mill, and the pH value is adjusted with ammonia and hydrochloric acid. SiC was prepared by ball milling for 3 h at a rotational speed of 190 r/min. The zero number rotor was

IOP Conf. Series: Materials Science and Engineering **389** (2018) 012033 doi:10.1088/1757-899X/389/1/012033

selected, and the viscosity of SiC slurry was measured under the condition of 60 r/min speed, and the relative settlement height of SiC slurry was observed by loading a 50 mL measuring cylinder. The slurry is coated on the SiC support body suface. The spraying rule is sprayed on the surface of the SiC support by the self-made spraying equipment in the laboratory. The SiC support body after the membrane is put into a dry blast oven to dry 4 h, and the dried green body is put into the SiC sintering furnace to burn to 1950 °C, and the heat preservation is 1 h. The SiC ceramic microfiltration membrane is obtained with the cooling out of the furnace.

3. Results and discussion

Table. 1 shows the influence of various factors on the viscosity of the slurry. The primary relationship of the viscosity of SiC is a>d>c>b, that is, the change of solid content has the greatest impact on the viscosity of the slurry, followed by the pH value and CMC-Na, and the smallest is TMAH. The optimum scheme is solid content of 6wt%, TMAH volume of 1.5wt%, CMC-Na volume of 0.2wt% and pH value of 9. Under this scheme, slurry viscosity is the lowest.

Table. 1 The influence of various factors on the viscosity of the slurry

Number	Solid content (wt %)	TMAH (wt %)	CMC-Na (wt %)	pH	Viscosity (mPa.s)
1	6	1	0.2	8	3.98
2	6	1.5	0.6	9	4.39
3	6	2	1.0	10	4.85
4	10	1	0.6	10	5.40
5	10	1.5	1.0	8	6.58
6	10	2	0.2	9	5.05
7	14	1	1.0	9	8.56
8	14	1.5	0.2	10	6.05
9	14	2	0.6	8	7.28
K1	13.22	17.94	15.08	17.84	
K2	17.03	17.02	17.07	18	
K3	21.89	17.18	19.99	16.3	

Number	Solid content (wt %)	TMAH (wt %)	CMC-Na (wt %)	pH
k1	4.41	5.98	5.03	5.95
k2	5.68	5.67	5.69	6
k3	7.30	5.73	6.36	5.43
R	2.89	0.92	1.64	1.7

Table. 2 shows the influence of various factors on the sedimentation height of the slurry. The primary relationship of the sedimentation height of SiC is a>c>d>b, that is, solid content is the most important, followed by CMC-Na and pH, and TMAH is the least affected. The preferred scheme is a1b2c1d2 or a1b3c1d2, that is, the solid content is 6wt%, the TMAH is 1.5 wt% or 2 wt%, CMC-Na is 0.2 wt%, and the pH value is 8. Under this scheme, the height of the slurry is the lowest and the most stable.

Table. 2 The influence of various factors on the sedimentation height of the slurry

Number	Solid content (wt %)	TMAH (wt %)	CMC-Na (wt %)	pH	Highth (mL)
1	6	1	0.2	8	1.5
2	6	1.5	0.6	9	2.0
3	6	2	1.0	10	2.5

4	10	1	0.6	10	4
5	10	1.5	1.0	8	4
6	10	2	0.2	9	2.5
7	14	1	1.0	9	5.8
8	14	1.5	0.2	10	5
9	14	2	0.6	8	6
K1	6	11.3	9	11.5	
K2	10.5	11	12	10.3	
K3	16.8	11	12.3	11.5	

Number	Solid content (wt %)	TMAH (wt %)	CMC-Na (wt %)	pH
k1	2	3.77	3	3.83
k2	3.5	3.67	4	3.43
k3	5.6	3.67	4.1	3.83
R	3.6	0.1	1.1	0.4

Figure. 1 shows SEM images of SiC membranes. After sintering at 1950 C and holding 1 h, the SiC ceramic membrane was obtained. The sample particles of the ceramic film became more rounded than before the burning, and the sintered necks formed between the particles were formed between the particles, and there were no separate SiC particles. This would help to increase the ceramic film. The intensity and the distribution of the pore size of the diaphragm are uniform.

Figure. 1 SEM images of SiC membranes(a) 6% solid content by dip-coating; (b) unsintered

Figure. 2 shows the XRD patterns of the SiC ceramic membrane. The ceramic film is mainly composed of the SiC of the space point group of P63mc (186), and also contains a small amount of 3C-SiC and 6H-SiC. By comparing the XRD diagram of the burning before burning, the ceramic film is sintered at 1950 °C, and the phase is no change but the diffraction peak becomes sharper. It shows that the crystalline degree of the crystal becomes higher at 1950 °C, and the result of Figure. 2 is also proved.

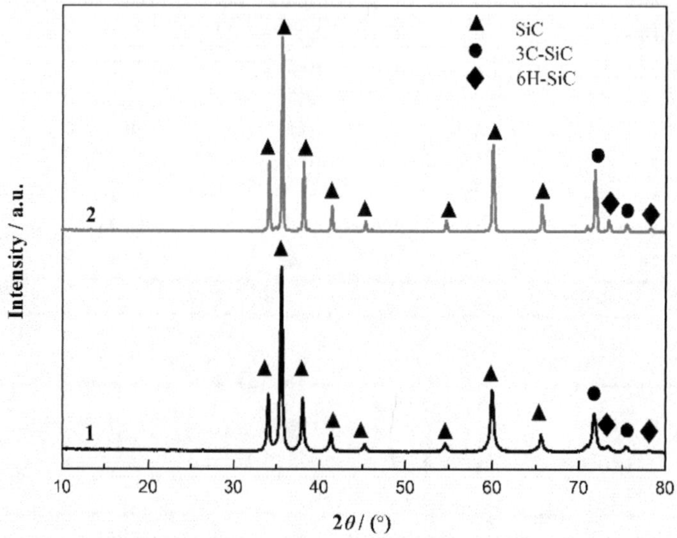

Figure. 2 XRD patters of membrane unsintered (1) and sintered at 1950 °C (2) for 1 h

Figure. 3 shows pore-size distribution of SiC membranes. The pore size distribution in 0.75-1.60 μm, the average pore size is 1.10 μm.

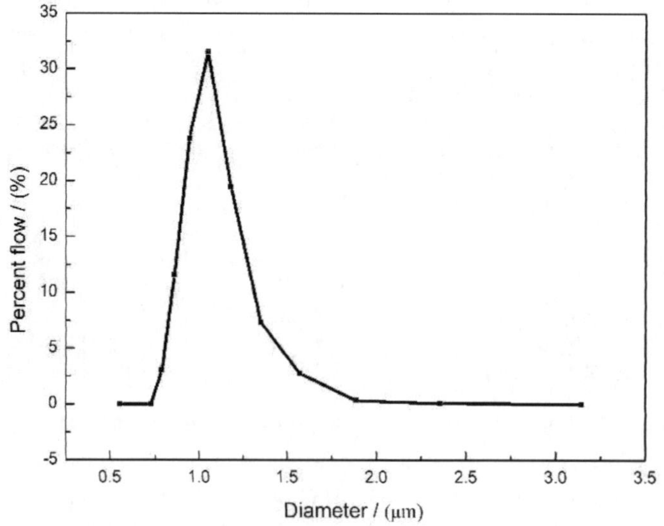

Figure. 3 Pore-size distribution of SiC membranes

4. Conclusions

Through the orthogonal experiment, the solid content has the greatest influence on the viscosity and stability of the slurry, when the solid content is 6 wt%, the amount of TMAH is 1.5 wt%, the CMC-Na is 0.2 wt%, the pH value is 9, and the viscosity of the slurry is the lowest. When solid content is 6 wt%, TMAH is 1.5wt% or 2wt%, CMC-Na is 0.2 wt%, pH value is 8, sedimentation height of the slurry is the lowest. After sintering, the grains around the silicon carbide are round and uniformly stacked, the phase is single, and the pore size distribution is narrow. The average pore size is about 1 μm.

Acknowledgments

This work was supported by the National Natural Science Foundation of China (51374155), the Hubei Province Key Technology R&D Program (2014BCB034), Major Projects of Technical Innovation in Hubei Province (2016ACA161) and the National Natural Science Foundation of Hubei province of China (2014CFB796).

References

[1] SUN G H,DENG X Y,SUN Y S,et al. 2015, Preparation and properties of silicon carbide porous ceramic filter membrane [J]. Key Engineering Materials, 655:36-40.
[2] SHQAU K,MOTTERN M L,DI Y,et al. 2006, Preparation and properties of porous α-Al2O3 membrane supports [J]. Journal of the American Ceramic Society, 89(6):1790-1794.
[3] HYUN S,KIM G. 1997, Synthesis of ceramic microfiltration membranes for oil/water separation [J]. Separation Science, 32(18):2927-2943.
[4] AHMAD N A,LEO C P,AHMAD A L. 2013, Superhydrophobic alumina membrane by steam impingement: Minimum resistance in microfiltration [J]. Separation & Purification Technology,107(4):187-194.
[5] TANAKA H. 2011, Silicon carbide powder and sintered materials [J]. Journal of the Ceramic Society of Japan, 119(1387):218-233.
[6] VALES F,REZAKHANLOU R,OLAGNON C. 1999, Determination of the fracture mechanical parameters of porous ceramics from microstructure parameters measured by quantitative image analysis [J]. Journal of Materials Science, 34(16):4081-4088.
[7] SHE J,YANG J F,KONDO N,et al. 2002, High-strength porous silicon carbide ceramics by an oxidation-bonding technique [J]. Journal of the American Ceramic Society, 85(11):2852-2854.
[8] ELYASSI B,SAHIMI M,TSOTSIS T T. 2007, Silicon carbide membranes for gas separation applications [J]. Journal of Membrane Science, 288(1/2):290-297.

IOP Conf. Series: Materials Science and Engineering **389** (2018) 012034 doi:10.1088/1757-899X/389/1/012034

Study on Intercalation of Organic Mica with $CnH_{37}(CH_3)_3NCl$ (n=12, 14, 16, 18) and Its Independent Interlayer Spacing

YAO Donghui[1], WEI Yu[1], HUANG Zhiliang*[1], LEI Hong[1], MA Dong[1] and PENG Weikang[1]

[1]School of Materials Science and Engineering, Wuhan Institute of Technology, Wuhan 430074, People's Republic of China

* correspondence: hzl6455@126.com

Abstract. We have prepared a special organic mica composite by inserting $CnH_{37}(CH_3)_3NCl$ (n=12, 14, 16, 18) surfactant into previously prepared sericite. This complex exhibits a unique interlayer spacing feature. As the surfactant concentration increases, the interlayer spacing will remain unchanged. So far, similar reports haven't been reported.In addition, FT-IR showed that the organic molecules in this material will adopt different structural arrangements at different concentrations.

1.Introduction
Lamellar organosilicate complexes possess special adsorption properties, which is a hot research topic in fields like industrial catalysis [1-2], membrane [3-4], and waste-water absorbent [5-7].

The height and stability of the layer space are two key factors that determine its application: the space determines the number of molecules and its corresponding stable efficiency. According to reports, the filling position of organic molecules and the extension of layer spacing are related to the concentration of organic substances. Organic molecules usually adopt a single-layer oblique ortho-arrangement at low concentrations, and turn to two-layer or even pseudo-three-layer arrangement at high concentrations.

Besides, due to the hydrophobic nature of organic matter, the interlayer space is also very hydrophobic. This means that his application will be limited, for example as an inorganic catalytic material.

This article describes the preparation of an organic mica composite by intercalating the surfactant $CnH_{37}(CH_3)_3NCl$ (n=12, 14, 16, 18). We have unexpectedly found that increasing the inter-layer space has nothing to do with the concentration, but it is related to the length of the carbon chain. This particular property implies his potential application.

2.Experiments

2.1Method
The Ba-mica (Ba-M) prepared by Ba-K ion exchange method was used, and its basic interlayer space was expanded to 1.206 nm[8]. Each time 0.2 g of Ba-M was weighed into a clean flask, the corresponding amount of $CnH_{37}(CH_3)_3NCl$ (n=12, 14, 16, 18) (≥95%, Aladdin) was added to the flask. Mix with 40.0 ml distilled water. The flask was then sealed and placed in water at 60 °C, and heated in a water bath while stirring with magnetic force for 5 hours. At the end of the reaction, filtration was

immediately carried out. The filter residue was washed with a mixture of alcohol and water (1:1) until it was neutral. The washed filter residue was then washed three times with distilled water and dried in a vacuum drying oven at 95 °C for 12 hours. Finally, the dried organic mica was stored in a sealed vacuum box.

The surfactant concentration was set to: 0.05 CEC, 0.15 CEC, 0.20 CEC, 0.25 CEC, 0.50 CEC, 0.75 CEC, 1.00 CEC, and 1.50 CEC. And the organic mica (O-M) samples obtained were named as M-0.05, M-0.15, M-0.20, M-0.25, M-0.50, M-0.75, M-1.00, and M-1.50 with different carbon chain length (n = 12, 14, 16, 18).

2.2. Characterizations

A powder X-ray diffractometer (D8 ADVANCE, Bruker, Germany) was used to characterize the basal spacing of crystalline planes of the samples. The acceleration voltage was 40 kV and the acceleration current was 40 mA. The X-ray wave length was 0.1540 nm. The scanning angle is from 2 to 50 degree, the scanning velocity was 0.01023 degree per step.

A thermo gravimetric analyzer (STA449F3, NETZSCH, Germany) was used to find out the organic quantity swallowed into interlayers. The TGA tests were under oxygen environment with a heating rate of 5 °C /min and from room temperature to 1000 °C.

A fourier transform infrared spectrometer (Nicolet Impact 6700) with a resolution of 4 K was used to examine the interaction between the organic groups and the aluminosilicate sheets. The sample was amounted on a KBr disc, and the sample used was measured as 2% wt. of the KBr.

3. Results and Discussion

3.1. Slight change of enlarged interlayer space

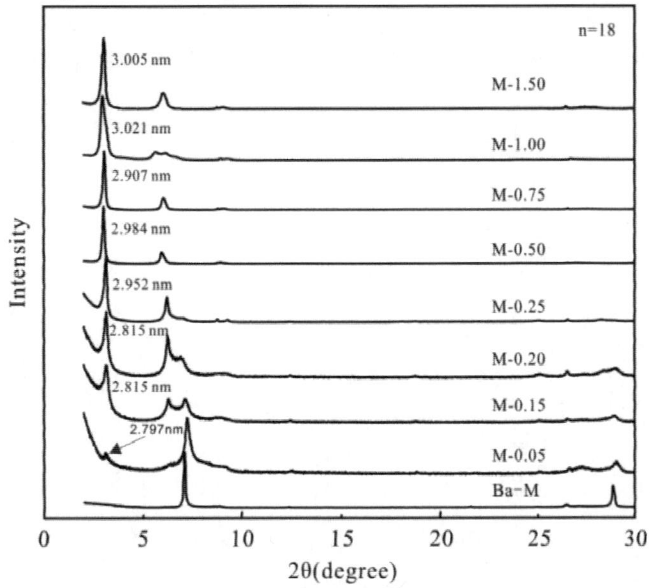

Figure 1. XRD patterns of $C_nH_{37}(CH_3)_3NCl$ (n=18) intercalated O-M.

As shown in Figure 1, the new peak appears gradually, and the original sericite peak gradually disappears, indicating that the sericite has been successfully intercalated. When the concentration of surfactant (n = 18) ranged from 0.05 CEC to 1.50 CEC., the size of d_{001} increased from 2.797 to 3.005 nm. According to reports, as the surfactant concentration increases, the substrate spacing also increases from about 1.4 nm to 4.0 nm. Therefore, the d_{001} for preparing organic mica in this paper seems to be independent of the surfactant concentration.

The space height between the interlayers is equal to the thickness of the sheet d_{001} minus 0.96 nm, and the increase range is 1.837-2.045 nm. The difference is 0.208 nm, and the thickness of a monolayer of water molecules is about 0.2 nm.Therefore, the slight difference in d_{001} may be caused by water molecules. We will discuss this further in FTIR analysis.

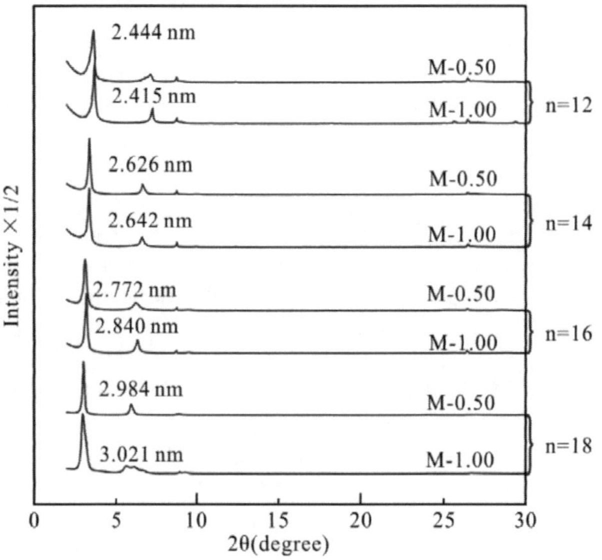

Figure 2. XRD patterns of $CnH_{37}(CH_3)_3NCl$ (n=12, 14, 16, 18) intercalated O-M

The concentration independence can also be observed in O-M pillared with different carbon lengths (n=12, 14, 16) (Figure 2). At 0.50 CEC and 1.00 CEC of same carbon length, the d_{001} are nearly same that the height of the basal spacing floated only 0.17-0.2 nm. And, the length of the long carbon chain shortens about 0.2 nm as every two carbon atoms cuts, which is in accordance with our observation.

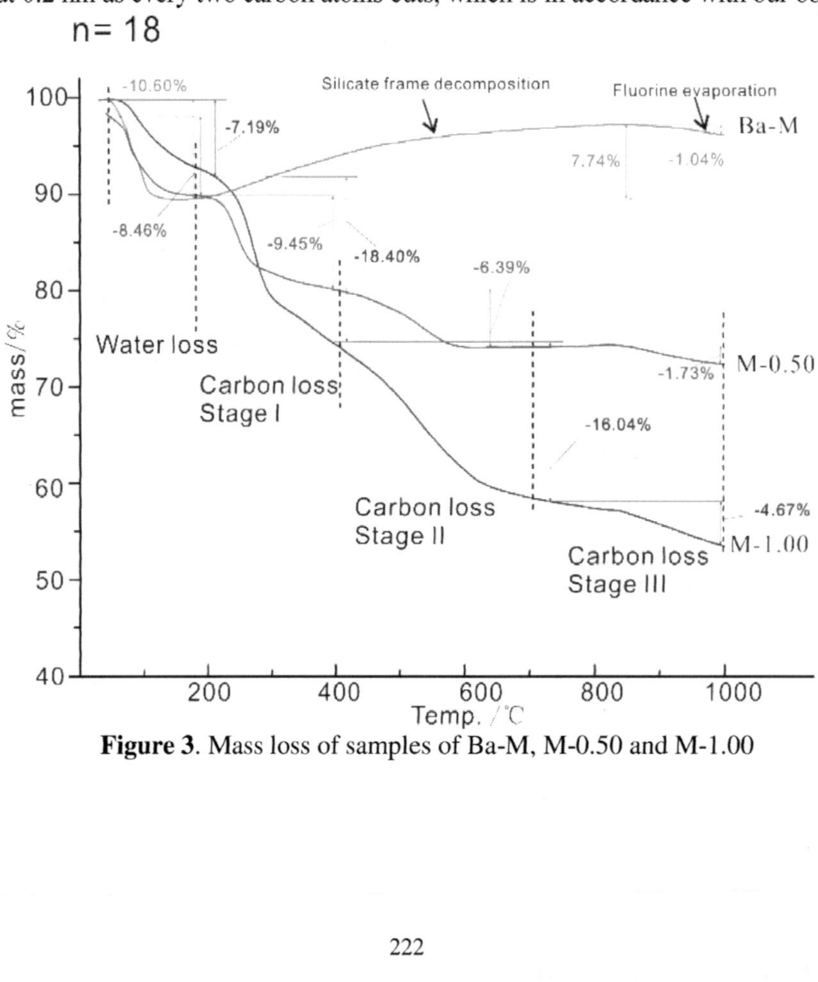

Figure 3. Mass loss of samples of Ba-M, M-0.50 and M-1.00

TGA tests were to conform if the concentration independence was caused by the saturation of surfactant molecules in the gallery space, as exampled $C_{18}H_{37}(CH_3)_3$ NCl (STAC) pillared O-M (Figure 3). In oxygen environment, the surfactant will burn out when heating. The mass loss between the 80-200 °C accounts for the structural water. The water loss is about 10.60% in Ba-M, 8.46% in M-0.50, and 7.19% in M-1.00. And the mass loss between the 200-1000 °C is due to the silicate frame decomposition in Ba-M, and in O-M it is the combination of organics and silicate frame decomposition.

The organics content in O-M, which is equal to the mass loss between 200-1000 °Cminus the corresponding mass loss in Ba-M, was 24.27% in M-0.50 and 45.81% in M-1.00. And ideally, the maximum of organics content was about 33.33% in M-0.50 and 50.00% in M-1.00. Therefore, the gallery space hasn't been completely soaked by the organic molecules below 1.00 CEC. It is then fair to say that the d001 of the O-M prepared is independent to the surfactant concentration.

3.2. Identical Conformational Arrangement of Surfactant Molecules

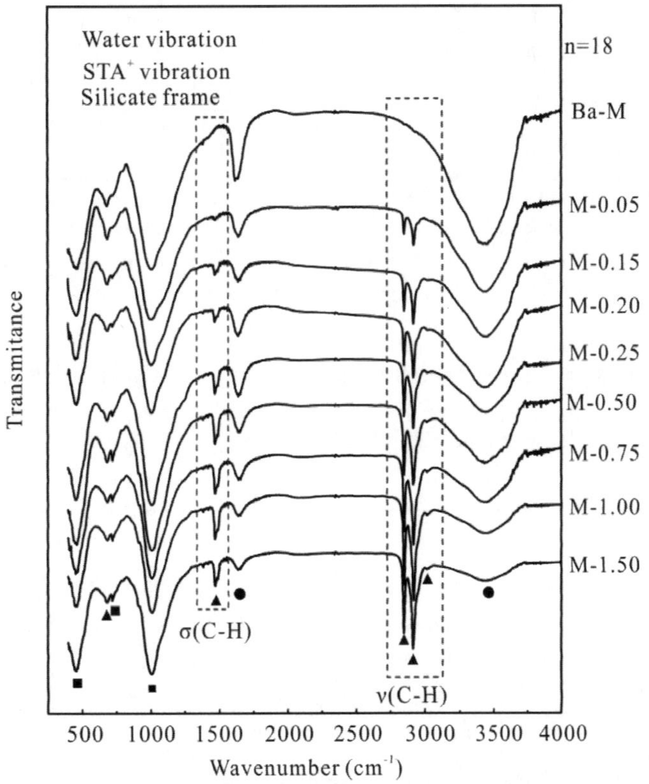

Figure 4. Mid-FTIR spectrums of Ba-mica and O-M intercalated with STAC.

The FTIR specturms are shown in figure 4. The intensity of water stretching band (3442 cm^{-1}) and bending band (1618 cm^{-1}) gradually weakened with the increase of STAC quantity. This is due to the hydrophobic effect conferred by the carbon chains. As more surfactant molecules intercalated into the gallery space, the water repellency grew stronger, resulting in the reduction of water content. TGA results also prove this point.

IOP Conf. Series: Materials Science and Engineering **389** (2018) 012034 doi:10.1088/1757-899X/389/1/012034

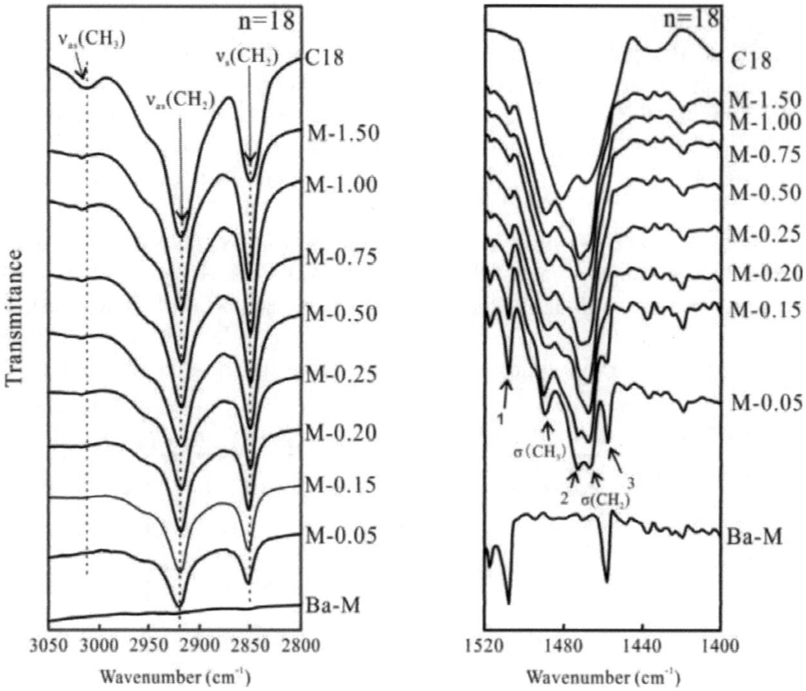

Figure 5. FTIR spectrums in the 3050-2800cm^{-1}and 1520-1400cm^{-1} regions of Ba-mica and O-M intercalated with STAC

Compared with crystalline state of STAC, the band frequencies of STAC in O-M associated with methylene (CH_2) were almost same, while the band frequencies associated with the methyl in the head ammonium group ($N(CH_3)_3$) shifted towards a higher wave number (Figure 5) though the peaks were weak. The exact data are shown in table 1.

Table 1. Vibration frequencies related to surfactant in O-M (Frequency unit: cm^{-1})

Assignment of group *	M-0.05	M-0.15	M-0.20	M-0.25	M-0.50	M-0.75	M-1.00	M-1.50	STAC
$v_{as}(N(CH_3)_3)$	~	~	~	3018.23	3017.96	3017.6	3017.44	3017.81	3012.58
$\sigma(N(CH_3)_3)$	1489.85	1490.52	1487.65	1488.22	1489.00	1488.53	1488.08	1489.38	1481.53
$v_{as}(CH_2)$	2919.94	2918.43	2917.75	2918.09	2918.16	2918.1	2918.09	2918.03	2918.06
$v_s(CH_2)$	2850.29	2850.27	2849.89	2850.00	2850.08	2849.96	2849.87	2850.50	2849.70

* This is referred to Barman et al. 2003, Venkataraman et al. 2000, Venkataraman et al. 2001.

At different STAC loadings, the four important vibrations of STA$^+$, including $v_{as}(N(CH_3)_3)$ (Asymmetric stretching vibration of the methyl in the head ammonium group) and $\sigma(N(CH_3)_3)$ (the corresponding bending vibration), $v_{as}(CH_2)$ (Asymmetric stretching vibration of the methylene) and $v_s(CH_2)$ (symmetric stretching vibration), altered insignificantly, indicating that STA$^+$ was in a similar environment once it was accommodated between sheets regardless its concentration.

Compared with pure STAC, the $v_{as}(CH_2)$ and $v_s(CH_2)$ in O-M fluctuated very little, demonstrating the carbon chains adopted the same conformational mode as the crystalline STAC in which the carbon chains are fully stretched (Venkataraman et al. 2001), i.e., the STAC dose had no influence on its conformational model in O-M.

However, compared with the crystal state of STAC, the $v_{as}(N(CH_3)_3)$ and $\sigma(N(CH_3)_3)$ in O-M moved to higher frequency, meaning the head ammonium group was immobilized. This was caused by the positively charged head ammonium group interacting with negatively charged aluminosilicate sheets.

It is also noticeable that the scissor vibration of the methylene $\sigma(CH_2)$, which reflects the interaction between carbon chains, was split (Figure 5b). The splitting was clear in M-0.05 and M-0.15, and blurred at higher STAC quantity, but it still could be resolved comparing with it in crystalline state, representing that the STA^+ had two different chain interactions. This was due to the uneven distribution of structural water. The STA^+ in water concentrated terrain were hydrated stronger. And the water was repelled out gradually as more STA^+ moved in, which therefore weakened its interaction with STA^+.

4. Conclusion

Organic mica composites were prepared with pre-prepared Ba-mica by intercalating $C_nH_{37}(CH_3)_3NCl$ (N=12, 14, 16, 18). The evidences in FTIR showed that the STA^+ in O-M adopted an identical conformational arrangement at different concentrations. Combining with XRD analysis, it is believed that this organic mica composite has a unique property that its basal space is regardless to the organics concentration. Even in a very low loading, the organic molecules are still strong enough to support the whole interspace from crashing that was believed to happen, i.e., the organics structure is robust so that it will not change easily even though a certain number of organics dissolve out. In addition, the remaining water molecule is helpful to adjust the hydrophilic-hydrophobic property on the premise of retaining this structure. According to this character, potentially, demanded layered silicate composite with certain height or hydrophilic feature can be precisely prepared.

Acknowledgments

This work was supported by the National Natural Science Foundation of China (51374155), the Hubei Province Key Technology R&D Program (2014BCB034) and the National Natural Science Foundation of Hubei province of China (2014CFB796).

References

[1] Ballini R., Bigi F., Conforti M.L., De Santis D., Maggi R., Oppici G., and Sartori G. 2000. Multicomponent reactions under clay catalysis. Catalysis today, 60(3), 305-309.
[2] Guo J. and Al-Dahhan M. 2003. Catalytic wet oxidation of phenol by hydrogen peroxide over pillared clay catalyst. Industrial & engineering chemistry research, 42(12), 2450-2460.
[3] Darder M., Colilla M., and Ruiz-Hitzky E. 2005. Chitosan–clay nanocomposites: application as electrochemical sensors. Applied Clay Science, 28(1), 199-208.
[4] Guth U., Brosda S., and Schomburg J. 1996. Applications of clay minerals in sensor techniques. Applied clay science, 11(2), 229-236.
[5] Babel S. and Kurniawan T.A. 2003. Low-cost adsorbents for heavy metals uptake from contaminated water: a review. Journal of hazardous materials, 97(1), 219-243.
[6] Jiang J.Q. and Zeng Z. 2003. Comparison of modified montmorillonite adsorbents: Part II: The effects of the type of raw clays and modification conditions on the adsorption performance.
[7] Oliveira L.C., Rios R.V., Fabris J.D., Sapag K., Garg V.K., and Lago R.M. 2003. Clay–iron oxide magnetic composites for the adsorption of contaminants in water. Applied Clay Science, 22(4), 169-177.
[8] Meng P, Li Z, Huang Z, et al.2016. Extraction of potassium from biotite by Ba^{2+} /K^+, ion exchange and the structural transformation[J]. Physics & Chemistry of Minerals, 43(6):1-7.

IOP Conf. Series: Materials Science and Engineering **389** (2018) 012035 doi:10.1088/1757-899X/389/1/012035

Experimental Study on Impermeability of Impervious Slurry in Landfill

Yanmin Sheng[1,2], Guozhong Dai[1,a] and Shujin Li[1]

[1]School of Civil Engineering and Architecture, Changzhou Institute of Technology, 213032, Jiangsu, Changzhou, China
[2]School of Environmental & Safety Engineering, Changzhou University, 213164, Jiangsu, Changzhou, China

[a] Corresponding author: daigz@e-mail.org

Abstract. The quantity and harmful ingredients of municipal solid waste are increased with the acceleration of urbanization which sets a higher request to the impervious performance of landfill. This paper takes the method of orthogonal experiment to prepare impervious material based on the modified bentonite. The effects of main components on the impervious properties are also researched. The experiment shows that critical aperture of the impervious slurry will reduce with the increase of cement and bentonite. This will enhance the impermeability of the slurry. The influence of cement on the permeability is more significant than that of bentonite. The experiment also finds that the permeability of the pulp reduces with the increase of Na_2CO_3 and polyvinyl alcohol (PVA). In the microstructure, the number of macropores and the critical aperture increases with the increase of Na_2CO_3 and PVA. This results in the improvement of permeability for impervious slurry.

1 Introduce

Municipal solid waste will increase dramatically with the development of urbanization. According to statistics, the output of municipal solid waste is 1.3 billion tons so far and expected to reach 2.2 billion tons by 2050 in the world [1, 2]. The landfill is widely used at home and abroad because of its characteristics of simple and low running cost [3]. However, a great quantity of leachate with high pollution will be produced in the progress of landfill. This would bring secondary pollution on surrounding soil and water if it cannot be handled properly [4]. The measures and materials of anti-seepage are the key to preventing pollution. It has been proved that taking curtain grouting technology to build anti-seepage system around the landfill site is one of the most effective way to prevent harmful ingredients migrating from leachate to outside [5].

Plee [6] has done the research of microstructure, osmotic coefficient and flowing property on the cement-bentonite slurry. He finds that the large amount of Ca^{2+} in cement and high pH value of mixed liquor will affect the macroscopic property of slurry. Schoenfelder [7] takes the NMR method to analyze the hardening characteristics and permeability of the cement-bentonite slurry with different mixture ratios. Rafalski [8] and Jefferis [9] have also studied the anti-seepage performance of the mud wall with cement, bentonite and fly ash used in European and American.

Raw bentonite can only effectively prevent the migration of low tension leachate[10, 11]. It will

IOP Conf. Series: Materials Science and Engineering **389** (2018) 012035 doi:10.1088/1757-899X/389/1/012035

flocculate and shrink under the action of high concentration organic leachate. This will result in the rapid decline of its ability to adsorb organic pollutant [12]. It can use organic modifier to improve microstructure and particle surface properties of bentonite because of its interlayer structure and physical properties [13, 14]. It can effectively control pollution at the situation of low permeability by this way. This will greatly improve the efficiency of bentonite in the practical engineering [15]. Therefore, it is of great practical significance to do research on the impervious properties of the antiseepage slurry based on the modified bentonite.

2. Experiment

2.1. Raw materials

In this work, the raw materials are as follows: the bentonite should take the treatment of sodium, the type of cement is 42.5 grade ordinary portland cement, the grade of fly ash which is produced in heat-engine plant is I level, the type of polycarboxylate superplasticizer is PCA-300P, the type of PVA is 1788, the Na_2CO_3 and $AlCl_3$ is analytically pure.

2.2. The method of experiment

Firstly, the experiment uses PVA, Na_2CO_3 and $AlCl_3$ to modified the bentonite. Secondly, it mixes the modified bentonite, cement, fly ash and water reducer to prepare the samples which the experiment needs. Lastly, the osmotic coefficient of the samples is tested by the variable water level method to evaluate the permeability of the slurry.

3. Results and Discussion

3.1. Orthogonal experiment

The factors and levels of orthogonal experiment are determined by contrastive experiments and the osmotic coefficient for 7 days and 28 days is tested (Table 1 and Table 2).

Table 1. The experimental factors and levels of slurry.

Level	Factors (g)			
	Cement	Bentonite	Na_2CO_3	PVA
1	180	180	1	1
2	200	190	2	2
3	220	200	3	3

Table 2. The arrangement and test results of orthogonal experiment for slurry. [a]

Number	The amount of various factors (g)				Osmotic coefficient (cm/s)	
	Cement	Bentonite	Na_2CO_3	PVA	7d	28d
					10^{-6}	10^{-8}
1	180	180	3	2	8.1	2.1
2	200	180	1	1	4.7	1.3
3	220	180	2	3	2.7	1.8
4	180	190	2	1	5.1	1.6
5	200	190	3	3	1.4	3.1
6	220	190	1	2	4.4	3.9
7	180	200	1	3	4.5	8.0
8	200	200	2	2	5.6	3.5
9	220	200	3	1	0.6	1.0

[a] Other admixtures: fly ash 180 g, water reducer 0.5 g, $AlCl_3$ 1.5 g.

It shows that the osmotic coefficient is from10^{-6} to10^{-7} cm/s for 7 days (Table 2). However, the osmotic coefficient of the slurry reaches 10^{-8} cm/s for 28 days which meets the Standard for Pollution

IOP Conf. Series: Materials Science and Engineering **389** (2018) 012035 doi:10.1088/1757-899X/389/1/012035

Control on the Landfill Site of Municipal Solid Waste.

This paper takes the range method to analyze the results of the orthogonal experiment (Table 3 and Table 4).

Table 3. The results of the analysis of osmotic coefficient for 7 days.

Factors		Cement	Bentonite	Na_2CO_3	PVA
Osmotic coefficient 10^{-5}cm/s	K1	2.0	1.6	1.4	1.0
	K2	2.0	2.4	1.3	1.8
	K3	0.8	1.1	2.3	2.1
	Range	1.2	1.3	1.0	1.1

Table 4. The results of the analysis of osmotic coefficient for 28 days.

Factors		Cement	Bentonite	Na_2CO_3	PVA
Osmotic coefficient 10^{-8}cm/s	K1	10.0	5.2	13.2	3.9
	K2	8.0	8.6	6.9	9.5
	K3	3.0	1.3	6.2	12.9
	Range	3.0	7.3	7.0	9.0

On the one hand, the osmotic coefficient decreases with the increase of cement and bentonite and increases with the increase of PVA. On the other hand, the osmotic coefficient increases for 7days while it decreases for 28 days with the increase of Na_2CO_3. For 7 days, the order of the effect of various factors on osmotic coefficient is bentonite, cement, PVA and Na_2CO_3. For 28 days, the order is PVA, bentonite and Na_2CO_3 and cement. The effect of cement on the osmotic coefficient is more significant than that of bentonite.

3.2. Single factor experiment

3.2.1. The effect of cement and bentonite on the osmotic coefficient of slurry. According to the results of the orthogonal experiment, the experiment controls the cement and bentonite from 180 g to 220 g and the other components are the same (Table 5).

Table 5. The results of osmotic coefficient with different amount of cement and bentonite.

Number	The amount of various factors (g)		Osmotic coefficient (cm/s)		
	Cement	Bentonite	7d	14d	28d
			10^{-6}	10^{-7}	10^{-8}
A1	180	190	11.0	8.2	3.1
A2	190	190	8.4	6.7	2.9
A3	200	190	7.2	5.5	2.3
A4	210	190	4.2	5.1	1.8
A5	220	190	2.5	4.5	1.1
B1	200	180	9.4	9.1	5.1
B2	200	190	9.2	7.0	3.5
B3	200	200	9.0	6.2	2.4
B4	200	210	7.2	5.5	2.3
B5	200	220	6.6	2.6	1.8

IOP Conf. Series: Materials Science and Engineering **389** (2018) 012035 doi:10.1088/1757-899X/389/1/012035

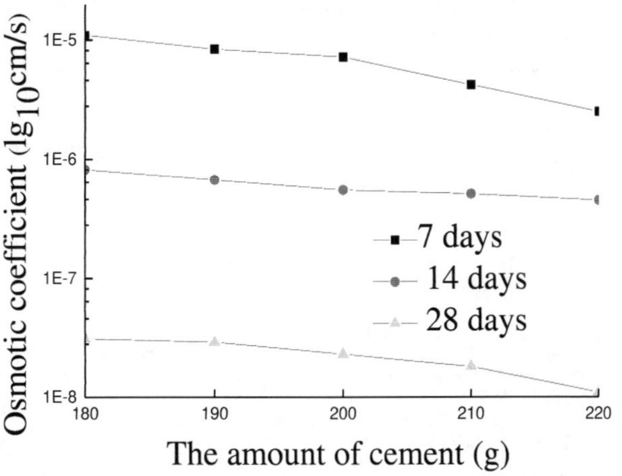

Figure 1. Relation curve of osmotic coefficient and the amount of cement.

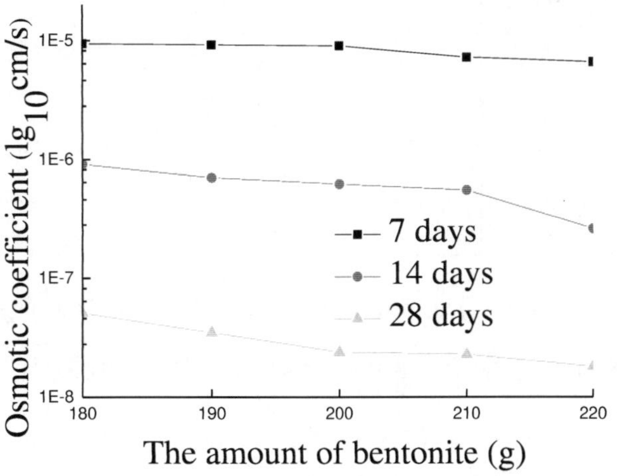

Figure 2. Relation curve of osmotic coefficient and the amount of bentonite.

The osmotic coefficient of impervious slurry decreases with the increase of cement and bentonite at all time points (Figure 1 and Figure 2). It is known from the study of Chao Xu [16] that the permeability of slurry is related to the critical aperture which reflects the degree of connectivity of pores and the tortuosity of infiltration paths inside the impervious material. This is the essence of permeability and critical aperture. The hydration products of cement produce more on structure of consolidation is more complete with the increase of cement. The bentonite will swell after reacting with water. The macropores decrease and the micropores increase when the bentonite is increased. This results in the further reducing of critical aperture and the number of connected pore which explains the decrease of osmotic coefficient with the increase of cement and bentonite.

3.2.2 The effect of PVA and Na2CO3 on the osmotic coefficient of slurry. The white floc will be producted after reacting between excessive Na_2CO_3 and PVA. The modification on bentonite is changed in that situation. The best ratio of Na_2CO_3 and PVA which is summarized from a lot of experiments is 1:1. According to the results of the orthogonal experiment, the experiment controls the Na_2CO_3 and PVA from 1.0g to 3.0g and the other components are the same (Table 6).

Table 6. The results of osmotic coefficient with different amount of Na_2CO_3 and PVA.

Number	The amount of various factors (g)		Osmotic coefficient (cm/s)		
	Na_2CO_3	PVA	7d	14d	28d
			10^{-6}	10^{-7}	10^{-8}
C1	1.0	1.0	2.8	2.3	1.2
C2	1.5	1.5	7.0	3.7	1.9
C3	2.0	2.0	7.2	5.5	2.3
C4	2.5	2.5	10.0	7.9	2.6
C5	3.0	3.0	13.0	8.5	3.1

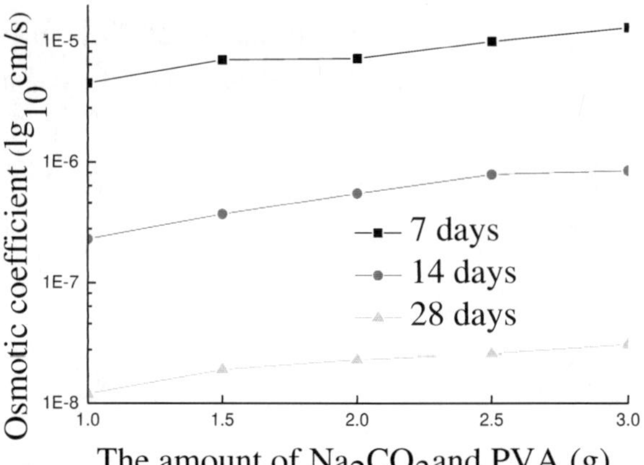

Figure 3. Relation curve of osmotic coefficient and the amount of Na_2CO_3 and PVA.

The osmotic coefficient of the consolidation increases with the increase of Na_2CO_3 and PVA (Figure 3). It is known from Weian Kong[17] that the increase of Na_2CO_3 concentration will reduce the expansion volume of bentonite which increases the permeability of the slurry.

4 Microstructure of the Slurry

The experiment selects three samples of C1, C2 and C3 to further explore the influence of Na_2CO_3 and PVA on the permeability of slurry. The three samples are tested under scanning electron microscope after drying, crushing and sampling.

It shows the morphology of particles in the enlarged 500 times (Figure 4). Inside the C1 sample, the crystal particles are close, the structural densification is well, the macropores are few and the critical aperture is small which determine that the permeability of the slurry is excellent. With the increase of Na_2CO_3 and PVA, the crystal particles are disperse, the structural densification is bad, the macropores are more and the critical aperture is bigger which determine that the permeability of the slurry is poor.

IOP Conf. Series: Materials Science and Engineering **389** (2018) 012035 doi:10.1088/1757-899X/389/1/012035

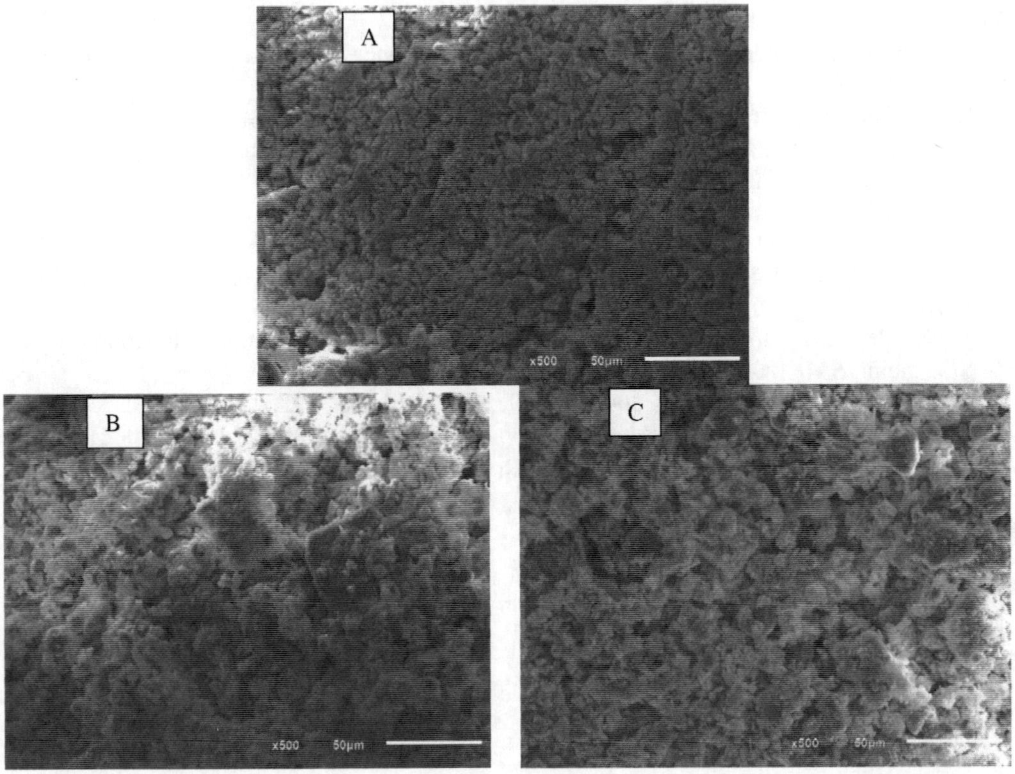

Figure 4. (A) SEM diagrams of C1 sample; (B) SEM diagrams of C2 sample; (C) SEM diagrams of C3 sample

The Na_2CO_3 concentration is low when the amount of Na_2CO_3 is small. In that situation, the bentonite will swell and compact the inter structure of the slurry after absorbing water which results in the low permeability of the impervious slurry. The Na_2CO_3 concentration will increase when the amount of Na_2CO_3 increases which represses the water imbibition and expansion of bentonite. The PVA will form a dense polymer film on the surface of bentonite particles and the bentonite particles present hydrophobicity which results in the weakness of impermeability of the impervious slurry.

5 Conclusions

This paper analyzes the relationship between the permeability and each component of the slurry which is based on the influence of components on critical aperture. The conclusions are drawn as follows:

The critical aperture of the impervious slurry will gradually decrease with the increase of cement and bentonite. The effect of cement on the osmotic coefficient is more significant than that of bentonite. The increase of Na_2CO_3 will repress the water imbibition and expansion of bentonite. The PVA will form a dense polymer film on the surface of bentonite particles and the bentonite particles present hydrophobicity. These reasons cause the decrease of the impermeability of the impervious slurry with the increase of Na_2CO_3 and PVA. The impermeability of the antiseepage slurry meets the requirements of the current standard and provides the basis for the landfill engineering.

Acknowledgments

The national natural science foundation of China (51678083) is gratefully acknowledged. The project is also supported by Postgraduate Research & Practice Innovation Program of Jiangsu Province (SJCX17_0734).

IOP Conf. Series: Materials Science and Engineering **389** (2018) 012035 doi:10.1088/1757-899X/389/1/012035

References

[1] TE. Butt, AA. Javadi, MA. Nunns and CD. Beal, Science of the Total Environment, **S**, 569-70: 815-29 (2016)
[2] K. Kieckhafer, A. Breitenstein, TS Spengler, Waste Management, **60**,768 (2016)
[3] RB. Brennan, MG. Healy and L. Morrison, et al, Waste Management, **55**,355 (2015)
[4] QQ. Zhang, BH. Tian and X. Zhang, et al, Waste Management, **33**,2277-86 (2013)
[5] PF. Hudak, Environmental Monitoring & Assessment, **90**,113-20 (2004)
[6] D. Plee, F. Lebedenko and F. Obrecht, et al, Cement & Concrete Research, **20**, 45-61 (1990)
[7] W. Schoenfelder, J. Dietrich and A. Märten, et al, Magnetic Resonance Imaging, **25**, 582 (2007)
[8] L. Rafalsk, Studies in Surface Science & Catalysis, **41**, 7-23 (1994).
[9] A. JS, Proc 10th Internatonal Conference for Soil Mechanics and Foundation Engineering, Stockholm: **SMFE**: 425-40 (1981)
[10] Y. Gao, Y. Guo and H. Zhang, Journal of Hazardous Materials, 302:105 (2016)
[11] X. Liu, P. Hicher and B. Muresan, et al, Applied Clay Science, **119**, 365-74 (2016)
[12] J. Wiszniowski, J. Surmaczgórska and D. Robert, et al, Journal of Environmental Management, **85**, 59-68 (2007)
[13] F. Haghseresht, S. Wang and DD. Do, Applied Clay Science, **46**, 369-375 (2009)
[14] A. Rashid, HN. Bhatti and M. Iqbal, et al, Ecological Engineering, **91**, 459-71 (2016)
[15] R. Yu, G. Hu, Environmental Sanitation Engineering (2003)
[16] X. Chao, L. Dan and H. Liang, Journal of Tongji University (Natural Science), **39**, 819-23 (2011)
[17] WA. Kong, SL. Zhang and CH. Bai, et al, Non-Metallic Mines, **33**, 42-4 (2010)

IOP Conf. Series: Materials Science and Engineering **389** (2018) 012036 doi:10.1088/1757-899X/389/1/012036

Microwave Absorbing Property of the Nanoparticle Concrete

Guang Peng[1], Jin-yu Xu[1,2], Er-lei Bai[1] and Jing-sai Zhu[1]

[1] Department of Airfield and Building Engineering, Air Force Engineering University, Xi'an 710038, China
[2] College of Mechanics and Civil Architecture, Northwest Polytechnic University, Xi'an 710072, China

E-mail: PengGuang122333@163.com

Abstract. Nano concrete samples with different sizes were prepared by incorporating particles of Nano ferric oxide, Nano silicon dioxide and Nano alumina with good dispersion. The compressive strength, flexural strength and radar reflected wave were tested. Mechanism of Nano concrete enhancement and mechanism of Nano concrete wave elimination were discussed. Result showed that Nano particles of iron oxide and alumina on the compressive strength of concrete had a different degree of resistance increase. The compressive strength of Nano alumina (NA) concrete was the highest, which was increased by 44.4% compared with that of ordinary concrete (PC). The flexural strength of Nano iron oxide (NF) concrete was the highest, which was 44.4% higher than that of ordinary concrete. As for the radar wave reflectivity, different Nano materials on the radar wave reflection effect were different. The full reflectivity of NSF compounded two kinds of Nano particles was lower than -8dB. With the increase of the content of Nano iron oxide particles, the reflectivity of radar wave decreases gradually. The minimum reflectivity of -16.7dB was achieved at 2.5GHz.

1. Introduction

With the continuous improvement of radar technology, radar stealth technology was also in constant development. A large number of structure materials and stealth coating constantly come out, which was low reflection and high absorption to radar wave. Stealth fighter warships continue to innovate. And to important ground of facilities, underground works et al belong the radar stealth design did not catch up with the pace of the times. With the universal application of wireless communication technology as well as the convenience of life, people are also facing the problem of communication security, electromagnetic radiation pollution, signal interference and other issues, and some even threaten the health of people. However hospitals, research institutes, communications relay and scheduling and other unites are facing the greatest threat. So it was an important research direction to study the wave absorbing of radar wave in the field of building materials.

Most of the research was mainly carbon black, ferrites, carbon fiber, steel fiber and other traditional wave absorbing agent in cement-based materials. Their properties of microwave-absorbing was good [1-3], but they often have a narrow absorption band, severe reduction in strength or other serious shortcomings. Nano concrete has excellent performance in many ways [4-5], but people never learned much about the absorption propertied of Nano concrete [6-7]. In theory, the application of Nano materials should have great development potential in the field of concrete absorbing. In this paper, a variety of Nano concrete specimens were prepared by using well dispersed Nano silica and alumina particles. Mechanical properties of the sample and the radar reflectivity were tested and the strengthening mechanism and wave absorbing mechanism were discussed. The research and application of Nano concrete wave absorbing will have a certain role in the future.

2. Experiment method

2.1 Test materials

The cement used was Portland cement (P•O32.5R) of Qinling Brand (Shaanxi China), main properties are shown in Table 1; Low calcium fly ash of Hancheng Power Plant; limestone Rubble (5 to 10 mm, 15%; 10 to 20 mm, 85%), river sand (bulk density of 1500 kg/m^3 with a fineness modulus of 2.78 and mud content was 1.1%). Superplasticizer liquor (Solid content of 40%) was from Zhongyi Chemical Plant (Shaanxi, China), as Nano-materials dispersant and concrete water reducing agent. Nano-SiO_2, Nano-Fe_2O_3 and Nano-Al_2O_3 were obtained from Veking New Material Co. Ltd (Zhejiang, China), appearance as shown in Figure 1, main properties are shown in Table 1.

(a) (b) (c)

Figure 1. Particles of Nano ferric oxide, Nano silicon dioxide and Nano alumina

Table 1. Properties of Nano particles

Item	Model	Appearance	Purity/%	Surface-volume Ratio(m^2/g)	Diameter/nm	Water%
SiO_2	VK-SP15	White powder	≥99.8	250±30	15±5	≤0.1
Fe_2O_3	VK-E01	Cinnamon	≥99.8	80-90	50	≤0.1
Al_2O_3	VK-L20Y	White powder	≥99.9	150-200	10~20	≤0.01

2.2 Test technique

Quasi static mechanical test refer to "ordinary concrete mechanics performance test method standard" (GB / T50081. 2002) in the provisions of the method. The polycarboxylate superplasticizer with 3/4 water were stirred into the solution, then adding Nano-particles in solution and stirring 30 s at a low velocity and stirring 30 s at high velocity, which were scattered for 15 minutes by using the ultrasonic dispersion method to ensure Nano particles dispersed evenly. Concrete prepared by "wrapped sand method", the specific process was as follows: Firstly, adding fly ash and half of the cement and mixing together into a mixing ash (30 s); secondly, adding Nano mixture and mixing into mortar (30 s); thirdly, adding to the sand and beat up (30 s); and then putting in the stones and stirring (30 s); lastly, adding 1/4 water washing container and the residual cement and stirring (120 s). A mixture of fresh concrete was uniform.

The radar reflectivity measured by sweep frequency measurement system arc method, as shown in Figure 2, whose testing standards are in line with the "radar reflectivity test method" (GJB2038A). The test frequency was in the range of 2GHz~18GHz and the test results take of 3 specimens of group average. The specimen of Radar absorbing material reflectivity test cast by wood formwork whose size was 200 mm x 200 mm x 100 mm. After pouring the specimens were covered with plastic film to prevent excessive evaporation of water. After standing the room for 24 h, the mold was carried out. Then put the test-pieces in the standard curing room for 28 d, which was in the temperature of 20±2℃ and whose relative humidity was more than 95%. It was polished accurately using a SHM-200 type double section grinding machine, whose surface irregularities must be controlled in the 0.1 mm.

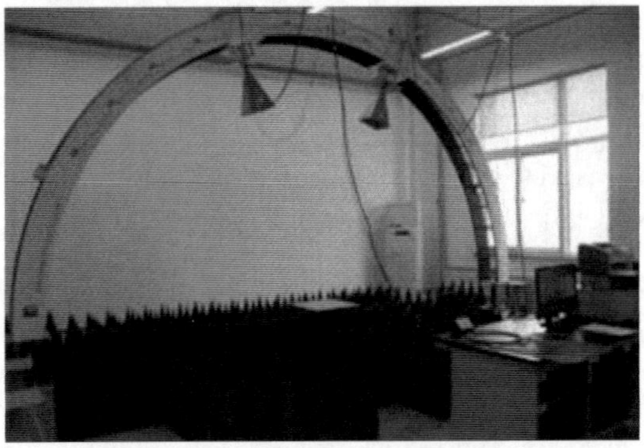

Figure 2. H62 brass pulse shapers

3 Experimental results and analysis

3.1 physical and mechanical properties

The physical and mechanical properties of Nano concrete is shown in Table 2. The water cement ratio of all samples was 0.36. Polycarboxylate superplasticizer was used as Nano-particles dispersing agent and water reducing agent, ensuring that the Nano-particles were uniformly dispersed in the concrete and keep the concrete working better than concrete in low water cement.

Table 2. Physical and mechanical properties of Nano concrete

Specimens	Nano materials	Volume of Nano concrete/%	Volume of superplasticizer/%	Strength of 28 d/MPa	
				Compression	Flexural
PC	0	0	0.4	34.2	4.3
NA	Al_2O_3	2.0	0.4	49.4	4.9
NS	SiO_2	2.0	1	33.8	3.6
NF-a		1.0		38.3	4.8
NF-b(NF)	Fe_2O_3	2.0	0.4	42.4	5.2
NF-c		5.0		40.5	4.7
NSF	Fe_2O_3, SiO_2	2.0	1	40.2	5.1

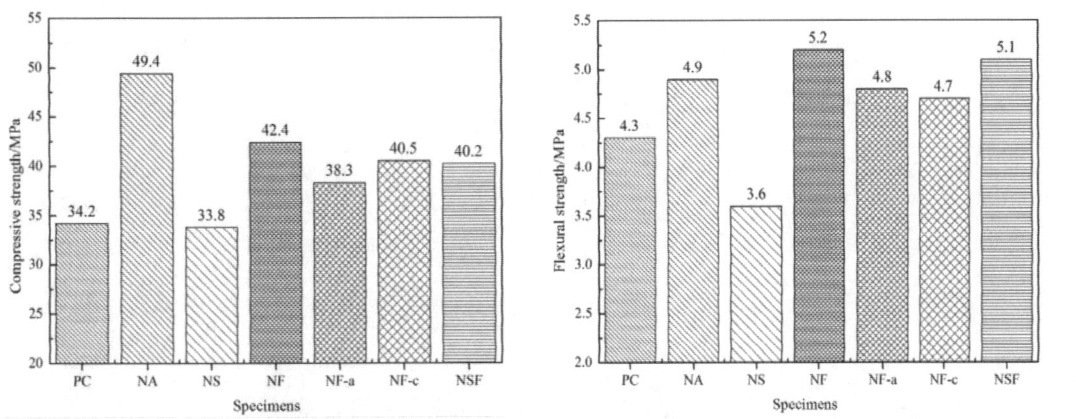

(a) (b)

Figure 3. Mechanical properties of Nano concrete

The 28 d compressive strength and flexural strength of Nano concrete in each group samples were shown in Figure 3. All groups except NS of Nano concrete compressive, flexural strength had different degrees of improvement compared with the blank group PC. NA compressive strength was highest and increased by 44.4% compared with PC among them. NF-b flexural strength was (the same ratio of NF) highest and more 20.9% than PC. However, the NS doped with Nano-SiO_2 has no advantage and even weakened compared with PC in compressive strength and flexural strength, which should be noticed in the following research. The NSF doped with two kinds of Nano materials has a certain advantage compared with PC in the two aspects of compression and bending, increased respectively by a factor of 17.6% and 18.6% and was basically the same as that of the NF. The compressive strength and flexural strength of NF-b was the highest in the NF group, explaining that the compressive and flexural strength had a first increased and then decreased in the 0~5.0% content range of Nano particles. So there should be the best volume to make the concrete strength reaches the maximum.

The different Nano particles had different effects on the strength of concrete. The formation of hydrated calcium silicate was due to the reaction of Nano-SiO_2 and $Ca(OH)_2$ in the early hydration of cement, which reduced the calcium hydroxide crystals. But with the increase of strength, a large number of Nano SiO_2 was in the reaction and water consumption. The cement hydration reaction was not sufficient, affecting the further development of NS late strength. Therefore, NS strength was weakened compared with PC [8]. When Nano-Al_2O_3 is in concrete, the process of concrete hydration was greatly promoted due to the great surface area and surface polarity charge. A part of Nano Al_2O_3 reacts with the alkaline substance in concrete in order to reduce the aggregation of harmful crystal. The other part filled in concrete pores in the form of decreasing the harmful hole, which had an influence on the strength of NA [9]. Nano-Fe_2O_3 particles dispersed uniformly were added into the concrete. On the one hand, due to the nucleation effect, hydration reaction takes Nano-Fe_2O_3 particles as the core and strengthens gradually for getting more uniform crystal. On the other hand, new three dimensional crystal structure produced by reaction continue to improve the strength of test-pieces owing to the original three-dimensional structure of cement stone [10]. Different Nano materials have different reaction mechanism, and the reaction activity of cement stone was different. So the influence on the macro mechanical properties of concrete was quite different.

3.2 effect of different Nano materials

Different Nano particles were added into the concrete, which has different effects on the radar reflectivity of the sample. The measured results were shown in Figure 4. NS, NF and NA were respectively on behalf of Nano-SiO_2, Nano-Fe_2O_3 and Nano-Al_2O_3 concrete, the sample's thickness was all 50 mm.

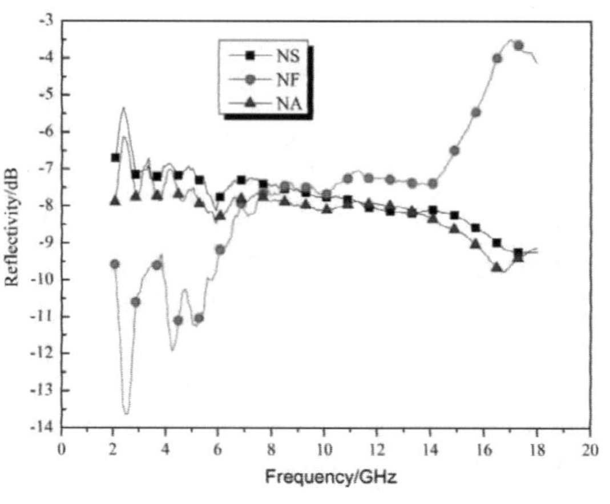

Figure 4. Stress versus strain curves by

The results of three kinds of sample had significant difference. The radar reflectivity of NF was smaller in the lower frequency range, larger in higher frequency range. NA and NS were opposite of NF. The reflectivity was larger in the low frequency range but lower in the high frequency range. NA's reflectivity was less than NS as a whole. The minimum of NF was -13.56dB near the 2.56GHz radar

IOP Conf. Series: Materials Science and Engineering **389** (2018) 012036 doi:10.1088/1757-899X/389/1/012036

frequency. When the wave frequency was 2GHz~7GHz Radar reflectivity were all less than -8dB. When the frequency is above 10GHz, NA and NS show more obvious advantages. The frequency was near 14.2GHz or better and the radar reflectivity would be less than -8dB.

It is difficult to achieve the dual requirements of impedance matching and strong absorption, for a single absorption complex. Composite material was conductive to improve electromagnetic loss ability as well as complementary electromagnetic parameters. So we can imagine mixing two kinds of Nano material, so that the sample achieved the purpose of low reflectivity in a wider frequency range

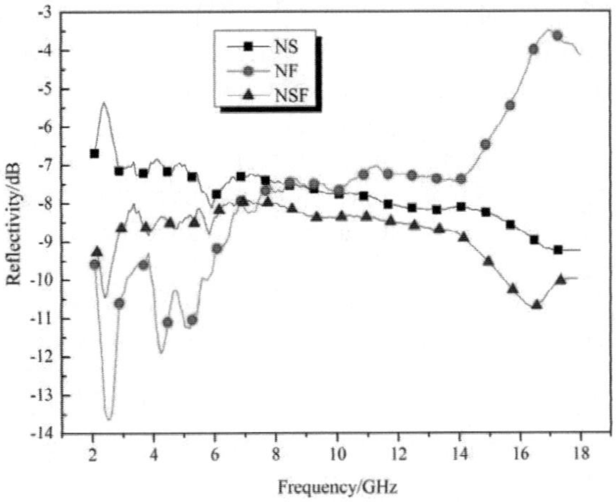

Figure 5. Stress versus strain curves

NSF was the Results of radar reflectivity that shown in Figure 5, which indicated that Nano-SiO_2 particles were added into concrete together with Nano-Fe_2O_3 particles. The figure shows that NSF can be seized of the common advantages of NS and NF. Radar reflectivity was almost less than -8dB in the full band range (2GHz~18GHz), but its effect was not a single linear superposition of them. The lowest reflectivity of NSF appeared near 2.56GHz and was consistent with the lowest frequency of NF whose reflectivity was -10.48dB. The second was near 16.42GHz and also synchronous with the lowest point of NS. Thus, it was feasible to use multiple absorbing materials to complement each other within a certain range.

3.3 The influence of different proportion

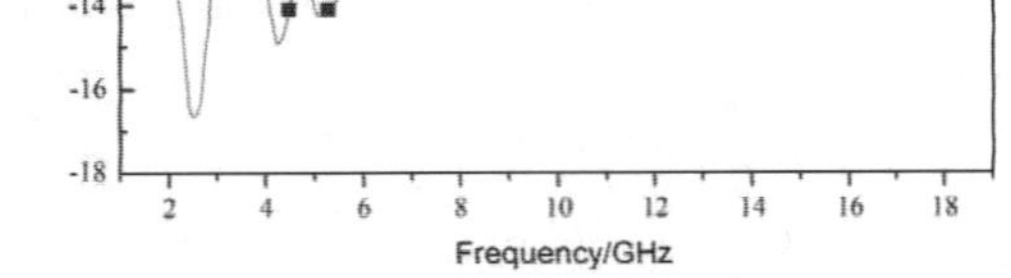

Figure 6. Stress versus strain curves

As shown in Figure 6, the reflectivity of concrete slab was under three kinds of different Nano-Fe_2O_3 content, including NF-a, NF-b and NF-c. And the mix proportion was respectively corresponding to the 3, 4 and 5 group in the table 2. The content of Nano-Fe_2O_3 respectively was 1%, 2% and 5% of cement content. It shows obvious regularity that three sets of curves were consistent with each other among the range of whole full frequency. This might be interpreted that the peaks and troughs of reflectivity curve keep synchronized. The three groups of curves reached the minimum at the frequency of 2.56GHz, which were respectively -11.2dB, -13.6dB and -16.7dB. Therefore, in the range of 1% to 5%, the larger the amount of Nano-Fe_2O_3 was, the smaller the radar reflectivity of concrete slab was. And the reflectivity got smaller at lower frequencies.

At present, great progress has been made in the research of novel Nano absorbing materials, but its mechanism has not yet formed a unified conclusion. Nanoparticles have the quantum size effect and macroscopic quantum tunneling effect because of extremely small size, which were not available in general materials. So it was difficult to explain with conventional macro mechanism.

Nano-Fe_2O_3 was a kind of double loss material with dielectric loss and magnetic loss, whose magnetism was consist of many domains and different from conventional polycrystalline and amorphous materials. The domain wall was in the middle of each domain. Every grain of Nano-Fe_2O_3 was a single magnetic domain. The different magnetic structures make it have new magnetic properties. The Nano-Fe_2O_3 at room temperature has super paramagnetic. There are multiple absorption peaks in the Mossbauer spectrum, which was a kind of reflection on the macroscopic quantum tunneling effect. The difference of this magnetic structure affects the complex permeability of NF when the radar reflectivity was measured. Nano-Fe_2O_3 was used as additives in the experiment of radar reflectivity because of those losses of unique domain wall resonance and natural resonance in an alternative power plant.

Nano-Al_2O_3 particles can be explained in terms of electrical loss mechanism. Dielectric constant can be expressed in the plural.

$$\varepsilon^*(\omega) = \varepsilon'(\omega) - i\varepsilon''(\omega) \tag{1}$$

"ε'" said the real part and "ε''" said the imaginary part, which respectively said diclectric constant and under electrostatic field and alternating electric field:

$$\tan\theta = \frac{\varepsilon''}{\varepsilon'} \tag{2}$$

This was the dielectric loss. The Phase-angle Difference existed between them, if the response of the electric displacement lagged behind the change of electric field. This was the dielectric loss. And the bigger the Phase-angle was, the more serious the loss was. The response of electric displacement was relative to polarization process. There were a large number of ionic bonds, approximately 63%, in the Nano-Al_2O_3 particles. So the arrangement of atoms in the huge interface was confusing in which Nano particles had many defects such as strong lattice distortion, vacancy and so on. And there are a large number of vacancies of oxygen irons in those Nano particles. The vacancy was equivalent to with positive charge. This kind of positive charge will form inherent electric distance with the negatively charged oxygen ions and turning-direction polarization in foreign electromagnetic field when the direction was changed. Turning-direction polarization was an important factor to Nano-Al_2O_3 with higher dielectric constant.

It can be inferred that the different electromagnetic parameters of different Nano concrete was due to these electromagnetic effects at the micro level and then the radar reflectivity had disparate impact. If the lattice structure and the molecular structure of Nano materials were further explored and to master the action parameters of different Nano materials to the radar wave, composite material can be used to improve and control the electromagnetic parameters. Radar stealth and electromagnetic shielding can be smoothly done or easily solved with the extension of the absorption bandwidth and absorption intensity to electromagnetic wave.

4 Conclusions

Polycarboxylate superplasticizer Liquor was be used as water reducer and the dispersant of Nano particle. Three kinds of Nano-concrete specimens were prepared by high-speed stirring and ultrasonic dispersion. And the static mechanical properties and the radar reflectivity were tested. The conclusion obtained was as follows:

(1) Nano-Al_2O_3 and Nano-Fe_2O_3 particles can significantly improve the compressive strength and flexural strength of concrete at 28 d. The c of NA increased by 44.4% compared with PC. The flexural strength of NF was 20.9% higher than that of PC. The effects of Nano-SiO_2 particles to improve compressive strength and flexural strength on the concrete was not obvious, even a certain weakening effect. The compressive strength and flexural strength of concrete in which Nano-SiO_2 and Nano-Fe_2O_3 particles were mixed was the same as that of NF.

(2) The effect of three kinds of Nano particles on the radar reflectivity of concrete was different. The reflectivity of NF was up to -8dB when wave frequency was in the range of 2GHz to 7GHz and the minimum value was -13.56dB when reflectivity was at 2.56GHz. The reflectivity of NA and NS was smaller in the range of 10GHz to 18GHz and below -8dB when wave frequency was above 14.2GHz. NSF in which Nano-SiO_2 and Nano-Fe_2O_3 were mixed had the advantages of NS and NF combined and the reflectivity of radar wave was all less than -8dB in the full range of frequency.

(3) It was very effective to reduce the radar reflectivity of the sample by adding Nano materials in a certain amount of content. The lowest frequency of NF's radar reflectivity was all near 2.56GH when the content of Nano-Fe_2O_3 were namely1%, 2% and 5% and the reflectivity were -11.2 dB, -13.6dB and -16.7dB respectively.

References

[1] Yue-fang Zhang, Wan-jun Hao, et al. Research progress of cement based electromagnetic absorbing materials [J]. *Bulletin of the Chinese Ceramic Society*. 2014, 33(11): 2908-2912.

[2] Jin-ping Ou, Xue-song Gao, Bao-guo Han. The absorbing properties of carbon fiber reinforced cement based materials and stealthy effectiveness [J]. *Journal of the Chinese Ceramic Society*, 2006, 34(8): 901-907.

[3] Xing-wen Jia, Ya-jie Zhang, Jue-shi Qian. Study on absorbing properties of foamed graphite concrete [J]. *Journal of Functional Materials*. 2012, 43 (17): 2397-2400.

[4] Di Gao. Study on the performance of Nano carbon fiber reinforced concrete with new building materials [D]. Changsha: Southwest Jiaotong University, 2011.

[5] Gu-hua Li. Effect of Nano materials on durability of concrete [D]. Changsha: Southwest Jiaotong University, 2006.

[6] Guo-xuan Xiong, Min Deng, Ling-ling Xu, et al. Microwave absorbing properties of cement matrix composites [J]. *Journal of the Chinese Ceramic Society*, 2004, 32(10): 1281-1284.

[7] Bao-min Wang, Zhi-qiang Guo, Yu Han, et al. Carbon nanotube cement based composites for electromagnetic wave absorption [J]. *Journal of Functional Materials*, 2013, 44 (9): 239-1248.

[8] Xue-biao Hou, Zheng-zheng Wang, Wei Wang. Progress of Nano SiO_2 high performance concrete[J]. *Concrete*, 2013, 3: 5-9.

[9] Ali Nazari, Shadi Riahi. Al_2O_3 nanoparticles in concrete and different curing media [J]. *Energy and Buildings*, 2011 (43): 1480-1488.

[10] Hui Li, Hui-gang Xiao, Jie Yuan. Microstructure of cement mortar with Nano-particles [J]. *Composites Part B: engineering*, 2004 (35): 185~189.

IOP Conf. Series: Materials Science and Engineering **389** (2018) 012037 doi:10.1088/1757-899X/389/1/012037

A Kind of Song Brocade Fabric with NFC Data Masking Function Used for Making Purse

ZHANG Xiuling[1] and JIN Zimin[1]

[1]College of Materials and Textiles, Zhejiang Sci-Tech University, Hangzhou 310018, China

15757121979@163.com

Abstract. This paper designed a kind of Song brocade fabric that can prevent NFC data of bank cards from being read. It is based on the Faraday cage principle, special metal fiber and Mulberry silk were designed in the fabric. In order to maintain the original texture and gentle luster of the Song brocade fabric , the traditional mulberry silk was designed as the surface layer and the other intertwined silver fiber or metal fiber were designed at the back of the fabric through special binding weave. It can prevent NFC data of bank cards from being read when the fabric was applied to the wallet, which ensure the bank card information's secure.

1. Introduction

NFC (Near Field Communication), also known as short-range wireless communication, is a short-range high-frequency wireless communication technology that allows non-contact point-to-point data transmission between electronic devices to exchange data. It is widely used in Alipay treasure wallet, the user simply need to attach the bank card (or bus card) on the back of the phone, pay treasure wallet can read the bank card (or bus card) information, and complete consumption transfer and other operations. There is a big risk in NFC function, the phone can read the information inside the card and complete the payment operation without direct contact when the conditions of hardware and software were satisfied. This can lead to a serious consequence , the bank card in the wallet is likely to be illegal hacking in the public places in a flash, then the user will not only disclose the privacy information but also may lose a lot of money in this case.

2.Pattern design

According to the characteristics of the wallet, designed the pattern of the product. Firstly, auspicious symbol with representatives of the Swastika grain, Chinese traditional auspicious patterns for the design of the country. Then decorated the pattern with moire, floral patterns and geometric flowers. Finally designed three programs as a jacquard pattern to meet the different aesthetic needs of different consumers. Bring Chinese traditional decorative grain type into contemporary designing will push forward the development of traditional Chinese art and culture. Patterns are shown in figures1a, 1b and 1c.

IOP Conf. Series: Materials Science and Engineering **389** (2018) 012037 doi:10.1088/1757-899X/389/1/012037

Figures 1a. Diamond pattern and Ruyi pattern

Figures 1b. Swastika grain

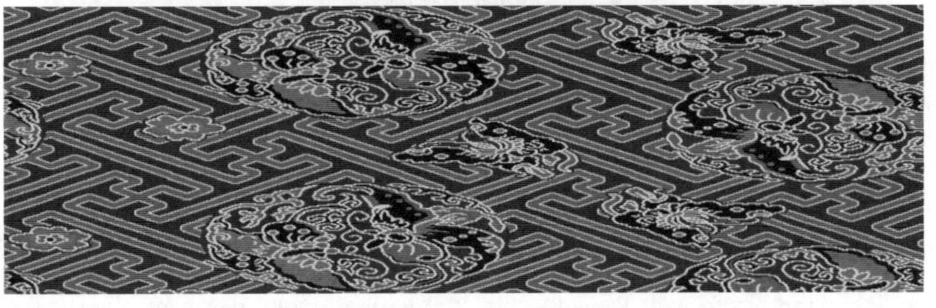

Figures 1c. Butterfly and flower pattern

3.The raw material selection and weaving process design

In order to overcome the limitation of current fabric technologies, this paper designed a kind of Song brocade fabric with NFC data masking function used for making purse [1]. There were two groups of wrap , ground wrap and face wrap. Raw material selection of ground wrap was mulberry silk while the face wrap used sliver ion fiber[2-3]. Two groups of warp were designed with mulberry silk and silver ion fiber at the ration of 3 : 1. There were N groups of weft, and N groups of weft ratio were 1 : 1, the front N-1 group of weft were designed with mulberry silk, the last group of weft (N group) used the silver fiber or metal fiber[4]. In order to maintain the original texture and gentle luster of the Song brocade fabric, the traditional mulberry silk was designed as the surface layer, the the face wrap (wrap B and weft N were silver ion fiber intertwined at the back of the fabric through special binding weave. This paper was designed with N=4, weft A : weft B : weft C : weft D =1 : 1 : 1 : 1. The number of ground warp density was set as 90 roots/cm, the number of face warp density was set as 30 roots/ cm, the number of weft density range from 30 roots/cm to 50 roots/cm.

4.Weave structure design

IOP Conf. Series: Materials Science and Engineering **389** (2018) 012037 doi:10.1088/1757-899X/389/1/012037

The song brocade fabric with NFC data masking function was weaved by two groups of wrap and N groups of weft[5]. The surface layer used mulberry silk, the the face wrap (wrap B and weft N were silver ion fiber intertwined at the back of the fabric through special binding weave[6]. This paper was designed with N=4 , the following offered a more thorough explanation of weave design.

This Song brocade jacquard fabric had three classes and 7 kinds of weaves. The three classes weaves were shown in Figures 2, and 3a, 3b, 3c, and4a, 4b, 4c. As seen in the following pattern drafts, ■means the wrap A over weft A, ○means wrap A over weft B, △means the wrap A over weft C, ☆ means the wrap A over weft D, ☒means wrap B over weft A, ▨means wrap B over weft B, ⊞means wrap B over weft C, ⧈ means wrap B over weft D. The arrangement of the warp were wrap A, wrap A, wrap A, and wrap B from left to right, the arrangement of the weft were weft A, weft B, weft C and weft D from top to bottom.

The first class of weaves were shown in Figure 2. The upper layer was interwoven by wrap A and weft A with two-down wrap twill weave, and the surface of fabric showed the color of wrap A. The middle layer was interwoven by wrap A and weft B, weft C with one- two down weft twill weave. The lower layer was interwoven by wrap B and weft D with two-down wrap twill weave.

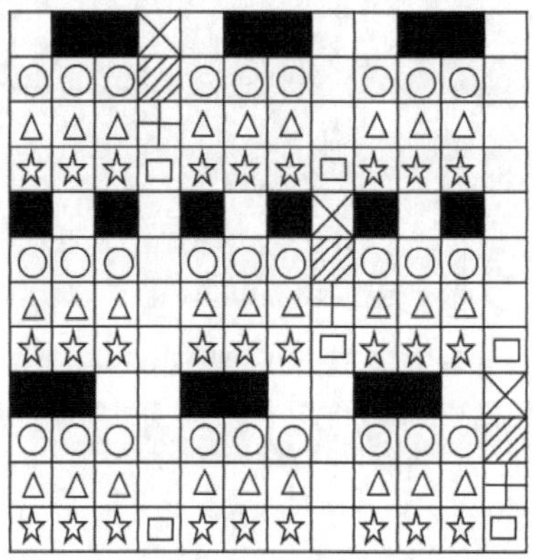

Figure 2.The first class of weaves

The second class of weaves were shown in figures3a and 3b and 3c. There were three kinds of weaves in this class. The first kind of weave is one-two down weft A twill. The upper layer were interwoven by wrap B and weft A with one-two down weft twill weave, wrap A and weft A with all weft float line, wrap A and weft B, weft C, weft D with all wrap over weft, the surface of fabric showed the color of weft A. The middle layer was interwoven by wrap B and weft B, weft C with one-two down weft twill weave. The lower layer was interwoven by wrap B and weft D with two-down wrap twill weave. The second kind of weave is one-two down weft B twill. The upper layer was interwoven by wrap B and weft B with one-two down weft twill weave, wrap A and weft B with all weft float line, wrap A and weft A, weft C, weft D with all wrap over weft, the surface of fabric showed the color of weft B. The middle layer was interwoven by wrap B and weft A, weft C with one-two down weft twill weave. The lower layer was interwoven by wrap B and weft D with two-down wrap twill weave. The third kind of weave is one-two down weft C twill. The upper layer was interwoven by wrap B and weft B with one-two down weft twill weave, wrap A and weft C with all weft float line, wrap A and weft A, weft B, weft D with all wrap over weft, the surface of fabric showed the color of weft C. The middle layer was interwoven by wrap B and weft A, weft B with one-two down weft twill weave. The lower layer was interwoven by wrap B and weft D with two-down wrap twill weave.

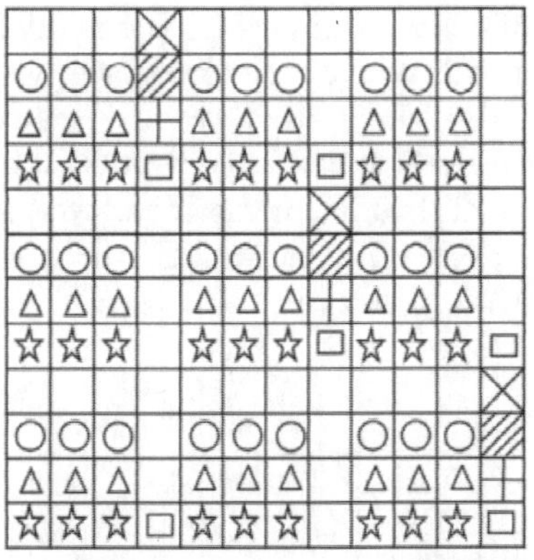

Figure 3a. The second class of first weaves

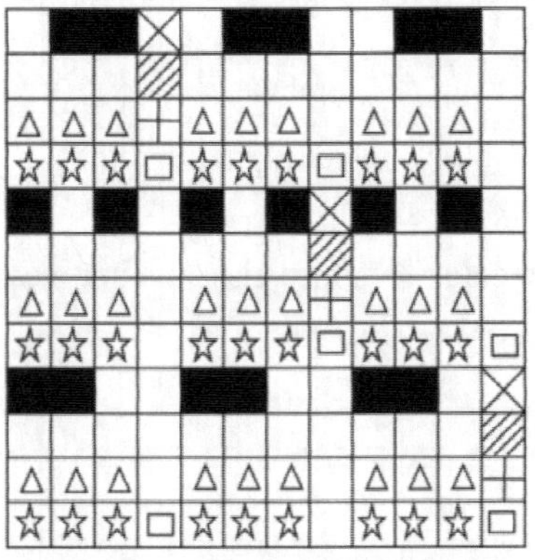

Figure 3b. The second class of second weaves

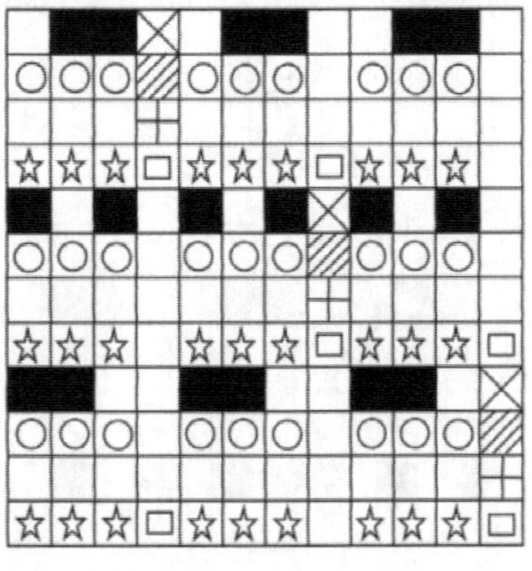

Figure 3c. The second class of third weaves

The third class of weaves were shown in figures 4a and 4b and 4c. There were three kinds of weaves in this class. The first kind of weave is one-two down weft A and weft B twill. The upper layer was interwoven by wrap B and weft A, weft B with one-two down weft twill weave, wrap A and weft A, weft B with all weft float line, wrap A and weft C, weft D with all wrap over weft, the surface of fabric showed the color of weft A and weft B. The middle layer was interwoven by wrap B and weft C with one-two down weft twill weave. The lower layer was interwoven by wrap B and weft D with two-down wrap twill weave. The second kind of weave is one-two down weft A and weft C twill. The upper layer was interwoven by wrap B and weft A, weft C with one-two down weft twill weave, wrap A and weft A, weft C with all weft float line, wrap A and weft B, weft D with all wrap over weft, the

IOP Conf. Series: Materials Science and Engineering **389** (2018) 012037 doi:10.1088/1757-899X/389/1/012037

surface of fabric showed the color of weft A and weft C. The middle layer was interwoven by wrap B and weft B with one-two down weft twill weave. The lower layer was interwoven by wrap B and weft D with two-down wrap twill weave. The third kind of weave is one-two down weft B and weft C twill. The upper layer was interwoven by wrap B and weft B, weft C with one-two down weft twill weave, wrap A and weft B, weft C with all weft float line, wrap A and weft A, weft D with all wrap over weft, the surface of fabric showed the color of weft B and weft C. The middle layer was interwoven by wrap B and weft A with one-two down weft twill weave. The lower layer was interwoven by wrap B and weft D with two-down wrap twill weave.

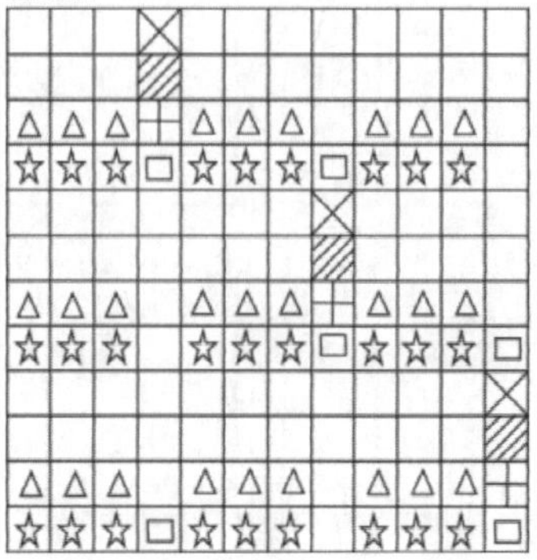

Figure 4a. The third class of first weaves

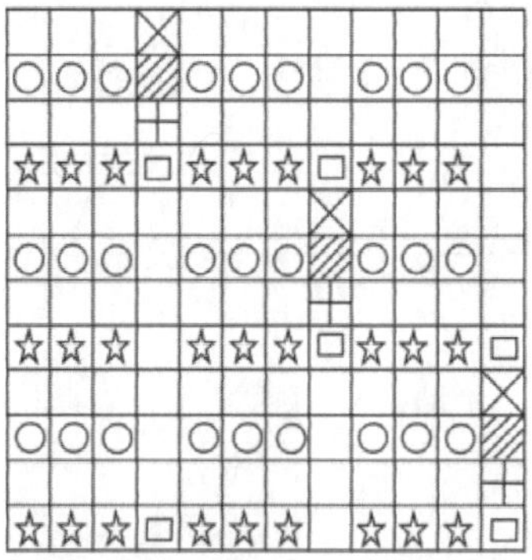

Figure 4b. The third class of second weaves

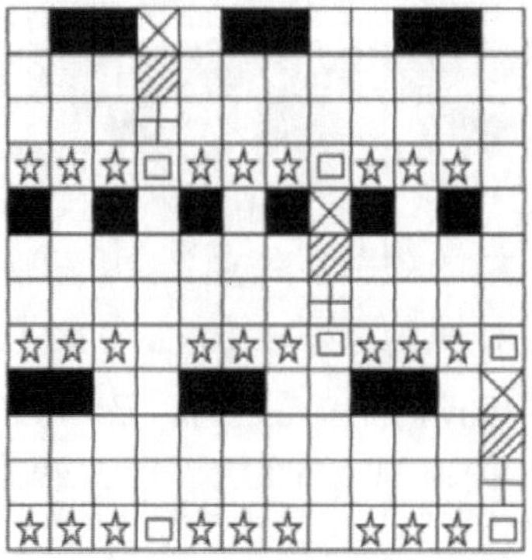

Figure 4c. The third class of third weaves

Finally, weaving the fabric on electronic jacquard machine after completion of jacquard cards in the jacquard weaving cad software[7-8].

5.Summary

Using textile CAD digital design technology, electronic jacquard weaving technology and fabric weave design technology, this paper designed a kind of Song brocade fabric that can prevent NFC data from being read. Two groups of wrap and N groups of weft(N = 2, 3, 4, 5, etc.) were designed on the fabric. Two groups of wrap ratio were ground warp(wrap A) : face warp (wrap B) = 3 : 1. Ground warp use silk, and the face wrap used mental silver fiber or metal fiber. N groups of weft ratio were 1 : 1, the front N-1 groups of weft for the silk, the last group of weft (N group) for the silver fiber or metal fiber. In order to maintain the original texture and gentle luster of the Song brocade fabric, the traditional mulberry silk was designed as the surface layer, the face wrap (wrap B and weft N were silver ion fiber intertwined at the back of the fabric through special binding weave. Compared with the existing technology, this fabric has the following advantages. Firstly, it can prevent NFC data from being read when the fabric is applied to the wallet, which effectively ensure the bank card information security. Secondly, the use of silver fiber can also effectively block the respiration and reproduction of bacterial cells, so the fabric has anti-bacterial deodorization ability[9].

References

[1] Qian X P. Analysis on characteristics of Shu brocade, Song Brocade and Yun Brocade[J]. Silk, 2011.

[2] Xiaoping Q. Research and Inheritance of Song Brocade[J]. Journal of Silk, 2015, 52(1) : 1-7.

[3] Miao Q. The Design of Fabric Structure and Patterns in Song Brocade[J]. Decoration, 2010 (1) : 120-121.

[4] Xiao H, Zhanghong T, Qun W. et al. Research on conductive grid structure and general influence factors to shielding effectiveness of electromagnetic shielding fabrics[J]. Journal of Textile Research, 2015, 52(1) : 1-7.

[5] Shibasaki A, Irimoto Y, Min K, et al. Selective binding of docosahexaenoic acid ethyl ester to a silver-ion-loaded porous hollow-fiber membrane[J]. Journal of the American Oil Chemists Society, 1999, 76(7) : 771-775.

[6] Jin Z, Zhou D, Zhu Y, et al. Weave design of two-sided velvet based on cut double layered and weft pulling[J]. Journal of Textile Research, 2014.

[7] Yingjie Z, Jin Z, Xinglong R,et al. Development of jacquard fabric of mental wire[J]. Journal of Silk, 2009(8) : 10-11.

[8] Wang X, Miao S, Cai Y, et al. Didactical Methods of Computer Aided Instruction in Home Textile Design[C] // Third International Workshop on Education Technology and Computer Science. IEEE Computer Society, 2011 : 94-97.

[9] Liu H. Development of Silver Fiber Knitted Antibacterial Functional Fabric[J]. Knitting Industries, 2015.